Ethics In Technical Communication

Shades of Gray

Lori Allen
Dan Voss

Wiley Computer Publishing

John Wiley & Sons, Inc.

New York • Chichester • Brisbane • Toronto • Singapore • Weinheim

Executive Publisher: Katherine Schowalter

Editor: Theresa Hudson

Managing Editor: Mark Hayden

Text Design & Composition: SunCliff Graphic Productions

This text is printed on acid-free paper.

Library of Congress Cataloging-in-Publication Data

Allen, Lori (Lori A.).
Ethics in technical communication: shades of gray / Lori Allen, Dan Voss
p. cm.
Includes index.
ISBN 0-471-15328-1 (pbk. : alk. paper)
1. Communication of technical information--Moral and ethical aspects. 2. Communication of technical information--Moral and ethical aspects--Case studies. I. Voss, Dan. II. Title.
T10-5.A44 1997
174 .98--dc20 96-26312
CIP

Printed in the United States of America
10 9 8 7 6 5 4 3 2 1

To our parents,
who taught us how to convert our dreams into reality . . .

. . . And to our families,
who shared the sacrifices of so doing.

With love and appreciation to:

Bill and Evelyn Zollinger
Russ Beitzel

Fred and Lorraine Voss
Ria Voss
Jonathan and Aaryn Burke
Rick Voss

Contents

Chapter 12 Professional Growth: Staying Ahead of the Information Avalanche 281

Acknowledgments

Thank goodness for all of the assistance we had with this book—and for the opportunity to publicly thank the people and organizations who provided it.

Special thanks to the reviewers whose perceptive comments improved the book's readability and ensured that its content reflects recent thinking on ethics. Dan Jones, Chair of the Department of Technical Communication at the University of Central Florida, and Michael Keene, Professor of English at the University of Tennessee, Knoxville, generously took time from their own work to read this manuscript and offer their expertise.

We greatly appreciate the talent and hard work of Nancy Lund, the Denver artist who provided most of the illustrations and cartoons, and Billie Bowman, who compiled the index.

We are also indebted to the Society for Technical Communication (STC). Much of our knowledge base in the subject area is a direct result of our efforts and those of our colleagues on numerous STC committees and task forces. The Society graciously gave us permission to use excerpts from the articles and reports that culminated many of those efforts.

We acknowledge also the longstanding, valuable support of the Fire Control Systems Communications Team and the Communications Department at Lockheed Martin Electronics and Missiles, which supported Dan in publishing and presenting at a series of professional conferences. We are also grateful to the Lockheed Martin Corporation for granting permission to include excerpts from those papers and presentations in this book, and to its CEO and president, Norm Augustine, for allowing us to incorporate a key passage from his article "The Cost of Being Ethical."

Equally heartfelt thanks go to the Technical Communication faculty and other colleagues at Metropolitan State College of Denver for their encouragement, ad hoc reviews, snipets of information, and accommodations while Lori worked on this text. In particular, Professor J.O. Yunker, mentor and friend, provided valuable information and insight gleaned from a distinguished career in the discipline.

Speaking of mentors, thanks also to Dr. Howard Livingston, Professor of English at Pace University, who helped develop Dan's writing ability.

We'd also like to acknowledge the students in technical communication classes at Metropolitan State, who served as guinea pigs while we developed the case studies. Their often unpredictable responses helped show us the error of our ways; their enthusiastic feedback encouraged us.

Thanks as well to Theresa Hudson, our editor at John Wiley, who gave us a chance and never tired of our incessant questions.

Preface

This book is written to help professional technical communicators and students preparing for the profession to deal with complex ethical issues, both on the job and in the classroom. We saw a need to address these issues in terms that were well-grounded in theory yet equally well-versed in practice. To meet that need, we have emphasized practical examples: in the text discussions, in the application and exploration exercises at the back of each chapter, and, especially, in the case studies.

We have provided a template (Figure 14.1) to help readers use value analysis as a practical tool with which to define and resolve day-to-day ethical conflicts they might confront as technical communicators.

Recognizing that ethics builds upon the personal values of our readers and realizing that these values reflect a wide range of world views, we deliberately drew from a wide variety of sources. We were seeking both breadth and depth in our philosophical foundation, yet trying also to bring our points close to home.

We believe the study of ethics is of vital importance to our profession. But we also believe that just because something is important is no reason for it to be boring. In fact, that's a very good reason for it to be interesting! That's why our style is deliberately informal, and that's why we have provided a generous supply of cartoons and illustrations.

We wanted to keep our heads in the sky, so to speak, but our feet on the ground. We hope this book will help you, our readers, do the same.

A Note About Style

We made two stylistic choices we'd like to alert you to right off the bat in case they are unfamiliar. First, throughout the book, we have tried to emphasize the personal nature of ethics by quoting from a broad spectrum of sources, from Shakespeare to *Star Trek*, from the biblical to the secular. There was method to our madness in selecting this eclectic assortment of references—ethics and values are, in fact, so personal that we could not hope to touch every reader by relying on any single source or any one type of source. So we drew freely from literature and myth, comic strips and movies, folk wisdom and scripture—as wide a variety as we could uncover. Our intent was neither to endorse nor to exclude, but to illustrate the all-encompassing nature of ethics. If we, as technical communicators, are to have a chance to "do what is right to achieve what is good,"—often against significant odds—we're going to need all the help we can get!

We also elected to use what we call "gender toggling" to handle pronoun references—that is, we alternate between masculine and feminine usage. Gender toggling is a relatively new practice intended to replace the sexist use of "he" to represent the whole human race without becoming mired in *he/she* references. We chose gender toggling because it exemplifies the cultural sensitivity we espouse as an important value. We also found it forced us to recognize and overcome any preconceived notions we may have had concerning gender as it pertains to occupations, activities, and basic values.

Part I

Setting the Stage

- A Definition of Ethics
- Using Value Analysis to Resolve Ethical Conflicts
- Professional Codes

CHAPTER 1

Dilemmas, Values, and Shades of Gray

Grub first: then ethics.

—Bertolt Brecht,
The Three Penny Opera

In ethics, actions speak louder than words.

—Mary Guy,
Ethical Decision Making in Everyday Work Situations

Are We Clear about Business Ethics?

Americans seem to be schizophrenic about business ethics. On the one hand, we seem to agree with Charles Dickens' cynical observation in *Martin Chuzzlewit,* " 'Do other men, for they would do you.' That's the true business precept." Several hundred years later, the *Wall Street Journal* echoed Dickens' sentiments by suggesting that the term "business ethics" should be considered an oxymoron like "jumbo shrimp."[1]

Numerous surveys document what appears to be a deep and abiding societal distrust of business people and their ethics. A 1983 Gallup poll showed that 82 percent of Americans ranked the honesty of business executives average or less, trailing most other professions, including doctors, teachers, bankers, and law enforcement officers. In fact, one of the *few* groups whose honesty rating was lower than that of business executives was Congress! In the same year, a Harris survey showed that only 18 percent of the public had "great confidence" in corporate executives; in fact, public confidence in business had plunged more rapidly than it had in all other institutions since a similar Harris poll 10 years earlier.[2] In 1987, 73 percent of those who answered a survey by *Crain's Chicago Business* said that they were aware of unethical practices in their industries.[3]

Things didn't change much as we moved into the 1990s. A 1990 survey by *Working Woman* magazine confirmed the predominance of the belief that fair play is fading from the business scene.[4] Much of this no doubt reflected the infamous scandals of the last two decades—junk bond swindles; alleged cover-ups by cigarette manufacturers; Medicare scams; Ralph Nader's celebrated exposé on corporate conduct in the automobile industry; even the demoralizing exposure of corruption at the highest levels of America's venerable grassroots organization, United Way. Is it any wonder why the public mood toward business ethics has remained cynical?

As Albert Z. Carr, special consultant to President Truman, once wrote in the *Harvard Business Review*:

> *In the last third of the twentieth century even children are aware that if a man has become prosperous in business, he has sometimes departed from the strict truth in order to overcome obstacles or has practised [sic] the more subtle deceptions of the half-truth or the misleading omission.*[5]

Do technical communicators fit the public perception of business ethics?

All of this suggests that the public seems to feel that Mark Twain's observation about politicians could just as well have read, "Reader, suppose you are a *businessman*. Now suppose you are of ruthless and greedy character. But I repeat myself."[6]

On the other hand, we still accord adulation to those captains of American industry who built the great industrial fortunes of the past. Railroad magnates like Vanderbilt, Morgan, and Gould; oil giants like Rockefeller; and steel tycoons like Carnegie are widely acknowledged to have engaged in highly unethical activities to amass their fortunes. In fact, a number of the well-known novels by the muckrakers of the last century, such as Upton Sinclair's *The Jungle*, exposed the practices of these very men. Yet, the industrial barons are still envied by many who aspire to corporate greatness. Many people question their ethics; few question their integrity. So how do we explain this apparent contradiction? The explanation resides in a definition of terms.

What Do We Mean by Ethics?

Just as the technical communication profession has changed dramatically over the last decade, so too has our understanding of our ethical responsibilities. As Renee Horowitz points out, our historic "approach to ethics was a firm warning against plagiarism."[7] But, as we began offering a more comprehensive range of services and using more sophisticated technology, we faced increasingly complex ethical dilemmas.

In response, many of the experts and leaders in our profession began exploring and researching ethics more intensely—a process which, of necessity, began with a definition. A brief examination of definitions in technical communication textbooks illustrates what these textbook authors say. Carolyn Boiarski, for instance, says that to be ethical simply means we must follow the law and that "concern for the consumer and citizen should remain foremost in relation to everything you write."[8] Jimmie Killingsworth defines ethics as "the art of setting standards of conduct and making moral judgments."[9]

Mike Markel sees ethics as "the study of the principles of conduct that apply to an individual or a group," identifying the principles as rights, justice, and utility.[10] John Lannon uses three different terms in defining ethics: obligations, ideals, and consequences.[11] And these are just a few of the available definitions in our field!

Clearly, although the details vary, all of these definitions have a common theme, one that we have adopted as our own definition of ethics:

Ethics is doing what is right to achieve what is good.[12]

That's about as basic as it gets. The problem is that any definition of ethics, no matter how basic, is still subject to individual interpretation and application. The definition of what is "right" and what is "good" is bound to vary according to the beliefs of the individual and the culture. It is also apt to vary with the situation.

Thus, the ruthless industrialists at the turn of the century believed that what they did was "right" (applying the law of the jungle) to achieve what was "good" (building their corporate empires). If their core value, or belief, was to achieve success, then any means they took to reach that end could be justified. This is what is referred to as *situational ethics*.

Some of the same robber barons who wouldn't flinch at extracting the last ounce of labor at near-slave wages to wring out a profit or at greasing the palms of government officials to remove business obstacles were also pillars of their respective communities and world-renowned philanthropists who donated millions to worthy causes and established major centers of healing and cultural exchange.

Their ethics? Driven by the situation. Their integrity? A matter of opinion. Their *personal* honor and civic responsibility were often unimpeachable, separate, and apart from their tactics in the business world of Social Darwinism. In other words, the ethics of the Corporate Board Room on Tuesday were not the same as the ethics of the Board of Deacons on Sunday. Business ethics and personal ethics operated on separate frequencies—the signals weren't mixed, just different.

Is it possible, then, for someone to have integrity in general yet still behave unethically in certain circumstances? Is that what we're saying? Yes and no.

On the one hand, we would certainly like to see "right" and "good" defined in terms more altruistic than those that accompanied the unprincipled rise of the industrial titans. We would certainly hope that the same basic human decency and concern for others that governs our personal principles would carry over into business.

On the other hand, we would be naive if we failed to acknowledge that many business people still operate on the *separate frequency* ethical bandwidth of the early industrialists. In other words, they do whatever they have to do to ensure success for their company and to advance their own careers; they have no qualms about taking actions in business that they wouldn't dream of taking in their personal lives.

Personal Ethics in the Business Arena

So the issue, as far as we're concerned, is not establishing separate ethics for the business world but fostering the extension of personal ethics into the business arena. Businesses are, after all, comprised of many individuals, all of whom bring their personal standards to the office each day. Unless you're the personal and professional equivalent of Dr. Jekyll and Mr. Hyde, it's impossible to divorce ethical behavior on the job from ethical behavior in all other parts of your life.

As Mary Guy, Professor of Political Science and Public Affairs at the University of Alabama, notes:

> *Unethical acts are not usually so much a product of greed or immorality as they are of ethical naïveté . . . When upper management is not clear about standards, priorities, or limits, the manager is left to his or her assumptions about what actions to take. Or, employees hidden in the anonymity of a large organization think that their actions will remain undiscovered regardless of the rightness or wrongness of their behavior.*[13]

We would add that it may not be so much the anonymity of an organization, but the culture that has been fostered by management (or allowed to grow on its own, uninhibited by management oversight) that provokes unethical actions by otherwise ethical people.

Let's say, for instance, that you work in the Public Relations and Corporate Communications Department of a Fortune 500 company. For the last month, the woman in the cubicle next to yours has had a most interesting assignment—orchestrating a large, *private* dinner party for the wife of the division president. Naturally, some members of the guest list were business associates, but most were friends and personal acquaintances.

"So, Mr. Bigman, you say your wife prefers lavender for the tablecloth?"

How do you respond? Well, other than feeling a little envious that she is schmoozing with caterers while you're haggling with printers, you don't much mind. In fact, it hasn't even really occurred to you that assigning such a task to a company employee might be breach of ethics. It is simply standard operating procedure—or at least it has been throughout your tenure with the firm.

But let's think about it for a moment. Isn't the division president, in fact, stealing from the company? After all, this employee is being paid good wages for work that is not contributing to the corporate bottom line. Carry the argument a little further. If it is indeed acceptable for the "pres" to steal, oh, let's say $15 an hour in this way, why isn't it be equally acceptable for another employee to take home $15 worth of office supplies? And if you look at the cost of the whole party planning effort—$15 an hour for 160 hours—well, hell, why not just take home that laptop computer the company purchased last year!

Now, it's not likely that any of you would walk off with thousands of dollars worth of computer hardware at the end of the workday, but you get our drift. If the corporate culture appears to condone *some* forms of unethical behavior, the perception is likely to be that *all* forms are equally acceptable. So in our occupations, we are often tempted to do things we would not ordinarily do, either because we discern that such behavior is accepted within the corporate culture or, sometimes, because we are instructed to do so.

Of course, not all ethical dilemmas are as black-and-white as "Should I steal this expensive laptop computer from my employer or shouldn't I?" There's no question, it probably wouldn't even cross your mind. But what about, "Should I take home this $20 print cartridge and a handful of diskettes?" More tempting? And how about, "Should I take home the 10 sheets of paper I need to finish my class report tonight, since I ran out of paper last night and I can't get to Office Depot?" Even more tempting.

Where do we draw the line—with one sheet of paper? That's the problem—black-and-white turns to *shades of gray*, and dealing with gray is rarely easy. Many situations are indeed gray because there are no clear-cut correct, or even acceptable, answers. Unfortunately, we sometimes call actions gray only because by doing so we can rationalize our decisions to produce a self-serving outcome. In other words, gray can be real—and quite confusing—or it can be just a disguise for self interest. It is the *real* gray that makes resolving ethical concerns in the workplace a trying proposition.

How do we cope with this apparent *bad influence* on our otherwise upstanding behavior? We reiterate that the issue is not so much business ethics as it is personal ethics. When faced with seemingly unethical influences or expectations in the workplace, we need to draw on the same ethics we invoke when we are not on the job.

The Key to Ethical Behavior

The key to ethical action is to behave with integrity that is based on a sound core of personal values. In this text, we present 10 basic values that we believe are central to the technical communication profession: honesty, legality, privacy, quality, teamwork, avoiding conflict of interest, cultural sensitivity, social responsibility, professional growth, and advancing the profession.

We do not presume to have these 10 values inscribed on stone tablets handed down from a mountain nor do we assert that they are any better than those put forth by other authors, whether individuals or organizations. We simply offer them as a starting point for your own value analysis, the six-step process we recommend as an effective approach to resolving the thorny ethical dilemmas you're likely to face in the workplace (refer to Chapter 2).

Two Approaches to Behaving Ethically

So how do we, as individuals, make decisions about ethical dilemmas in the workplace? How do we handle the unethical behavior of others? We suggest two approaches.

The first is simply to do your jobs in the best way you know how, using the 10 core values and applying value analysis along the way. Be honest in your use of language and graphics, for instance. Comply with the laws that govern communication. Respect others' privacy. Be sensitive to their cultures.

The second approach is less direct. In a number of places throughout the book we allude to our unique expertise as professional communicators—our ability to use the language. We argue that under normal circumstances, our ethical burden is no greater than that of our friends, family, or neighbors. But when it comes to language, our obligation is, in fact, greater. You'll hear this from us again, but it bears repetition, so here goes: Unusual capability carries with it higher responsibility. In circumstances in which we can't take direct action by virtue of our professional responsibilities, we can act indirectly by using our language skills to expose ethical issues to public scrutiny and debate.

Fortunately, we need not stand alone in the quest for ethical behavior in the workplace. More and more, the professional organizations to which we belong are adopting codes of ethics that can serve us in negotiating sticky situations for which we can find no relief through the appointed corporate channels.

How Can This Book Help Your Ethical Decision Making?

We offer no magic formulas, nor do we even purport to be in the business of *teaching* values. We trust that has already happened. Our goal is more modest: to help you sharpen your ability to apply your existing personal values in your role as a technical communicator within the rough-and-tumble world of business. It's not an easy task, to be sure, but it's an essential one if we're to maintain and enhance the integrity of our profession.

To accomplish this task, we've divided the book into three parts. Part I provides a theoretical and philosophical framework. In Chapter 1 we craft an operational definition of ethics that can be applied in the form of specific values to guide the professional practices of technical communicators. Chapter 2 presents value analysis, a logical and powerful six-step process you can use to resolve ethical dilemmas. Chapter 3 introduces the concept of codes of ethics as one way in which professions attempt to define their ethical position and assist their members in ethical decision making.

Part II, Chapters 4 through 13, presents the 10 core values mentioned earlier. In each chapter, we define the value, explore its permutations, and examine its implications. As you read through these chapters, you'll find dozens of recommendations, all guided by adherence to the core values. We present all manner of scenarios, examples, and exercises, both within the realm of business and technical communication and from a broader frame of reference.

Part III brings all the theory and values closer to home. Chapter 14 contains the basics of case studies—what they are, how they benefit you, and how you resolve them. Chapters 15 through 23 contain nine detailed case studies in technical communication. Each presents a realistic workplace situation, assembling an interesting cast of characters who represent various fields of expertise, and setting up ethical dilemmas that bring several different values into conflict. Chapter 24 gives you the opportunity to try your hand at developing an ethical case study in all its ethical shades of gray. Get your experience here and the real world of business ethics will be a much less intimidating place!

Throughout the book we don't mince words—we make our positions clear. You may not agree with us at every turn, but one thing is certain—it will be an enlightening trip! So if you dare, buckle your philosophical seat belts and step right up to ride *Shades of Gray*, the roller coaster of ethics and technical communication.

References

1. *Wall Street Journal*, 5 May 1983, 1.
2. Tad Tuleja, *Beyond the Bottom Line: How Business Leaders Are Turning Principles into Profit* (New York: Facts on File Publications, 1985), 4.

3. Mark Hornung, "Honesty's valued, but transgressions abound," *Crain's Chicago Business*, 10 (34):1 (1987): 61-63.

4. "How Ethical is American Business?" *Working Woman* (September 1990), 113.

5. Albert Z. Carr, "Is Business Bluffing Ethical?" *Harvard Business Review* 46 (January-February 1968): 143-153.

6. This modification to Twain's quotation appears in Laura Nash, "Why Business Ethics Now?" in *Managing Business Ethics*, eds. John Drummond and Bill Bain (Oxford: Butterworth-Heineman, 1994), 9.

7. Renee Horowitz, "Beyond Plagiarism: Ethical Issues in the Technical Communication Classroom," *Proceedings of the 41st Annual Conference in Minneapolis, MN. May 1994* (Washington D.C.: Society for Technical Communication), 134.

8. Carolyn Boiarski, *Technical Writing: Contexts, Audiences, and Communities* (Boston: Allyn & Bacon, 1996), 131.

9. M. Jimmie Killingsworth, *Information in Action: A Guide to Technical Communication* (Boston: Allyn & Bacon, 1996), 6.

10. Mike Markel, *Technical Communication: Situations and Strategies*, 4th ed. (New York: St. Martin's Press, 1996), 22.

11. John Lannon, *Technical Writing*, 6th ed. (New York: Harper Collins, 1994), 72.

12. Jules Harcourt, "Buddy" Krizan, and Patricia Merrier, *Business Communication*, 3d ed. (Cincinnati: Southwestern Educational Publishing, 1996), 83.

13. Mary Guy, *Ethical Decision Making in Everyday Work Situations* (New York: Quorum Books, 1990), 7.

Chapter 2 Overview

Value Analysis . . .

The Mission: **When confronted with thorny ethical conflicts in technical communication, we can apply a logical process to impose order, assess priorities, and make the best possible decisions. This process is called value analysis.**

Chapter Roadmap

Definition and Focus. This section defines *values* and *interests*, applies them to an ethical framework, then carries that framework into the world of business. It takes a realistic look at the two basic reasons to be ethical: prudence and virtue. The chapter provides a step-by-step treatment of value analysis and explains how to apply it to resolve value conflicts, both inside and outside the workplace and the profession.

Value Analysis Imposes Structure on Ethical Conflicts. This section helps you master and apply a sophisticated and powerful tool for ethical analysis by breaking it down into six basic steps. It then describes two models for value analysis and provides examples for each.

- *The Six Steps in Value Analysis*
  Define the *Issue* and Identify the *Stakeholders*
  Determine the Stakeholders' *Interests*
  Identify All Relevant *Values* That Bear on the Issue
  Determine the Values and Interests That Are in *Conflict*
  Apply a Model to Rank the Values According to Importance
  Resolve the Conflict in Favor of the Higher Value
- *Two Basic Approaches to Value Analysis*
  The Concentric-Ring Model
  The Hierarchical Model

Conflicts Vary: Black, White, and Gray. This section can help you deal not only with clear-cut ethical conflicts (black-and-white) but also with situations in which ethical values collide with other ethical values and with situations where none of the alternatives are ethically desirable.

- ❑ *The "Black-and-White" Scenario: Right versus Wrong*
- ❑ *The "Greatest Good" Scenario: Right versus Right*
- ❑ *The "Least of All Evils" Scenario: Wrong versus Wrong*

What Price Gray? This section probes the perils of *situational ethics*, when people bend the rules to fit their interests. It can help you detect and avoid pitfalls in ethical reasoning, such as:

- ❑ *The Diet Coke Theory of Morality*
- ❑ *The Relative Theory of Filth*

Applying Value Analysis to Technical Communication. This section brings value analysis into technical communication by defining 10 basic values pertinent to our profession. It also gives you four practical guidelines on how to deal with ethical conflicts in the workplace—and still keep your job!

- ❑ *Ten Basic Values in Technical Communication*
- ❑ *Four Guidelines for Dealing with Ethical Conflicts on the Job*

Summary, Practical Exercises, and References

. . . The Key to Unlocking Ethical Dilemmas

CHAPTER 2

Value Analysis Helps Us Resolve Ethical Conflicts in Technical Communication

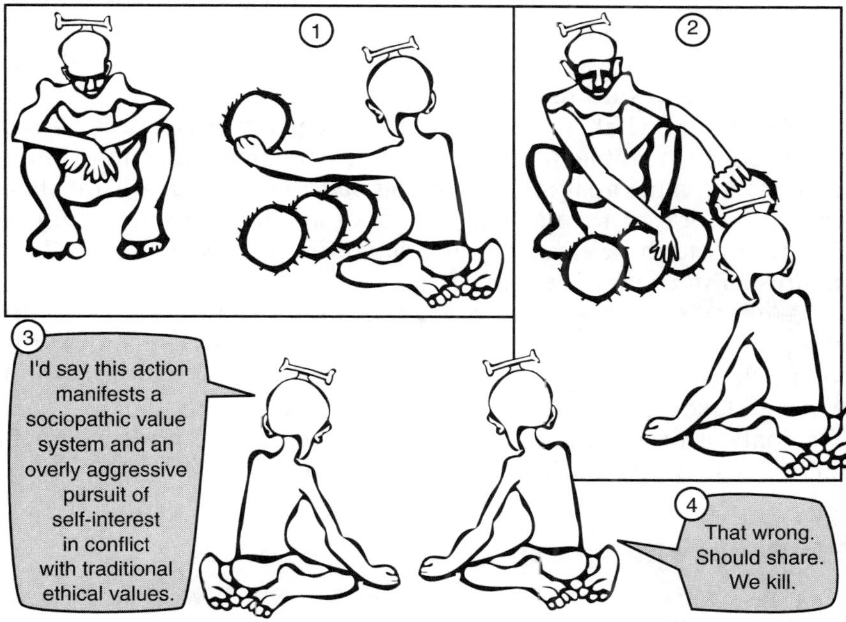

Who is the better ethicist in the cartoon, Aborigine #3 or Aborigine #4?
In today's political climate, #4 could carry many election districts!

Definition and Focus

What is value analysis? In simplest terms, it means examining ethical conflicts by breaking them down into the underlying values that are colliding, then deciding how those values stack up in importance, to determine the most appropriate course of action. Of course, it is not always that simple. Indeed, as we were researching this book we discovered a wide range of paradigms for conducting value analysis and resolving ethical issues, including the following:

- Hierarchical value analysis
- Concentric-ring value analysis
- Ethical algorithms
- Issues emergence and evolution

These are just a fraction of the models we studied. All are valuable. Each provides a different perspective, a unique *spin*. But after a while, try as we might, as we studied these complex models our eyes would begin to glaze over. We found the human brain can labor to impose structure on the ethical thought process for only so long before it begins to shut down. As you struggle with a complex and elegant paradigm intended to process the myriad factors impinging on ethical decisions, you have to resist the growing urge to throw up your hands in despair and listen instead to the little voice in the very back of your weary brain that keeps insisting, "Do I really have to understand all this stuff to know right from wrong?"

You don't. That was the insight that made this chapter—indeed, this whole book—possible. It's not that there isn't great value in exploring formal avenues for value analysis—there had *better* be, in that we are about to walk you down one of them!—but it's important never to lose sight of the fact that no matter how elegant the paradigm, *value analysis still boils down to the ability to distinguish right from wrong.*

Most of us are born with that. From a tenured professor with three Ph.D.s at a prestigious university to an uneducated primitive on a desert island, human beings know that it's good to share and it's bad to take—and it's even worse to take from someone who has just shared with you.

That's it in a coconut shell. Now all we have to do is apply that fundamental insight to technical communication.

We Begin with Values

First, let's work on values. In the context in which we're using it, *The American Heritage Electronic Dictionary*, 3d ed., defines the word "value" as:

> *A principle, standard, or quality considered worthwhile or desirable.*

In simplest terms, then, *a value is something we believe in.* As eminent ethicist Michael Josephson of the Josephson Institute of Ethics in Washington, D.C. points out: "An important thing to remember about values is that *not all values are ethical.*"[1]

I may place a value on being the most efficient burglar in the neighborhood, but it is doubtful my neighbors will share that value. Indeed, one of them may place a value on being the top markswoman in the National Rifle Association competition—in which case I may experience serious value conflict on my next burglary.

Next, Let's Look at Interests

Another important term to define is *interest.* Let's go back to the *American Heritage Electronic Dictionary*:

> *Interest: Regard for one's own benefit or advantage.*

Is the picture beginning to clear? Let's return to the aborigines. They both share an interest in having coconuts. The difference is that Aborigine #1 curbs his self-interest with a value that calls for sharing, whereas Aborigine #2 acts purely out of self-interest. Indeed, you could almost say that his self-interest has risen to a value of blind entitlement: "I want them. I deserve them. I'll take them."

Now, Let's Move to Ethics

In Chapter 1 we defined *ethics* as *doing what is right to achieve what is good.* The next step in the process is to differentiate ethical values from unethical values and from nonethical values.

> *Ethical Value*: A belief or principle rooted in moral behavior, based on a sense of what is right.
>
> *Unethical Value*: A belief or principle rooted in immoral or amoral behavior, based on a sense of what is wrong or, at least, of consciously disregarding what is right.
>
> *Nonethical Value*: A belief or preference that is not related to right and wrong.

By this definition then, we see Aborigine #1's selfless action as reflecting an ethical value ("it's good to share") and Aborigine #2's treacherous repayment of that kindness as a manifestation of an unethical value ("if you want it, take it"). To illustrate a nonethical value, say Aborigine #1 liked coconuts and #2 favored kumquats. They might place a high value on their preferences, indeed, fervently defend them, but this issue clearly doesn't cross into moral territory like sharing versus stealing.

Business Ethics = Personal Integrity + Social Values

The sense that we should share and cooperate, even if it means compromising our self-interest, is a basic and ethical *social value*. A person who adheres steadfastly to an ethical value system, a moral code of behavior, is said to possess *personal integrity*. This takes us to the trampoline for our leap from the desert island to the world of business and technical communication.

Before we can apply value analysis to business ethics, however, we first must explain what *business ethics* are and how they relate to personal values. As to the concern that the expression "business ethics" is a contradiction in terms, such as "military intelligence" or "happily married," we answer that we are not quite that cynical—well, not always!

Business ethics boil down to a combination of personal integrity and social values (see Figure 2.1).[2] A company's ethics can be no better than the individual conscience of its employees. No amount of ethical codes, no amount of ethical pronouncements from top management, will make an iota of difference if the rank and file of the organization are operating out of self-interest

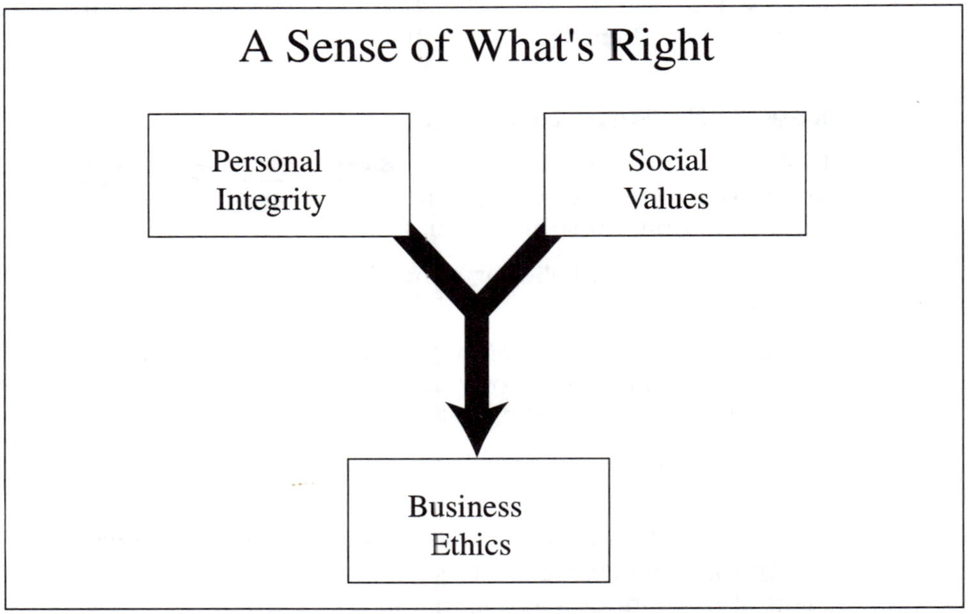

Figure 2.1 Business ethics are the vector of two component moral imperatives: social values and personal integrity.

rather than from a sense of personal integrity and social responsibility. In fairness, then, we should be no more cynical about corporate ethics than we are about individual goodness. If we believe most people are capable of behaving decently, then it follows that most companies can also behave decently.

It's very important not to view business ethics as an arcane and abstract discipline, embodied in weighty codes that sit on the shelf to impress corporate auditors but are never opened, let alone acted upon. Business ethics are nothing more than sound, moral, personal values exercised in a business environment. This concept can be very helpful in resolving ethical conflicts.

For example, how many of you would reach into another person's wallet or purse and steal $25? Compare this with the employee making $12.50 an hour who conducts two hours of personal business on the company clock.

And how many of you would shoplift a $5 item from a drugstore? Compare this with the employee who runs a ream of paper through the company copy machine on personal business. What's wrong in the personal environment is also wrong in the business environment.

Now suppose a 25-year employee runs one sheet of paper through the copy machine to make a copy of her grandchild's kindergarten diploma. Fire her? We don't think so. But how is that different than the 500 sheets in a ream? That's where the *shades of gray* come in. Common sense tells us that while copying the single sheet might technically be wrong, it's hardly in the same category as pumping a ream of paper through the copier. Who could possibly want to work in an environment where someone could be fired for such a minor infraction—if indeed, something that trivial could be considered an infraction at all?

But what if it were 250 sheets? 100? 50? 25? *Where do you draw the line*? That's the trap. That's where the gray comes in. Developing a viable scheme of ethics demands an appreciation of the *shades of gray* that can sometimes apply. However, it's important not to allow the sense of gray to become a rationalization for behavior that really isn't gray at all, rather flat-out wrong. If the individual employees in a company are ethical chameleons who adjust their values to meet their needs, then the company cannot be expected to be any more ethical.

Prudence and Virtue: Two Reasons to Be Ethical

Suspending our sense of moral righteousness for just a moment, let's ask the question: Why, after all, when you get right down to it, should businesses *be* ethical? *Are good ethics really good business*? Josephson says, "Yes." He offers two reasons:

Prudence*: It's the* smart *thing to do.*

Virtue*: It's the* right *thing to do.*[3]

The *virtue* rationale reflects the moral high road—doing what's right for its own sake, out of integrity and pride. According to the virtue theory, employees refrain from making any more than minimal personal use of business time and materials because they understand that doing so is a form of stealing from their employer, and they know that stealing is wrong. Bumping that up a notch, the company refrains from using false advertising to drive up lagging sales because it knows that misleading advertising is a form of dishonesty, and dishonesty is wrong.

The *prudence* rationale reflects the practical considerations—doing what's right for fear of the consequences of getting caught doing something wrong. Under this theory, employees refrain from making personal use of business time and materials because they fear they may be fired if they get caught. And the company refrains from false advertising because it knows it could receive severe sanctions from the Federal Communications Commission if its deception were to come to light.

Does it really matter *why* individuals or companies behave ethically, as long as they do? Well, it would be nice to think that Aborigine #2 might have a sudden attack of virtue, see the error of his ways, return his ill-gotten gains, and voluntarily refrain from stealing coconuts in the future. But if you're Aborigine #1, with a migraine headache and no coconuts, you'd be inclined to vote for Aborigine #4's solution.

That's why we have *laws* against stealing coconuts—as well as trade secrets, competition-sensitive proposal data, or software. And that brings us to value analysis.

Value Analysis Imposes Structure on Ethical Conflicts

Value analysis is an extremely effective approach to resolving ethical conflicts because it is sufficiently structured to impose order upon an often confusing morass of issues, yet sufficiently flexible to acknowledge the reality of *gray* when it comes to such issues.

The Six Steps in Value Analysis

When confronted with an ethical issue, whether it be in the workplace, in school, or in one's personal life, the six basic steps to follow are:

1. Define the *issue* and identify the *stakeholders*.
2. Determine the stakeholders' *interests*.
3. Identify the relevant *values* that bear on the issue.
4. Determine the values and interests that are in *conflict*.

5. *Apply a model* to rank values according to importance, to weigh the values and interests that are in conflict.
6. *Resolve* the conflict in favor of the higher (more important) value.

Define the Issue and Identify the Stakeholders

Defining the issue is usually closely linked to identifying the stakeholders. If the issue is teacher raises, the primary stakeholders are the teachers, the students, and the taxpayers. If the issue is logging original redwood forests in the Northwest, the primary stakeholders are the timber companies, environmental protection groups, the tax-paying public, and, in this case, future generations.

Determine the Stakeholders Interests

Once the basic issue is on the table and the stakeholders are seated on each side of the table, so to speak, the next step is to determine each stakeholder's interest in the issue: service or profit, dedication or advancement, exploitation or conservation. Watch out for hidden agendas. Be wary of the salesperson who has your best interests at heart; that may or may not be so, but it sure doesn't hurt to find out if the person is on commission! Don't let healthy skepticism cross the line into blanket cynicism, however. Just because a person has a legitimate interest in the outcome of an issue does not necessarily mean the person will behave unethically to serve that interest. In fact, despite all the stereotypes, for every dishonest auto repair or unnecessary surgical procedure, there are probably dozens of legitimate ones.

Identify the Relevant Values that Bear on the Issue

Next come the values. Take the issue of teacher raises. Typical American values involved would include the importance of educating our children (social justice, upward mobility), the time-honored principle of providing a fair day's wage for a fair day's labor (work ethic), and the need to be frugal with the public coffers (waste not, want not). Consider the logging issue. Underlying values would include free enterprise, resistance to government regulation, respect for nature and other species, and saving resources for our posterity.

Determine the Values and Interests that Are in Conflict

Sometimes determining the value conflict can be straightforward; other times it can be complex. In the case of teacher raises, it's not hard to see that the conflict boils down to the importance of education versus the use of taxpayer money, with the work ethic falling somewhere in between. In the logging scenario, the first two values listed obviously contend with the second two.

Apply a Model to Rank the Values According to Importance

As previously mentioned, numerous models exist to weigh values and interests that are in conflict, ranging from basic and straightforward to elegant and complex. We present two simple and useful value models later in this chapter: the concentric-ring model and the hierarchical model.

Resolve the Conflict in Favor of the Higher Value

The final step is to resolve the conflict in favor of the higher value. That is much easier said than done. For instance, with the teacher raises, resolving the conflict is far more difficult than defining it. Since the authors have been involved in teaching full-time or part-time for the past three decades, we must be wary of our own hidden agenda in this scenario! Probably the fairest thing to say is that neither value is a clear winner in this conflict; more data is needed. If teacher salaries in the school district in question are already among the highest in the state (and the state is well up the ladder on the national scale), and if the taxpayers' millage is already higher than in several surrounding counties, the scales may tip in favor of a lower wage increase. If the facts show the opposite trend, then perhaps it's time for the taxpayers to bear the cost of bringing their teachers' salaries more in line with fair market value. (How'd we do?)

As to the logging, it may be tempting—but manifestly unfair—to portray the logging interests as ruthless despoilers of the land and the environmentalists as noble defenders of nature. But what if we're talking about private land, not under public protection? Does protecting the environment mean every tree, in every forest? And are we all willing to curb our taste for lumber to serve the cause? Here again, we need all the facts.

Two basic methods for assigning relative priority to values, the concentric-ring approach and the hierarchical approach, are presented in the following sections. Sometimes the decisions are rather clear-cut; more often than not, however, there may be several *right* courses of action and several *wrong* courses of action, and maybe even some in between. In that event, you analyze the possible consequences of each action and choose the path that will serve *the greatest good*: That is, *help as many of the stakeholders as possible*, as much as possible; and *hurt as few of the stakeholders as possible*, and as little as possible.

A good rule of thumb in addressing ethical issues is to picture yourself as an impartial mediator who is paid to weigh all sides of an issue objectively and choose the best path of action. The analogy is not perfect, because mediators by definition are usually seeking to forge a compromise between contending parties. In value analysis, if a clearly ethical value is in conflict with a clearly unethical one, there is no room for compromise.

So if the logging company decides to clear-cut a couple hundred acres of federally owned land that happens to border its own, that's a clear violation of both ethics and law. The company's interests must be subjugated to the public good and the violation must be punished. What if the teachers go out on strike in defiance of public law? That's a little trickier, in that the logging company in the previous case has no title whatsoever to the land while the teachers, as public servants, might have a legitimate claim for an adjustment if the facts on salary show them indeed to be below fair market value. Still, as we shall see in both value models to be discussed, respect for law is a core value and one that we cannot abrogate lightly. So if the teachers *do* strike, even though they may feel morally justified, they have to be ready to accept the legal consequences of their action.

Even in the case of the desegregation movement (see Chapter 1), where the ethics of equal justice for all clearly superseded racist laws, the leader of the movement, Dr. Martin Luther King, Jr., advocated nonviolent *civil disobedience* on the model of Mahatma Ghandi. Right or wrong, if the law was on the books and you broke it, you did so with the expectation and acceptance of being arrested. The idea was to put the law itself on trial and, through sheer public outcry, ultimately have it changed to reflect the higher good. It took a while, but it worked.

Two Basic Approaches

Several models have been developed to define and prioritize values. Two of the most useful are the concentric-ring model and the hierarchical, or vertical, model.

The Concentric-Ring Model

In the concentric-ring model for value analysis, values are arranged in circles with core values at the center, peripheral values at the edge, and authority values in between (see Figure 2.2). An extremely important point is that *there is no one formula for emplacing values within this or any other model.* There are consensus values in our culture; there are consensus values within other cultures (not all the same as ours); there are even basic human values that reflect some consensus across cultures—but no two people, whatever their culture, are going to hold exactly the same values and in exactly the same order of precedence.

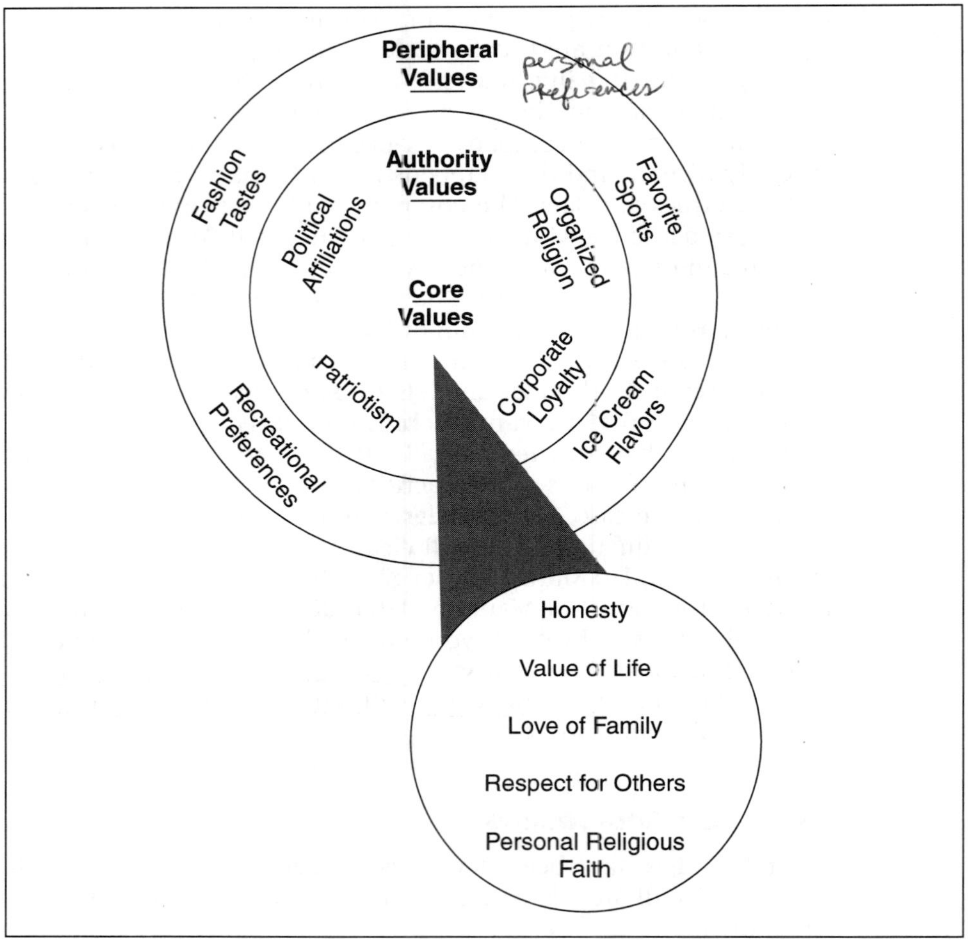

Figure 2.2 In the concentric-ring model, the innermost, or core, values generally take precedence.

With that caveat, let's go ahead and assign some consensus values to the concentric-ring model and take a look at how it works. Typical *core values* in our western culture include honesty, value of life, love of family, respect for others, commonly held axioms such as the "Golden Rule," and basic spiritual beliefs that are common to many religions. (This is not, of course, an exhaustive list.) Typical *authority values* include political affiliations, national allegiances (patriotism), corporate loyalty, and affiliations with organized religion. Typical *peripheral values* include personal-preference items such as fashion tastes, favored recreational pursuits, even favorite flavors of ice cream (*very* peripheral!).

When Values Collide, Core Values Prevail. Now let's examine a conflict situation and apply the concentric-ring model. Suppose a lifelong, dyed-in-the-wool Republican who is also a strong advocate of the right-to-life position on abortion must choose between a Republican candidate who is pro-choice and a Democrat who opposes abortion. The party affiliation, while well-entrenched, is an authority value, not a core value, whereas the right-to-life conviction springs from religious beliefs that are core values. In all likelihood, the person will vote Democratic rather than support a pro-choice candidate. The choice may be difficult, but *core values generally take precedence over authority values.*

Or suppose you're considering a life's mate. Would you be more apt to marry someone who differs in respect for others (for example, a closet racist), someone of a different religion, or someone who preferred a different recreational activity? The deeper the values that were in conflict, the less likely you would be to consider the match.

Using Value Analysis to Communicate Persuasively. Let's look at another example. Say a car–bike accident has seriously injured a child in your community (call her Amy Stephenson, daughter of Mr. and Mrs. Paul Stephenson of Azalea Circle). The child will recover, but the accident could easily have been fatal. Let's further suppose that there is a referendum coming up to install bike paths, which would greatly reduce the hazard. The problem is that your community is already chafing under federal, state, and county taxes over which it has little direct control; it is therefore likely to seize upon the bike path referendum as the only tax issue handy over which it *does* have control and therefore reject it—more in angry reaction to other taxes than on the merits of the instant case. If you were a supporter of the bike path initiative, how could you use value analysis to overcome the tax backlash and advance your cause?

First, you would analyze the issue to see which values are involved. The two that are central to this case are the high value western culture generally places on human life and the obligation of elected public officials to be wise stewards of public funds. The value placed on life is clearly a core value. However, frugality is also a core value, deeply rooted in the American culture. The question is, if two core values collide, which prevails? That depends on two factors: how the issue is presented to the stakeholders and whether one of the core values is more central than the other.

As a supporter of the bike path initiative, you would act on the assumption that the value of life would be the more central of the two; the key would be to present the issue in those terms to the stakeholders. Thus, your editorial in the local newspaper might start with a description of the accident, proceed through the arguments in favor of bike paths, openly acknowledge the price tag—in the process using well-documented facts like contractor quotes and projected impact on millage rates to dispel any exaggerated cost claims by the opposition—then conclude with the question, "We have before us, then, a referendum that will cost the average homeowner in our community $117 in new taxes to install bike paths. Is it worth it? Ask Mr. and Mrs. Paul Stephenson of Azalea Circle."

"That's pretty heavy stuff!" you might object. "Dramatizing an injury to a child to drum up support for bike paths? Come on! Is *that* ethical?"

We think so. In this case, the humanistic end—saving other children's lives—justifies the journalistic means (as opposed to a lurid tabloid interview with the bereaved parents of a young shark attack victim, where the only objective is to sell newspapers). An important point in our position is the fact that Amy Stephenson *will* recover; if she had been killed, using the dramatic final line of the editorial would become unethical. Basic human decency and respect for the parents' feelings would demand a different closing. Interestingly, publishing the dramatic line even in that tragic event *would* probably be *legal*, under freedom of speech—but so was the rush of photographers to snap close-ups of little John F. Kennedy, Jr. saluting as his father's coffin rolled by on the way to Arlington. That didn't make it right.

Cognitive Dissonance: When Core Values Clash. Here are two final examples before proceeding to the second model. Sometimes a direct clash of core values can produce a situation psychologists refer to as *cognitive dissonance*—a state of anxiety approaching panic. An example would be a fervent vegetarian in a foreign cultural setting where refusal to eat meat would constitute a grave insult to the host (value of life versus respect for other humans). Because of the pain involved with cognitive dissonance, the person in conflict is strongly motivated to find a compromise, an "escape hatch" of some kind to allow relief. The rationalization could go either way (for example, "The sheep has already been sacrificed; I might as well eat it and not offend my host," or "I don't like these people anyhow; why should I compromise my principles in order not to insult them?").

Now suppose you had the unenviable task of delivering a speech to an audience of disabled veterans to argue *against* the constitutional amendment that would ban the burning or desecration of the American flag. Impossible? Well, at the very least it's certainly challenging. About the only possible path you could take would be to set up a cognitive dissonance between patriotism and loyalty to comrades-in-arms versus freedom of speech, one of the American core values they had sacrificed to defend. Your closing could be:

> *In conclusion, then, it may be the ultimate irony that the very flag we all love, the flag for which you have sacrificed so dearly to defend, protects the free expression of* all *Americans—even these misguided souls who choose to use that freedom in an attempt to dishonor their birthright. But they shall fail. Dishonorable people with dishonorable actions can never dishonor*

what we all cherish. They may burn the flag, but they can never burn what it represents.

The Hierarchical Model

Another model for value analysis, developed by Dr. Lynn Payne at the University of Virginia, arranges values in hierarchical fashion with the most important values at the top and the least important at the bottom (see Figure 2.3).[4]

Critical Shared Values Are Backed by Law. The uppermost values are called critical shared values. These values are unique among the four clusters in that they alone, in western culture, are backed with the force of law. Within these values, honesty is considered to be the highest (corresponding to the innermost core value in the concentric-ring model), followed by respect for property and law, keeping commitments, and protecting the physical integrity of others. Here, as in the other model, *the selection of values and the precise succession of values is arbitrary*; in this case, they represent one expert's assessment on a consensus of the American conscience. A con artist, to use an extreme example, might place *dis*honesty in the highest rung on the value ladder; no-

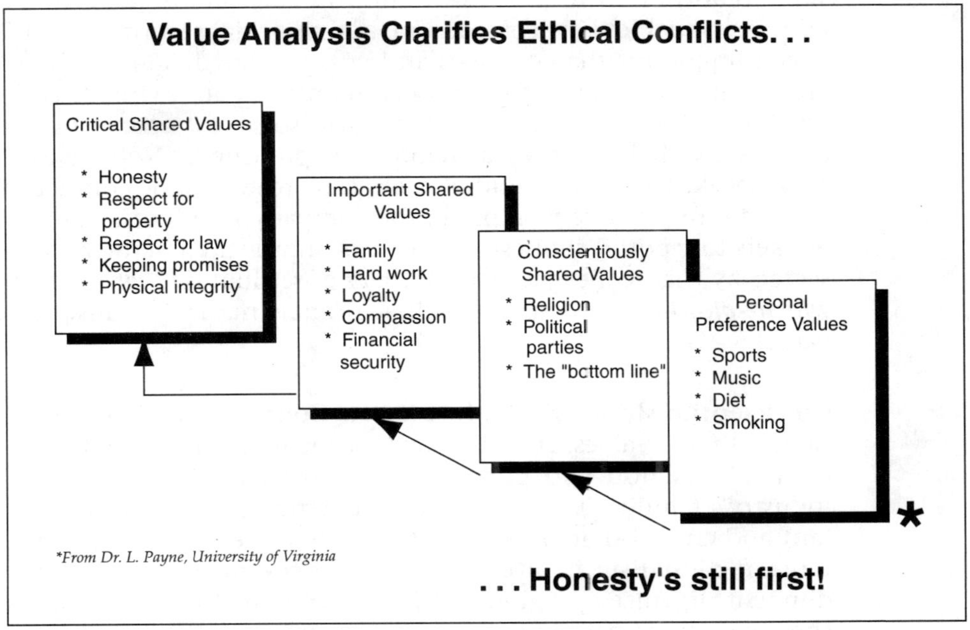

Figure 2.3 In the hierarchical model, the highest, or critical shared values, often, though not always, take precedence.

body can stop him from doing that, but the consensus of our society would condemn that value as unethical, since it serves self-interest at the expense of the social interest.

That brings us to the thread that runs through the critical shared values—the rights of the individual versus the rights of the society. To paraphrase British philosopher and political scientist John Stuart Mill, in his seminal work *On Liberty*, the rights of the individual extend only insofar as they do not infringe upon the rights of any other individual. Chief Justice Oliver Wendell Holmes made the same point in more homespun fashion when he said: "My right to swing my fist in the air ends where your nose begins." Mill goes on to argue that by our very membership in society, we sign an invisible "contract" to honor the rights of others; and that society has the right, through its laws and its enforcing agents, to hold us to that contract. On the other hand, society has no right to impose its will on those whose behavior injures only themselves; thus, Mill would probably have opposed laws against victimless crimes such as solitary drug use or riding a motorcycle without a helmet.

Important Shared Values Reflect General Consensus. The second level, important shared values, includes beliefs that enjoy a broad consensus of support in our society. These values include family allegiances, the work ethic, personal loyalty, compassion, and financial security. The difference between important shared values and critical shared values is that the former are driven not by laws but by social opinion. Thus, while it is not illegal to "rat" out a friend, doing so is apt to incur social ostracism unless the friend's deeds are so heinous they demand to be punished. "Nobody loves a 'narc,'" so to speak. This fact, in and of itself, can create a value conflict in maintaining the ethics of the workplace; the company ethics office must depend on workers to report dishonest acts such as improper charging or misuse of trade secrets by their colleagues without fear of retribution, yet the very process of *whistle-blowing* tends to go against human nature because of the loyalty value.

Conscientiously Shared Values often Conflict. The third level, conscientiously shared values, contains some of the most intensely defended values—after all, religion and politics are assigned to this category, along with industry's familiar *bottom line*. The difference between these and the important and critical shared values is that here there *is* no consensus; indeed, intense factionalism is often the order of the day. Consider the myriad of different organized religions—it has been argued that more blood has been shed over differences in religions than from any other cause in recorded history. And anyone who thinks there is consensus in the political arena need only to look back at the campaign tactics used in the most recent election. Within industry, there is some consensus on the importance of the bottom line, but for every statute-combing corporate lawyer working to minimize the impact of an environmental damage lawsuit, there is a placard-wielding environmentalist marching in protest outside the corporate headquarters.

Personal Preference Values Are Outside the Ethical Framework. The fourth level consists of personal preference values. These are akin to the peripheral values in the concentric-ring model and include such preferences as smoking, diet, music, and sports. The values within this level cannot be considered right or wrong; they are *non*ethical, as opposed to ethical or unethical. Rachmaninoff is not intrinsically "better" than rap—try as we baby boomers might to inculcate this concept in our young! The only instance in which these values can be viewed as right or wrong occurs when they come into conflict with values higher up in the hierarchy.

Smoking is a classic example. Dr. Payne's model, not to mention those of John Stuart Mill or Oliver Wendell Holmes, would clearly establish the right of the smoker to smoke—in private. But the minute that right infringes upon the rights of another, such as a nonsmoker seated next to the smoker on an airplane, it is forfeited. In Payne's model, that's a case of a critical shared value (the physical integrity of the nonsmoker's lungs) superseding a personal preference value (the choice of the smoker to destroy his own lungs).

That conflict is clear-cut and relatively easy to resolve, since it pits a Level 1 value against a Level 4. Pit a critical shared value against an important shared value, however, and the problem can become knottier.

Suppose, for example, you learn that your brother-in-law has been engaged in a low-level embezzlement scheme at his company for the past several years—nothing earth-shaking, say perhaps five thousand dollars a year, to stay below the threshold of detection. Suppose, further, that his company has passed him over repeatedly for well-deserved promotions, imposed a heavy amount of unreimbursed overtime, and otherwise abused him. And to complete the picture, he's using the embezzled funds to put his daughter (your niece) through college. Oh, and while we're at it, let's assume your sister has a disability, so there is no other significant source of household income beyond a small Social Security stipend.

How many of us would blow the whistle, turn him in to the company to be fired and possibly arrested? Yet if we apply the hierarchical model, we can see that it's a case of three critical shared values versus one important shared value. What your brother-in-law is doing is dishonest, it clearly doesn't respect the property of the company, and it is illegal. On the other side is his motive—taking care of family. That would presumably be your conflict also—family loyalty versus the knowledge of criminal activity.

We theorize that many people who in the same situation would not themselves embezzle would probably stop short of turning in a family member who did. They would tend to rationalize: "The company will never miss that small an amount. He's really only taking what he rightfully deserves . . . and so on." This is called *situational ethics*, which can be both seductive and dangerous. (There will be more on that later.)

Going back to the model, let's create another scenario using your brother-in-law. Instead of a small-time embezzler, let's make him an ax murderer who specializes in children on their way home from Sunday school. Now are there any whistle-blowers in the crowd?

Clearly, the physical integrity of innocent children is a critical shared value that would prompt nearly everyone to set aside family loyalty and turn in the deranged brother-in-law to authorities.

At the opposite end of the spectrum, let's say you're driving with your spouse on the interstate and he/she is 10 miles over the speed limit. Would you exercise your option as a citizen of your state to perform a citizen's arrest—direct your spouse to drive to the nearest police station, swear out an affidavit, and have a ticket drawn up? Right—and spend the rest of your life on the couch! No question here: family loyalty (Level 2) would overcome the letter of the law (Level 1).

At the extremes, then, the conflicts are easier to resolve. Nearly everyone would turn in the family ax murderer; virtually no one would arrest a spouse for speeding. The embezzling scenario falls closer to the middle, in the gray zone, but tilting toward family loyalty.

While these examples have been hypothetical, core value conflicts can be very real and very compelling. In fact, a real-life case occurred just as we were about to go to press. Federal law enforcement officers were led to the suspect in the infamous Unabomber case by his mother and brother, who reported his suspicious actions.[5]

The FBI agents to whom the suspect's family reported him said the brother's anguished value conflict was evident during the questioning. The brother clearly sensed that his greater duty was to help society put an end to a senseless series of sophisticated and cold-hearted letter bombings that had already claimed 3 lives and maimed 23 other people. Yet, he still agonized over taking an action that, assuming conviction, could wind up ending his brother's life. According to the wire service report:

> *David Kaczynski was torn between duty to country and loyalty to family before he pointed investigators toward his older brother as a possible suspect in the 18-year series of Unabomber attacks, federal agents said yesterday. . . .*
>
> *It took "many discussions, many interviews" with the lawyer to persuade David to meet face-to-face with FBI agents, one official said . . . .*

"He was torn, as anyone would be, between doing what is societally right and loyalty to his brother," said one agent.[6]

Consider another real-world example: A surgeon is operating on a child who hemorrhages and must receive an immediate blood transfusion or die. What if the parents belong to a religion that forbids blood transfusions and refuse to sign the consent form? This actually happened. The doctor ordered the transfusion, the child survived, and the parents sued the doctor. The court found in favor of the doctor, ruling, in essence, that the child's right to life (critical shared value, physical integrity) superseded the parents' religious beliefs (conscientiously shared value).

What if the patient were an adult who had personally refused to sign the consent form before the operation? That's just a hop, step, and a jump from the Dr. Kevorkian issue: Do human beings have the right to end their own lives? We won't attempt to resolve that one for you.

Conflicts Vary: Black, White, and Gray

When it comes to ethical conflicts, some areas are gray; they are simply not clear-cut. To apply value analysis meaningfully, one cannot ignore that fact. However, to insist that everything is gray can become a moral cop-out, leading to a rationalization process wherein values are *adjusted* to conform to personal interests (situational ethics). Let's consider three basic types of conflict:

- Conflict between clearly ethical and clearly unethical values.
- Conflict among two or more values, all ethical.
- Conflict among two or more values, all unethical.

The "Black-and-White" Scenario: Right versus Wrong

The first case is the "black-and-white" scenario. The great coconut heist in the chapter opener falls into this category. Similarly, if somebody breaks into your car and steals your stereo, there isn't much room for gray. The social contract has been violated, as has the law. The thief has no right to your property. The value of respecting others' property is clearly ethical; the value that one can take what one wants is clearly unethical. When value analysis can reduce a conflict to ethical versus unethical values, the course of action is clearly defined.

The "Greatest Good" Scenario: Right versus Right

The second case is sometimes referred to as "the greatest good." Let's say Clara is a newly hired attorney in a prestigious law firm who is also a single parent and a member of the Board of Deacons at her church. Three values intertwine here—professional advancement, family responsibility, and religious service. All are ethical. So what happens if they come into conflict?

Let's say Clara faces her first major trial and has a critical late-afternoon deposition. It comes time to pick up her daughter at day care. What gives? Or suppose last-minute, after-hours pretrial preparation collides with a commitment she made to her church, such as visiting shut-ins?

At first glance, it's tempting to relegate the professional responsibilities to second place—that somehow serving the church or being a good parent must be a greater good than serving one's employer. Indeed, this message comes across loud and clear in the film *Kramer vs Kramer* when Dustin Hoffman's character misses a key business meeting to take care of his son and loses his job over it. We are encouraged to view him as the ultimate selfless parent, willing to compromise his career advancement for his child's well-being.

But Clara's case may be more a matter of professional *survival* than it is advancement. How, realistically, can she possibly do anything less than her best on her first major trial? Consider, also, her ethical responsibilities to the person she is defending.

Life is filled with conflicts like this. The best Clara—or any of us—can do is to resolve the immediate conflict through a hasty compromise (for example, calling a friend or fellow parishioner to pick up her daughter at day care, getting another Deacon to cover this month's visits to shut-ins), then, over the longer term, make strategic decisions that minimize the incidence of such conflicts in the future. Perhaps Clara can use the excellent salary a prestigious law firm offers to hire a private afternoon child-care provider with more flexibility than a day-care center. Perhaps she can do extra church work during the quiet period between trials in return for the realization that she will sometimes be "out of pocket" due to professional commitments.

As a side note, in the two decades since *Kramer vs Kramer*, employers, on balance, have become much more tolerant of the contending values and priorities employees face in their professional and personal lives. It used to be that some Draconian employers viewed anything short of 24-hour commitment to the firm as a dereliction of duty. Now, we see the onset of corporate day-care centers, flextime, and other humanitarian concessions to life's value conflicts.

The "Least of All Evils" Scenario: Wrong versus Wrong

The third case is sometimes referred to as "the least of all evils." If we do not have an ethical value to choose from and yet a choice must be made, about all we can do is choose the path that does the least harm. We must perform moral "triage." Suppose it's raining, the roads are slippery, you have two children in the car, and a dog runs out in front of you. Do you risk skidding into a lane of onrushing traffic, jeopardizing your life and those of your children, to save the dog? No. You swallow hard and hit the dog. Your usual value to protect life in any form is superseded by the need to protect human life, including your own. Self-preservation falls more under instinct than value, but you get the idea.

In the *Titanic* disaster, competition for insufficient lifeboat space illustrated the venal corruption of the moral triage to serve the self-interest of the more powerful. Historically, when it comes to evacuating a vessel, women and children go first, then men. However, in the case of the *Titanic*, close comparison of the passenger register with the survivor list shows that lifeboat access boiled down more to status (that is, first-class passengers versus tourist-class) than to age or gender. Indeed, the owner of the cruise line, a man, was among those saved—but the captain, literally, went down with the ship.

The ultimate example of choosing between two evils was the real and horrible choice U.S. soldiers had to make in foxholes in Vietnam when the Viet Cong wired children with bombs and sent them scurrying toward the entrenched U.S. troops. The soldiers had a split second to decide: Shoot the child and save themselves, or not shoot the child and all perish (including the child). What a choice! Actually, there *was* no choice. But the forced violation of a deeply engrained value of human life, particularly a child's life, inflicted untold psychological harm upon those who had to make that choice—as well as upon the nation that looked on in horror on the six o'clock news.

In some instances of intense value conflict, the higher value on the hierarchy (or innermost value within the concentric rings) does *not* prevail. As previously mentioned, we believe that family members would most likely put loyalty to kin above the law if the law-breaking involved did not have any immediate and visible victim (as in the embezzling case). Even though respect for law appears under critical shared values in the hierarchical model, there are times when the law itself is unethical. In that case, it is entirely appropriate for a lower, but ethical value to supersede the law. Interracial marriages are a good example. They are still "illegal" in a number of states, but social acceptance (important shared values) has grown to the point that the law has become largely a dead letter.

Sexual preference is a another good example, but one where social acceptance has not yet progressed sufficiently to overturn the law. By one perspective, sexual preference would be a Level 4 on the hierarchy (personal preference values). However, when it comes to abrogating civil rights on the basis of sexual preference (or of any else, for that matter, short of felony), we are violating individual liberties guaranteed under "life, liberty, and pursuit of happiness." Laws denying equal treatment to those with same-sex preferences are clearly unethical, just as were the Jim Crow laws in the Old South.

That's where the principle of civil disobedience kicks in: the appropriate form of resistance is peaceful defiance, until the law itself is brought under the microscope of a higher authority, such as the Constitution; found to be inimical with the fundamental founding principles of the nation; and stricken from the books.

What Price Gray?

Before we apply the principles of value analysis to technical communication, let's go back to the *gray* again for a moment and probe the perils of situational ethics.

In the arena of ethics, gray comes in two packages:

1. *Solid, or Legitimate, Gray*. This refers to situations where basic ethical values are in conflict, where difficult choices must be made, where not all the stakeholders' interests can be served, where there is no one right answer.
2. *Chameleon, or Self-Serving, Gray*. This refers to situations where the stakeholder resolves an ethical conflict in a manner that satisfies self-interest, then rationalizes the process by altering his or her values to cover the decision. This is what is known as ***situational ethics***; namely, altering one's ethics to meet the situation.

So how do we determine if the *grayness* in an ethical conflict is legitimate or self-serving? We must be honest and follow our individual conscience. Most of us have a "little voice" that tells us, "That's not right. That's not right either. *That* is." As long as we are analyzing values, identifying conflicts, assessing priorities, projecting the amount of good or harm that will come from our actions *before* we make our moral decisions, we are on the turf of legitimate gray. But the minute we make our decisions *first*, to serve our self-interests, and then conduct our value analysis after the fact to support what we have already done, then we have become ethical chameleons; we are practicing situational ethics—and gray has become just a disguise for wrongdoing.

The Diet Coke Theory of Morality

Consider the employee who has been putting in hellacious hours to come up with a more efficient assembly process. She finally succeeds. The process is implemented throughout the plant, realizing an immediate savings of thousands of dollars a month. The employee gets a commendation for her personnel folder, but management does not bother to translate her productivity suggestion into a cash reward. Legitimately bitter over this oversight, the employee begins padding her overtime—just slightly, to avoid detection. She resolves to continue this practice for several years until she has accrued $5,000 in extra pay—an amount similar to the kind of productivity awards the company has bestowed in the past, which she has read about in the company newspaper. Are her actions ethically justified?

Understandable, yes. Ethically justified, no. Dishonesty is still dishonesty, even if the person or company being cheated is itself morally bankrupt. Josephson refers to this form of situational ethics as "The Diet Coke Theory of Morality"[7]: The weak-willed dieter reasons that since he has drunk one diet Coke, he has therefore earned the right to one jelly doughnut. Applying this theory, the employee with the productivity suggestion reasons that since her work ethic has not been sufficiently rewarded, she is justified taking matters into her own hands.

Applying the "Diet Coke" theory at a corporate level, a company that donates a large sum to the city council's "Save our Parks" drive only to be denied an easement it desperately needs to allow a factory expansion might feel justified in obtaining the easement by bribing a public official. After all, fair is fair, right?

Wrong—fairness and ethics are not the same thing. Just because someone else behaves unfairly does not give us the right to behave unethically in response. What if the company retaliates against the city council by cutting back its contribution to the parks campaign? Unlike a bribe, such an action would represent no legal breach; however, its ethical foundation is extremely shaky, because the primary victims of such a cutback would be those who enjoy the parks, not the bureaucrats who blocked the easement.

The Relative Theory of Filth

Another form of justification given for wrongdoing is the time-worn excuse: "Everybody's doing it. Everybody cheats on income taxes; everybody pads insurance claims to offset the deductible; it's the American way. I'm no worse than anybody else."

Josephson dryly refers to this form of ethical rationalization as "The Relative Filth Theory of Morality."[8] The operative principle is that no matter how bad your actions are, if you look far enough you can always find somebody worse. Thus, a Third World dictator with a long pedigree of human rights violations can point to history and claim that Adolf Hitler sent far more people to their deaths, in far more gruesome fashion.

Arthur Miller aficionados will recall the last line of *All My Sons*. It has come out, though not proven in a court of law, that to save his business, protagonist Joe Keller ordered his workers to weld over cracks in airplane cylinders, which wound up sending several pilots plunging to watery graves in the war-torn Pacific. Through dramatic irony, one of his two sons was onboard one of the planes. In the play's final scene, his surviving son asks Keller: "How could you do that?" Keller replies: "Everyone was doing it. I'm no worse than anybody else." To which the son replies quietly, "I know that, Dad. I just thought you were better." Curtain.

John G. Bryan, professor at the University of Cincinnati and manager of the Ethics Committee for the Society for Technical Communication, eloquently "warns against the dangers lurking behind small, seemingly innocuous ethical compromises."

> *Small ethical concessions on the job are the first steps on a gentle but slippery slope.*[9,10]

Applying Value Analysis to Technical Communication

So how does value analysis apply to technical communication? Value analysis can be applied to any specific discipline in the same way it was applied to the general cases set forth previously. Before we can use it, however, we must first have a set of values to apply.

Chapter 3 discusses how several different professions, in particular technical communication, went about defining sets of values, or ethical codes, to govern the practice of the profession. Just as there is no one absolute set of standards for moral behavior in general, there is no uniform code of behavior within any given profession or across professions.

Thus, the American Medical Association has one set of standards; the American Bar Association has another. Some of the differences pertain to the unique elements of the respective professions, but even the relative emphasis

Cartoon designed by Steve Otte.[10]

on basic moral principles (such as honesty and compassion) may vary from code to code.

Similarly, there are no stone tablets engraved with the immutable ethical principles of technical communication. The ethical guidelines for the Society for Technical Communication (STC) do not march lockstep with those of the American Medical Writers Association or the International Association of Business Communicators, nor should they. Each organization has made its best effort to define the principles governing its membership.

Ten Basic Values in Technical Communication

We propose 10 basic values that are central to the technical communication profession. If you notice a greater similarity in our 10 values to the ethical guidelines of the STC than to those of some of its sister organizations, you are correct—that's because we are active on STC's national ethics committee and were closely involved in the development of the organization's guidelines. In no way are we implying STC's guidelines are *better* than those of any other organization, merely that to some degree they reflect the same line of thinking that has produced this book.[11]

Although we introduced you to the 10 basic values in Chapter 1, definitions are appropriate at this point.

1. **Honesty**	Our duty to tell the truth
2. **Legality**	Our duty to obey the law
3. **Privacy**	Our duty to respect the rights of others
4. **Quality**	Our duty to provide quality products and services that will best serve the end user
5. **Teamwork**	Our duty to work together to meet mutual objectives
6. **Avoiding Conflict of Interest**	Our duty to be loyal and to observe fair play
7. **Cultural Sensitivity**	Our duty to reflect the growing diversity of the workplace in our technical communications
8. **Social Responsibility**	Our duty to preserve and protect the public good
9. **Professional Growth**	Our duty to maintain and develop our skills
10. **Advancing the Profession**	Our duty to respect and assist our colleagues and enhance the reputation of our profession

The rest of the book is organized around these 10 values. In Part II, we examine each of the values in detail. In Part III, we explore scenarios that involve value conflicts within technical communication.

Inspecting our list of values brings up an obvious question: Are the values ranked in hierarchical fashion? The answer is no. We tried that and found there is too much overlap to establish a meaningful, consistently applicable order of precedence. That is not to say, however, that all the values always apply equally. It depends on the situation. For example, under most circumstances technical communicators are duty-bound not to reveal sensitive business or personal information of employers or clients (privacy). An exception would be if we were subpoenaed to disclose information in court (legality); in that case our obligations under the law would supersede our commitment to observe the privacy of our employer or client.

As another example, our duty to respect and assist our colleagues (advancing the profession) does not extend to the point of allowing an inferior product to go out to our users (quality) or to being less than candid in a product review or personnel appraisal (honesty).

And cooperation with colleagues in the workplace (teamwork) does not mean that we should enter into collusion to cover up an environmental violation by our company; in this case, our commitment to the public good (social responsibility) demands we take the high road and insist that our colleagues do the same thing. What if one of those colleagues is also our boss? Now it becomes a conflict between our job security and our duty to protect the public good.

Conflicts of this nature are not easy to resolve. Sometimes there is a compromise that allows the individual to hold firm on values without paying a serious price, such as a company environmental office or a company ethics office that will leverage our boss to do what's right without exacting a penalty upon us. Sometimes, however, we find ourselves boxed into positions where upholding our values carries a price tag. Companies sometimes find themselves in the same position. No mistake about it, that can be tough. Chapter 4 explores the short- and long-term consequences of principled versus unprincipled behavior.

Firebrand baseball coach Leo Durocher is famous for his cynical observation: "Nice guys finish last." We can see how the cynicism comes about; unethical behavior often yields short-term gains and ethical behavior sometimes carries short-term penalties. But to take the full moral measure of the situation, you have to play out the piece to its conclusion. Most of the time, unethical behavior is eventually self-defeating whereas ethical behavior is ultimately self-rewarding. But you have to look down the road. At the time you take a moral stand and suffer an immediate consequence for it, Durocher's words tend to ring true. In the final analysis, however, we think "nice guys finish first"—but there's no denying we often take some bumps and bruises along the way!

Four Guidelines for Dealing with Value Conflict on the Job

When confronted with value conflicts in the workplace, technical communicators are well advised to follow these four guidelines:

1. Communicate
2. Analyze
3. Personalize
4. Escalate

Communicate . . . Don't Go It Alone

The first step is to communicate with other trusted professionals to bring their experience to bear on your problem. If you have colleagues on the job who can serve that function, work with them. If, on the other hand, your colleagues are part of the conflict, it may be more appropriate to communicate with other professionals outside your workplace, such as members of a professional association or a former professor.

Analyze . . . Impose Order on Conflict

The next step is to analyze the conflict. Ask questions such as who are the *stakeholders*? What are their *interests*? What fundamental *values* are involved? Which values should take *precedence*? What action provides the *greatest good* to the most parties; or, alternatively, the *least evil* to the fewest parties? In other words, put into practice the same processes we described throughout this chapter.

Personalize . . . Trust Your Judgment on What's Right and Wrong

In the process of communicating and analyzing, don't let the issue become an abstraction. Personalize it. Listen to the insistent little voice which says "That's right" and "That's wrong." If, for example, you want to know whether your colleagues are justified in their rationalization that there's no harm in your factory dumping "just a few chemicals" into the river—"after all, we provide the community with hundreds of jobs"—consider whether you would accept that line of thinking if your next-door neighbor decided to empty a vat of used auto oil into the drainage ditch in your back yard. If it's not okay personally, it's not okay for the company.

Escalate . . . Cover Your You-Know-What!

Suppose you do all of the first three steps and you're still between a rock and a hard place? You've consulted with other professionals, conducted a value analysis, and come to the conclusion that your company cannot be permitted to dump chemicals in the river and that you cannot personally participate in covering up that activity. Unfortunately, your boss still disagrees with you. It's time to escalate.

Every organization has a structure; every boss has a boss. If you've come to the conclusion that you cannot morally do what your boss tells you, then you don't have a whole lot to lose going up the line. An important point here is that unless your boss is part of the problem, he/she is the *first* party to whom you should escalate the problem. When it comes to ethical conflicts, especially those with potential legal ramifications (such as the environmental lawsuits which could ensue in the previous case), it's best not to play the ethical Lone Ranger!

Once you escalate to your supervisor, you are off the immediate hook. And even if your boss *is* part of the conflict, you will still generally improve your situation by going up the organization chart as soon as you reach moral impasse. In the extreme case that you confront conflict right on up the line (let's say your entire department management is arrayed against you), you should escalate to any other available resource within the company who will listen (the ethics officer, if there is one; human resources personnel; in an extreme case, even the senior executive of the company).

And if all that fails, you can always take your problem *outside* the company. In the previous scenario, if nobody in the company will listen to your plea to protect the environment (in today's culture this would be extremely unlikely, by the way), you can in good conscience blow the whistle to the cognizant government agency. What if the company then fires you? A long line of attorneys would form outside your door to take on *that* case!

The point is that you're not alone in facing ethical issues. Indeed, there can be great danger in facing them alone. Use the system. Rely upon the resources that are available to you. Ethical issues can be dark and thorny; if you get a thorn under your fingernail while groping in the dark, it helps to find someone with a flashlight to help you dig it out!

Summary

There are many different models for defining issues, interests, and values—some elegant, many complex. The authors believe a simple, straightforward approach works best.

A value is defined as "a principle, standard, or quality considered worthwhile or desirable"—something we believe in. An interest is defined as "regard for one's own benefit or advantage." The two often come into conflict, both with individuals and with companies.

Values come in three varieties: ethical, unethical, and nonethical. Ethical values are moral (right); unethical values are immoral or amoral (wrong); nonethical values are preferences that do not intrinsically involve right or wrong.

Business ethics grow out of personal integrity and social values. Like personal morals, business ethics boil down to a question of what's right and what's wrong.

Two factors motivate good business ethics: prudence and virtue. Virtue means being ethical because it's the *right* thing to do; prudence means being ethical because it's the *smart* thing to do. The two are not mutually exclusive.

Value analysis is a systematic process to determine the conflicts of interest and value that underlie a problem, to find the best course of corrective action. The six steps are:

1. Define the issue and identify the stakeholders.
2. Determine the stakeholders' interests.
3. Identify all the relevant values that bear on the issue.
4. Determine the values and interests that are in conflict.
5. Apply a model to rank values according to importance, to weigh the values and interests that are in conflict.
6. Resolve the conflict in favor of the higher (more important) value.

Two models for ranking values are the concentric-ring model, in which the most important values are at the core of a circle, and the hierarchical model, in which values are ranked in descending importance down a *staircase.* Value analysis explores which values within the model are in conflict and, wherever possible, applies the higher or innermost value to resolve the conflict.

Types of ethical conflict include ethical values versus unethical values ("black-and-white"); conflict between ethical values ("greatest good"); and conflict among unethical values ("least of all evils").

Not all ethical issues are clear-cut. There are legitimate areas of gray. However, the concept of gray can also be used as a cop-out to mold values to fit one's self-interests, a rationalization process known as situational ethics.

To apply value analysis to technical communication, one must begin with a set of values. This book uses 10 basic values: honesty, legality, privacy, quality, teamwork, avoiding conflict of interest, cultural sensitivity, social responsibility, professional development, and advancing the profession. These established, the same process of analysis exercised for general ethical conflicts can be applied within technical communication.

Maintaining ethical behavior in a conflict situation can be costly. To minimize that cost, technical communicators faced with such a conflict are urged to communicate, analyze, personalize, and escalate.

Application Exercises

1. What is the difference between a value and an interest? Give an example of a conflict between the two.
2. Are all values ethical? Give examples to support you answer.
3. Suppose you are looking for a marriage partner. A potential candidate attracts your fancy, but differs from you in five value areas. Rank them in

order of importance. What value areas would you consider critical to your decision (i.e., where would a different value cause you not to marry the person?):

Political affiliation

Favorite flavor of ice cream

Religious affiliation

Attitude toward different races

Preferred recreational activities

Explain how this exercise relates to the concentric-ring value model.

4. Draw upon your own values and shape them into a concentric-ring model. What are your core values? Your authority values? Your peripheral values?
5. For each of the four levels in the hierarchical value model proposed by Dr. Payne, identify two of your personal values which fit within that level.
6. Explain the following and give a brief example of each:

   "Black versus white" conflicts

   "Greatest good" conflicts

   "Least of all evils" conflicts

7. Define situational ethics and give one example.
8. Define Michael Josephson's "Diet Coke Theory of Morality" and give a brief example.
9. Define Michael Josephson's "Relative Filth Theory of Morality" and give a brief example.
10. A technical communicator attending a networking luncheon on ethics at a national conference recounted this horror story: A non-English-speaking family purchased a high-powered motorboat. They had no clue on how to operate it. On their first outing as they were clumsily maneuvering the vessel, still within the marina, disaster befell them. The mother, who was very short, had corkscrewed her swivel deck chair into the highest possible position, which is unstable during turns. The husband, not knowing how to steer the vessel, made a sharp turn, pitching the mother into the water. She was sucked into the propellers and mangled. She survived the accident but suffered a severe disability.

    You can imagine the litigations that ensued. The family sued the marina, the boat manufacturer, and—here's where we, as technical communicators come in—the company which had been contracted by the manufacturer to prepare the users' manual. The manual included clear verbal instructions spelling out the dangers of rotating the deck chairs into the highest position, but there was no illustration accompanying the warning. The plaintiffs contended that using verbal instructions alone constituted inadequate warning. The defense argued that in a primarily English-speaking nation it was unreasonable to expect the manual to include a picture for every possible danger.

What values in the area of technical communication apply to this situation? Consider the 10 values we set forth in this book. Factor in your own values. Do you feel the preparers of the users' manual for the boat can legitimately be held liable?

11. The ultimate horror story came from a technical publications manager who told the story—which occurred about 25 years ago—of a powerful gas-powered compactor whose cast iron extrusion was replaced with aluminum to save weight and cost. The aluminum was strong enough to permit safe operation, but it created a safety hazard during disassembly for maintenance. With the cast iron extrusion, the eight bolts holding the compactor together could be removed safely in any order. With the aluminum extrusion, each of the eight bolts had to be removed a quarter of the way, then each one another quarter of the way, and so on—to avoid sudden and explosive disassembly. As it turns out, if the user removed three bolts all the way, without loosening the others, removing the fourth would trigger catastrophic failure, causing the apparatus to blow up in the user's face.

    The technical publications manager insisted on providing a prominent warning in the maintenance documentation; the marketing manager said the warning had to go. The company president sided with the marketing manager. The technical publications manager was essentially given a choice between removing the warning or removing himself from the premises. The product was released without the warning. A few months later a **compactor did blow up in a user's face** during routine maintenance. Needless to say, litigation ensued. The technical publications manager had kept a rigorous audit trail of his memos, which limited his personal liability, but the company was correctly held to task and is no longer in business.

    What values in the area of technical communication apply to this situation? Do you feel the publications manager can legitimately be held liable? Should (and could) he have done more than go on record that the warning should be included?

Exploration Exercises

1. Develop an ethical scenario that involves conflict across levels in the hierarchical value model. Identify which values are in conflict and suggest

possible courses of action as well as potential consequences of those courses of action.

2. Develop an ethical scenario that involves conflict between an ethical value and an unethical value ("black-and-white" conflict). Identify the values that are in conflict and suggest possible courses of action as well as potential consequences of those courses of action.
3. Develop an ethical scenario that involves conflict among two or more values, all ethical ("greatest good" conflict). Identify which values are in conflict and suggest possible courses of action as well as potential consequences of those courses of action.
4. Develop an ethical scenario that involves conflict among two or more values, all unethical ("least of all evils" conflict). Identify which values are in conflict and suggest possible courses of action as well as potential consequences of those courses of action.
5. Describe a value model that is part of your personal value system. Consider your organizational affiliations (religious, political, professional) and the beliefs those organizations espouse.
6. Set up a situation of potential conflict within the value model you set forth in #5 and propose potential solutions.
7. Describe a situation in your life where you feel you would be tempted to apply (or have applied) situational ethics and explain why.
8. Consider Leo Durocher's cynical slogan: "Nice guys finish last." Give an example of something that happened in your life which might support such a philosophy. Give an example of something else that would refute such a philosophy. Do you fundamentally agree or disagree with Durocher? Why?

References

1. Michael Josephson, videotape *On Ethics*, Josephson Institute of Ethics, Washington, D.C., c. 1992. Presented within inservice training program on ethics at Martin Marietta Corporation, Orlando, FL (now Lockheed Martin Corporation).
2. Daniel W. Voss, "Sharpening the Focus: A Workshop on Ethics and Technical Communication," Martin Marietta Corporation, Orlando, FL, 1993 (now Lockheed Martin Corporation). Presented at 41st Annual Conference of the Society for Technical Communication, Minneapolis, MN, May 1994. Workshop description appears in *Proceedings*, p. 380.
3. Josephson, *On Ethics*.
4. Dr. Lynn Payne, "Hierarchical Model for Value Analysis," University of Virginia, c. 1991. Presented within inservice training program on ethics

at Martin Marietta Corporation, Orlando, FL (now Lockheed Martin Corporation).

5. *CBS Evening News*, April 3, 1996; *KCNC Morning News*, Denver, CO, April 4, 1996.
6. Michael J. Sniffen, "Brother Did Agonize Before Going to FBI," Associated Press wire story published in the *Denver Post*, 5 April 1996, p. 5A.
7. Josephson, *On Ethics*.
8. *Ibid.*
9. John G. Bryan, University of Cincinnati, "Responding to Ethical Dilemmas when the Mortgage is Still Due," lecture notes from workshop, 15th Annual Practical Conference on Communication (PCOC 15), Society of Technical Communication, East Tennessee Chapter, Oak Ridge, TN, October 1991.
10. Daniel W. Voss, "A Slippery Slope: Using Value Analysis to Resolve Ethical Conflicts in Technical Communication," Martin Marietta Corporation, Orlando, FL, 1992 (now Lockheed Martin Corporation). Presentation at 40th Annual Conference of the Society for Technical Communication, Dallas, TX, June 6-9, 1993. Paper by same title appears in *Proceedings*, pp. 288-291.
11. Shirley Andersen-Hancock, "STC Ethical Guidelines," *Intercom* (July/August, 1995), Arlington, VA: Society for Technical Communication, pp. 6-7.

Other Sources Consulted

Bigelow, Barbara, Liam Fahey, and John F. Mahon. "Political Strategies and Issues Evolution: A Framework for Analysis and Action." Chapter 1 of *Contemporary Issues in Business Ethics and Politics*, ed. Karen Paul. *Mellen Studies in Business* 7, Lewiston, PA: Edwin Mellen Press, pp. 1-23.

Brown, Marvin T. "Discovering the Right Decision." Chapter 3 in *Working Ethics: Strategies for Decision Making and Organization Responsibility.* San Francisco: Jossey-Bass, 1990. pp. 30-53.

Guy, Mary E., "Values, Ethics, and Personal Responsibility." Chapter 1 in *Ethical Decision Making in Everyday Work Situations*. New York: Quorum Books, 1990.

Henderson, Verne E., "The Ethical Algorithm." Section within Chapter 2 of *What's Ethical in Business*. New York: McGraw-Hill, 1992.

Hodgson, Kent, *A Rock and a Hard Place: How to Make Ethical Business Decisions when the Choices Are Tough*. New York: American Management Association, 1992.

Chapter 3 Overview

Codes of Ethics

The Mission: Codes of ethics express the beliefs and values of organizations, based on the ways of thinking and patterns of behavior of the members of those organizations.

Chapter Roadmap

Definition and Focus. This section examines the ways in which we both influence and are influenced by more than just our personal values in making decisions about thorny ethical issues.

Codes of Ethics Have Varying Purposes. This section defines the differences between aspirational codes—those intended to guide—and regulatory codes—those intended to police.

What the Codes Say. This section familiarizes you with the contents of professional codes of ethics, exploring such aspects as:

- *Common Threads among Codes.* This section identifies similarities among a number of codes, such as:

  A Commonality of Values among Codes

  A Commonality of Concepts among Codes

- *Codes Can't Be Flawless.* This section acquaints you with several weaknesses and limitations of codes of ethics

Summary, Practical Exercises, and References

CHAPTER 3

Codes of Ethics

There are intimate conceptual connections between the possible ways in which individuals may live their lives and the institutions of society.

—Peter Winch,
Ethics and Action

I know only that what is moral is what you feel good after and what is immoral is what you feel bad after.

—Ernest Hemingway,
Death in the Afternoon

Definition and Focus

So far, we've looked at ethics and values from the perspective of the individual. But each individual is a part of many larger wholes. Those larger communities may be families, religious groups, societies, or any of a variety of organizations. As working communicators, each of us is a member of several occupational communities, some of which we join by necessity and some by choice. Unless we inherit a fortune or win the lottery, we generally must earn a wage, for which we function as employees of a company, whether in-house or on contract. We are also members of a profession and as such may elect to join one or more organizations that formally represent the discipline. If you document new surgical procedures for a health sciences center, for instance, you belong to three occupational communities: employees of the center, professional communicators, and medical writers. Your professional affiliations may include the Society for Technical Communication (STC) and the American Medical Writers Association. If you author environmental impact statements for a consulting firm, you are at the same time an employee of the firm, a technical communicator, and an environmental writer. In this case, your professional affiliations may include STC and the National Association of Professional Environmental Communicators.

Within these multiple professional affiliations we both influence and are influenced by more than our personal values in making decisions about thorny ethical issues. As stated in Chapter 2, business ethics are only as good as the values employees bring to the company. Ethicist Eric Mount notes that "different professionals bring the traditions, codes, slants, and concerns of their professional contexts to the organizational contexts of corporations."[1] Gordon MacKenzie, creative Director of Hallmark Cards, cleverly demonstrated Mount's point in his 1994 keynote address at the 41st Annual STC Conference in Minneapolis. He demonstrated the different perspectives and priorities of the creative and financial arms of the company.

The creative types, led by MacKenzie, were forever running afoul of the financial types, the "bean counters." For example, the bean counters wanted to account for every labor hour, charging it to a specific account, whereas the creative types insisted on a period of incubation to work their magic—time that could not be compartmentalized to fit the accounting structure of the company.

MacKenzie refers to the corporate "hairball" of entangled bureaucratic policies and procedures, likening it to a dangerous but necessary mother planet for creative types who remain in orbit about it. Whenever the gravita-

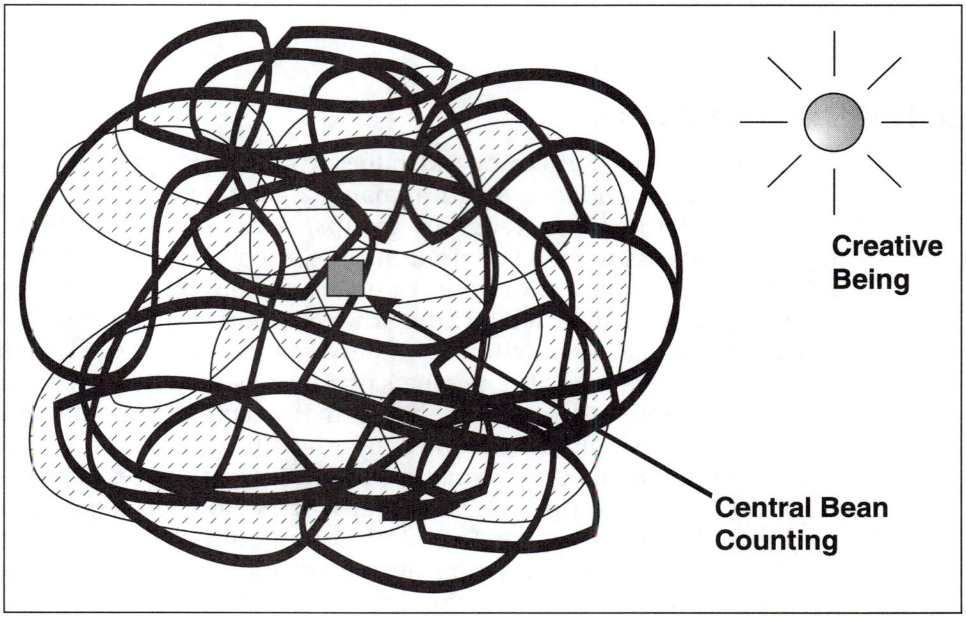

tional pull of the hairball threatens to pull the creative employee into its clutches, she must exert sufficient personal creative energy to achieve "corporate escape velocity"—yet must not exert so much energy that she leaves the gravitational field altogether and drifts off into nothingness. In other words, creative spirits and bean counters must coexist. To be productive in a corporate environment, creative energy must be harnessed and channeled without being regimented and stifled. It's a thin line. MacKenzie made his career walking that line.[2]

Although MacKenzie did not address ethics per se, his anecdotes exemplify the contrasting perspectives and priorities—Mount's codes, slants, and concerns—that different professions bring to a company.

Certainly, among the *codes* that professionals bring to corporations are values and ethics. This synthesis of values from varied professions imparts to corporations values that are "more than the sum of those individuals who make them up."[3] This influence works in reverse, too. Mount also states that we practice our professions—and ethical decision making is certainly part of practicing our professions—"within the expectations and standard of our peers."[4] We would add that we also operate within the expectations of the companies that pay our salaries.

Given this perspective, then, we need to examine the codes and rules of ethics of the organizations in which we practice our occupations. Such an examination provides the philosophical framework for the core values discussed later. In particular, we want to explore the values and ethical statements of the professional associations that serve technical communicators. Certainly the ethical policies of our employers exert a significant influence on our lives; however, they differ from those of professional organizations in two important ways.

First, we "worker bees" rarely have an opportunity to participate in the formulation of the ethical policies of corporations. Instead, they are generally handed down from the executive offices, so they reflect managerial philosophy regarding ethics. When professional organizations devise an ethical code or guidelines, on the other hand, they most frequently reach a consensus of membership opinion, or at least of the opinions of representative members. For that reason, the codes of professional societies generally reflect the personal values of their membership more directly than corporate codes reflect the values of their workforce.

Second, membership in professional organizations is totally voluntary, so we assume that none of us would pay annual dues to belong to an outfit whose ethics conflicted with our own. Inheritance or lottery windfalls aside, most of us are not as free to choose our employers, given our geographical constraints, salary requirements, and a frequently tight job market.

Before we look at professional codes of ethics and the values they represent in more detail, let's be sure we agree on what we mean by the term *code of ethics*. We believe that *a code of ethics expresses the beliefs and values of an organization, based on the ways of thinking and patterns of behavior of the members of that organization.*

Having defined ethical codes, we need to point out that there is more than a single viewpoint on their intent and nature. Before examining the codes and values of specific organizations, then, let's look briefly at two approaches that occupy the extreme ends of the continuum.

Codes of Ethics Have Varying Purposes

The primary distinction between approaches to ethical codes is what people believe to be the intent. One camp sees a professional code as a set of goals that members should strive to attain. From this perspective, codes provide guidelines that remind professionals to look beyond what is prudent to what is virtuous. The standards set forth in the codes articulate the shared values of the members of the organization, who are believed to be basically ethical people. Mark Frankel calls this stance, which offers a "moral anchor" to deter unethical behavior, *aspirational*. "An aspirational code is a statement of ideals to which practitioners should strive. Instead of focusing on right and wrong, the emphasis is on the fullest realization of human achievements."[5]

The opposing camp views a code of ethics as a list of specific obligations that govern the behavior of the profession's members, who are perceived as generally inclined to pursue their own self-interest and lollygag along the path of reluctant prudence rather than toe the straight and narrow line of virtue. From this perspective, codes express strict requirements for which members are accountable. Frankel calls this stance *regulatory*.[6] R.T. DeGeorge says that to be effective, regulatory codes must be both policeable and policed.[7]

Of course, other views occupy the continuum between these two extremes. But the extremes serve to demonstrate our point.

What the Codes Say

As we have seen, and will continue to see in the scenarios presented throughout this book, the unfortunate truth is that individual and corporate values *can* conflict. Suppose a conscientious and ethical engineer finds himself in conflict with an employer who uses inferior construction methods and materials. If, as we suggest in Chapter 2, the engineer attempts to resolve the problem through appropriate corporate channels without success, he should seek outside assistance. He could escalate his concerns to the local building inspector. Sometimes, though, people may feel uncomfortable taking what they perceive to be such a drastic step; in other cases, there may be no outside authoritative entity to which they can appeal.

DeGeorge cites an example of an engineer for a tire company who blew the whistle on an unsafe design practice that had been swept under the rug by management to cut costs. The engineer was fired.[8] In instances like this, the ethical codes of professional organizations can counterbalance the concerns of individuals and companies. This is what Frankel means when he says codes can be used both to help adjudicate disputes and to defend professionals against unwarranted attack.[9] Let's explore the attributes of professional codes of ethics and the common values they reflect from one organization to the next.

Common Threads among Codes

Because codes of ethics reflect the unique traditions, perspectives, and needs of a profession, no single code can adequately serve all professions. Nor could a uniform code even begin to scratch the surface of all the ethical problems that are unique to each profession.

The wide range of purposes for professional codes and the equally wide variation in their approaches are evident in Gorlin's *Codes of Professional Responsibility*, a compilation of codes of ethics of major professions in the United States. Even a cursory review of the two dozen professional codes contained in Gorlin's compilation drives home the extraordinary range of ethical concerns that must be addressed. Four examples are representative:[10]

- *Medicine*. The code of the American Medical Association, which is built on the cornerstone of the Hippocratic Oath, addresses a wide range of issues that are currently volatile, from abortion to genetic manipulation to capital punishment.
- *Social Work*. Codes listed from the National Association of Social Workers and the National Federation of Societies for Clinical Social Work address relationships with clients, colleagues, professional associations, and society.
- *Engineering*. The code of the National Society of Professional Engineers stresses safety and health, competence, objectivity and truth, loyalty to client, and economic fairness on the job market.
- *Accounting*. The code of the American Institute of Certified Public Accountants emphasizes competence as well as standards.

To illustrate the broad spectrum of ethical conflicts that can arise within the various professions, let's briefly consider one potential conflict within several of the representative professions previously discussed.

In *medicine*, how does the commitment to relieve human suffering square with the oath to preserve life when it comes to the controversial question of euthanasia? Does *active* euthanasia of the type practiced by Dr. Kevorkian violate or fall within the ethics of the profession? How about *passive* euthanasia? Is a physician ethically bound to provide the last full measure of technological life support if it extends the extreme suffering of a terminal patient whose only wish is to die with dignity? These are tough questions—questions that drive toward individual core values. No uniform code can possibly encompass all moral viewpoints within the profession, let alone the public.

Let's move to *social work*. Suppose a clinical social worker is assigned to an elderly couple whose economic pressures are being exacerbated by a manifestly unfair wrinkle in the Medicaid system. The social worker knows how to "skirt" the problem by shuffling some papers; in essence the letter of the law will be breached, but the spirit of the law will be sustained. Situational ethics at their finest! (See Chapter 2.) Which takes precedence—duty to follow the law (which bears in large measure on keeping one's professional license) or commitment to the welfare of the elderly clients? Again, this is a tough conflict.

Suffice it to say, it is well-nigh impossible to sever professional performance from ethical standards. At any moment, the two are likely to overlap and collide. That's where the tough choices lie. That's where we find gray. And that's why we need codes and guidelines for ethical behavior.

Obviously, a single code will not serve diverse professions. A single code may not even be adequate within a profession, particularly when that profession crosses as many disciplines as does technical communication. Indeed, the range of ethical conflicts encountered within technical communication mirrors the range illustrated in the representative professions we've already looked at, precisely because technical communication *does* cross so many fields—from *a*ccounting software to *z*oology.

The revised ethical guidelines adopted by the Society for Technical Communication are intended to be "sufficiently broad to apply to different working environments." The many other professional organizations that serve specific disciplines within technical communication also have codes of ethics tailored to the unique needs and problems of highly specialized fields. However, they share many common threads with each other and with the more general STC code.

A Commonality of Values among Codes

Eight values turned up in one form or another in the majority of the professional codes we reviewed:

- Honesty
- Integrity
- Fairness
- Objectivity
- Competence

- ❑ Respect for privacy
- ❑ Loyalty
- ❑ Freedom from discrimination

These values closely parallel the 10-value structure we adopted for this text. In addition, we felt that both legality and professional development merited separate treatment—the former because technical communicators often brush up against legal concerns; the latter because we consider professional growth indispensable to survival in a field that is inextricably wedded to on-rushing technology.

A Commonality of Concepts among Codes

In addition to common values, the various professional codes also tend to *cluster* conceptually. Six key concepts stood out:

1. It is important to avoid not only impropriety but also the *appearance* of impropriety (refer to the discussion of sexual harassment in Chapter 10).
2. It is critical to avoid, or at least fully disclose, conflicts of interest (see Chapter 9).
3. A central value in any field is to meet the quality standards of the profession (see Chapter 7).
4. It behooves professionals to be upright and fair with others in the profession (refer to Chapter 13).
5. Moral conscience demands a profession address social issues surrounding its practice (see Chapter 11).
6. To be meaningful, a code of ethics must be enforced. (This is not presently a strong suit for the technical communication field; refer to the discussion of certification in Chapter 13.)

Codes Can't Be Flawless

Of course, no code can be a magic bullet—flaws exist. For instance, codes rarely, if ever, prioritize the values they present. It is up to the individual to do so. As noted in Chapter 2, not all values apply equally in a given situation. Thus, if a code delineates several core values, it is possible for those values to come into conflict. For example, most professional codes call for both loyalty to one's employer and commitment to produce the highest possible quality for the end user. What happens if the employer directs the professional to deliberately cut corners on quality to wring out the last nickel of profit? It's all well and good to say the professional must put quality first, but that doesn't help when the mortgage payment comes due!

Another weakness of codes is that they cannot cover all of the emerging ethical dilemmas or potential *shades of gray*. As industry dashes ahead with ever-increasing speed, new issues and predicaments crop up. Almost weekly we face new temptations, new possibilities for imprudent and unvirtuous behavior. By its very bureaucratic nature, a professional organization cannot hope to revise its code frequently enough to keep up with the dizzying pace. Moreover, in a profession as diverse as technical communication, with the myriad of industries we serve, it would be impossible for a code to address every potential moral quandary.

Finally, no code—from the most punitively regulatory to the most altruistically aspirational—can force its members to internalize all of its values. Those who disagree with any of the stated beliefs will continue to follow their own convictions in spite of a published code. The architect who believes that foreigners should not be allowed to take jobs away from Americans is not likely to suddenly display cultural sensitivity toward Latins on the construction site simply because his professional society incorporates that value in its code of ethics. The marketeer who believes that all is fair in love, war, and the pursuit of commissions probably won't remove slightly dishonest claims from her sales pitch because her professional association codifies its belief in telling the truth. Nonetheless, even with their flaws, codes of ethics provide practitioners with guidelines that can ease the difficulty of ethical decision making.

Summary

As technical communicators, we are likely to join one or more professional organizations that represent our discipline. Many of these organizations have expressed their ethical standards in the form of codes or guidelines.

Two basic types of codes are aspirational and regulatory. Aspirational codes present ideals for which practitioners should strive, acting out of virtue. Regulatory codes prescribe and proscribe behavior, and practitioners are expected to follow the rules out of prudence if not virtue.

Two significant aspects of professional codes are their attributes and the common values they represent. Aspirational codes tend to be general and philosophical; regulatory codes are more specific and directive.

No single code can serve all professions. In fact, no single code can cover all situations within *one* profession—particularly not in technical communication, which spans many disciplines. Certain common threads run through a majority of professional codes. Among them are the values of honesty, integrity, fairness, objectivity, competence, privacy, loyalty, and tolerance.

No organization's code can force all members to espouse and act upon its values. Furthermore, there are times when one value conflicts with another.

Application Exercises

1. Explain the difference between aspirational and regulatory codes of ethics. Which is a better inducement for governing the ethics of a professional organization? Why?
2. We provided representative ethical conflicts for the fields of medicine and social work. Reread the guidelines for engineering and accounting, then provide similar representative ethical conflicts for those fields.
3. Should a professional code of ethics provide general guidelines, detailed direction, or some combination thereof? Why?
4. List three professions other than technical communication. In each case, suggest some ethical concerns the professions might have that are unique and others that are common to those faced by technical communicators.

Exploration Exercises

1. Apply Gordon MacKenzie's "corporate hairball" metaphor to an organization (professional or other) with which you have been involved. Focus on the need to strike a balance between creative energy and administrative structure.
2. Read Gorlin's *Codes of Professional Responsibility* and prepare a brief review (2-3 page written report or 3-minute oral presentation). Focus on commonality, and differences in conduct and approach.
3. Using the resources in the library's reference section and searching the World Wide Web, find the codes of ethics of the following organizations that serve technical communicators: Society for Technical Communication, American Medical Writers Association, Association of Professional Writing Consultants, and International Association of Business Communicators. Prepare a brief review (2-3 page written report or 4-minute oral presentation). Focus on commonality and differences in the codes.

References

1. Eric Mount, *Professional Ethics in Context: Institutions, Images, and Empathy* (Louisville, KY: Westminster/John Knox Press, 1990), p. 43.
2. Gordon MacKenzie, keynote address, 41st Annual Conference, Society for Technical Communication, Minneapolis, MN, 1994.

3. Mount, *Professional Ethics*, p. 31.

4. *Ibid.*, p. 39.

5. Mark Frankel, "Professional Codes: Why, How, and with What Impact?" *Journal of Business Ethics* 8 (1989): 110.

6. *Ibid.*

7. R.T. DeGeorge, "Professions, Business, and Ethical Codes of Behavior." *Business Ethics*, 3rd ed (New York: MacMillan, 1990), 381-398.

8. *Ibid.*

9. Frankel, *Professional Codes*, p. 112.

10. R. Gorlin, *Codes of Professional Responsibility*. (Washington, D.C.: Bureau of National Affairs, 1986.)

Part II

Ten Values in Technical Communication

- Honesty
- Legality
- Privacy
- Quality
- Teamwork
- Avoiding Conflict of Interest
- Cultural Sensitivity
- Social Responsibility
- Professional Development
- Advancing the Profession

Chapter 4 Overview

Honesty . . .

The Mission: **Honesty in technical communication means making our best effort to provide honest, clear, and accurate communication. It means neither falsifying, omitting, nor slanting information with the intent to deceive our audience. It means being candid in our intellectual and professional assessments, whether they be personnel evaluations, cost estimates, or research conclusions.**

Chapter Roadmap

Definition and Focus. This section provides a philosophical framework for honesty as a core value. The chapter is based on the belief that unusual capability carries higher responsibility, and it is designed to help you understand and fulfill this important ethical charter.

- ❑ *Our Unique Responsibility*
- ❑ *What Price Honesty?*
- ❑ *What Price Dishonesty?*

Communicating Honestly Means Avoiding Deception. This section arms you with insight into nine forms of deception in communication—it's your responsibility to avoid them!

- ❑ *Lying by Commission*
- ❑ *Lying by Omission*
- ❑ *Lying with Language*
- ❑ *Lying with Statistics*
- ❑ *Lying with Graphics*
- ❑ *Lying with Photographs*
- ❑ *Lying with Multimedia*

- *Lying with Logical Fallacies*
- *Lying with Propaganda Techniques*

How to Be Honest in Technical Communication. This section helps you understand what, specifically, it means to be honest in technical communication.

- *Honesty in Research*
  - Never Falsify Information
  - If Evidence Conflicts, Report It as Is
  - Openly Acknowledge Flaws in Research
  - Always Provide Proper Attribution
- *Honesty in Professional Assessments*
  - Assess Communication Requirements Accurately
  - Prepare Cost Estimates Realistically
  - Evaluate Communication Products Objectively
  - Conduct Personnel Appraisals Candidly
- *Honesty in Persuasive Communication*
- *Resolving Conflicts*
  - Dealing with the "Publish or Perish" Phenomenon
  - Toeing the "Bottom Line" without Toppling Over!

Summary, Practical Exercises, and References

. . . Our Duty to Tell the Truth

CHAPTER 4

Honesty May Be the Best Policy, But It Often Comes at a Price

Thou shalt not bear false witness against thy neighbour. —Exodus 20:16

Do you swear to tell the truth, the whole truth,
and nothing but the truth, so help you God? —Any courtroom in the United States

This above all: To thine own self be true,
And it must follow as the night the day,
Thou canst not then be false to any man. —*Hamlet,* Act I, Sc. III

Definition and Focus

The question of honesty and truth has been around since the dawn of time—the subject of philosophers, priests, politicians, poets. We could begin by defining truth, but that's more appropriate for Philosophy 101 than for ethics and technical communication. Besides, it's too difficult. Instead of waxing eloquent on the relative merits of Platonic idealism versus ontological realism, then, let's get right down to basics—values that even children grasp intuitively from their first communication:

- Honesty is telling the truth. It's right.
- Dishonesty is lying. It's wrong.

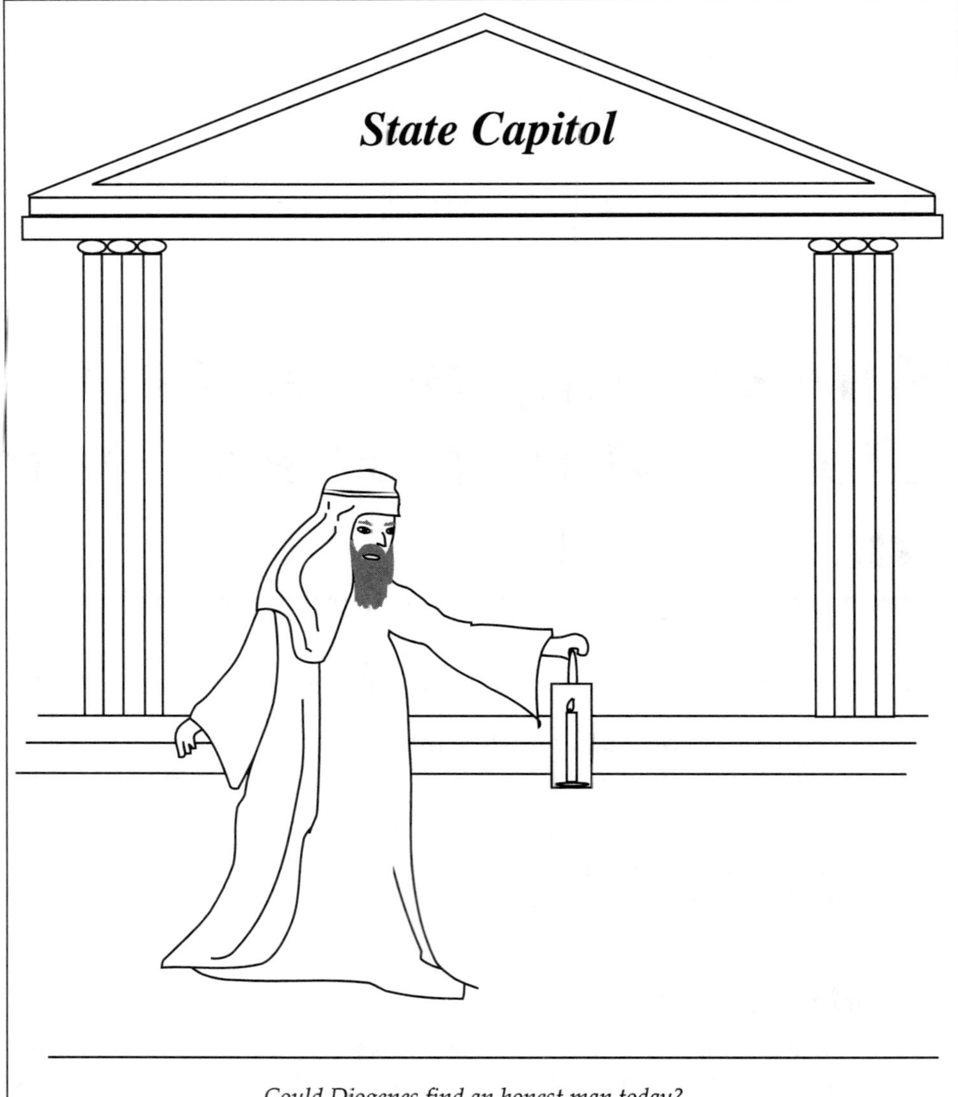

Could Diogenes find an honest man today?

We wish it were that simple. We realize instinctively that it isn't. Once again, *shades of gray* enter the picture. To suggest that honesty and dishonesty are merely opposite ends of a continuum with no meaningful reference points in between is to deny one of the fundamental human values. The area between the extremes is fraught with gray. Most value systems place honesty at or near the top or the core (see Chapter 2). Honesty is basic to human relationships. Tell the truth a hundred times and you slowly build trust. Lie once, and you tear it down—maybe forever.

But we're drifting back to Shakespeare. What we need here is not an appreciation of the importance of honesty to human relations—although that might not be such a bad fringe benefit to take from this book!—it's a good, solid operational definition of honesty as it pertains to technical communication.

> *Honesty in technical communication means making our best effort to provide honest, clear, and accurate communication. It means neither falsifying, omitting, nor slanting information with the intent to deceive our audience. It means being candid in our intellectual and professional assessments, whether they be personnel evaluations, cost estimates, or research conclusions.*

Our Unique Responsibility

In a number of places in this chapter, we will allude to the linguistic expertise of technical communicators and suggest that we therefore bear a commensurately greater responsibility to use our skills honestly. Why should our ethical burden be greater than the next person's? In general, it shouldn't. But when it comes to language, it should.

The situation is analogous to that of a person with a 10th-degree black belt in karate. If you or I are forced to defend ourselves against an assailant, we are morally free to use our hands to fend off the attack in any way, even to counter-attack to protect ourselves. But if a martial arts expert is confronted with a situation of that nature, he or she is morally obligated to control the situation without resorting to lethal force—if possible, without even inflicting serious injury. *Unusual capability carries with it higher responsibility.*

So, communication experts, remember—your cursors are deadly weapons. This chapter will help you use them honestly and responsibly.

What Price Honesty?

Michael Josephson (see Chapter 2) has conducted numerous industry surveys involving values and ethics in the workplace.[1] The surveys invariably include the question, "Are you honest?" Overwhelmingly, the responses boil down to, "Well, yes, I'm basically honest." The operative term here is *basically*. Josephson interprets this to mean, "I'm basically honest unless it costs too much." This, he contends, is not "basically honest" at all, but "basically dishonest"; that is, left to our own devices, we tend to manipulate the truth as necessary to serve our own agenda unless it's convenient to be honest without jeopardizing our self-interest. The clear implication is that when it comes to honesty, nearly everyone has his or her price.

A cleaning lady in New York, unobserved, found $5,000 in unmarked bills in a public bathroom. She turned in the money to authorities. We've all heard similar stories on the news. Sure, the woman got great reviews in the press, but she still doesn't have a penny to her name. There wasn't even a reward. Think about it. Was her action noble or foolish? Couldn't a person who is capable of leaving $5,000 in a bathroom afford to lose it more than this poor, honest soul could afford to give it up? For the loser, the money was probably pocket change. For the finder, it may have represented basic necessities.

Put yourself in the cleaning lady's place. What would you have done? (You don't have to answer that, by the way. We're merely asking you to focus on values and consequences in a down-to-earth example.)

For the cleaning lady, the price tag for honesty was $5,000. What might it cost you as a technical communicator? What might it cost your employer? Suppose you're up for a promotion on the job and your chief rival for the position is falsely blamed for a serious error that will eliminate her from consideration. Suppose, further, that the mistake is actually yours, but there is no way to pin it on you unless you come forward. In this case, the price tag for honesty is a lost opportunity, maybe even a lost job.

Suppose your employer is a major defense contractor involved in a multimillion-dollar procurement. Just before the proposal is due to be submitted, your CEO learns that the marketing department has inadvertently stumbled upon the opposition's confidential pricing information—information that can spell the difference between victory and failure, between a thousand jobs and a thousand layoffs. The two courses of action for the CEO are fairly clear:

1. Involve as few employees as possible, swear them to secrecy, use the information to win the proposal, and hope that it never comes out.
2. Voluntarily turn the information over to the government and trust for it to deal equitably with the situation.

In the case of Martin Marietta (now Lockheed Martin), a situation just like this actually happened. President and CEO Norm Augustine explains:

> *Martin Marietta, the firm I serve, is very concerned with ethics. Some time ago, we were in competition for a major government contract. The day before we were to submit our proposal, we received in the mail a copy of our competitor's price sheet, presumably sent by a disgruntled employee of our competitor. Once realizing what we had, [we promptly handed] the package to our attorneys, who informed the government and the competitor what happened. We did not change our bid price. Incidentally, we lost the contract, and some of our employees lost their jobs due to lack of work. By any*

measure we paid a heavy price for acting ethically. But I am convinced that it does "pay off" in the long run to be ethical. Martin Marietta has been asked to participate as a partner in many projects, not only because we have the engineering talent and the technological expertise but also because we are known as an ethical company.[2]

What Price Dishonesty?

When you think about it, the previous scenarios are rather depressing, aren't they? But we've looked at only half the picture. Let's revisit the scenarios from the opposite perspective. In each case honesty was indeed very costly—at least in the near term. But what about in the long term?

Consider the defense procurement, for instance. True, the company's ethical behavior resulted in a significant business setback. And true, also, that the company may have gotten away with using the sensitive financial data to win the proposal. But consider the corporate consequences if such an action were subsequently to surface under a routine government audit or through a disgruntled, whistleblowing employee. In this instance, severe civil penalties for the company—maybe even criminal penalties for its executives—would be a foregone conclusion. And more damning still would be the irreparable damage to corporate credibility. What customer would trust the company on a future procurement? Ultimately, dishonesty could cost the company much more than a single contract. It could close the doors. As Norm Augustine pointed out, *the right thing to do was also the smart thing to do.*

How about the situation in which you allow a colleague to take the rap for a mistake in order to capture a coveted promotion? True, coming forward would, in all likelihood, cost you the promotion. But you'd probably keep your job. Your credibility, at least, would be preserved; and, assuming even a modicum of ethical awareness on the part of management, perhaps your courage would be acknowledged as well. In the long run, you might be well served by such forthright behavior. And if you *were* fired, then you'd certainly have to ask yourself the question: Do I really want to work for a company that fires people for owning up to an honest mistake?

Now let's look at the flip side. Suppose you were to let your rival take the rap, enabling you to nail down the position you covet. What happens if your treachery subsequently comes to light? You can bet that not only the promotion and your job would be lost, but your professional reputation as well. You might well become unemployable. Once again, *the long-term price tag for dishonesty is far greater than the short-term price tag for honesty.* In other words, *good ethics is good business.*

For this chapter, that brings us full circle to the time-worn homily: "Honesty is the best policy." Unless, some would argue, you're a cleaning lady in New York. But we never said we were going to give you all the answers.

Communicating Honestly Means avoiding Deception

Communicating honestly is more than just telling the truth. It's more than not telling a lie. It demands a genuine commitment to convey the message to the audience in the most forthright, comprehensible manner possible. The key words are *message* and *audience*. The message part of honesty often—but not always—pivots on the use of words or numbers. In today's multimedia communication environment, truthfulness in the message may hinge instead on the use of graphics, photos, sounds, or video imagery.

The audience part of honesty refers to the technical communicator's ethical obligation to consider the audience while framing the message, so as to maximize clarity and minimize the possibility of misunderstanding. Thus, the deliberate use of a sophisticated vocabulary with a less verbally adept audience may constitute deception. As a minimum, it constitutes arrogance and disrespect, which are also objectionable, if not unethical. What if the use of terms unfamiliar to the audience is inadvertent? Well, that's not an honesty issue, but it certainly raises questions of competence on the part of the communicator. That, in turn, falls under different values—quality and social responsibility (see Chapters 7 and 11).

A useful approach to ensuring honest communication is to recognize and avoid dishonest communication. In other words, learn to tell the truth by learning the many ways of lying so as to *not* use them, even inadvertently. The flip side of this strategy, of course, is that an increased understanding of how to communicate deceptively also equips the learner to do precisely that. We feel safe, however, because obviously none of our readers would ever do such a thing. And if they do, we will, of course, disavow any knowledge of their actions. With that disclaimer, then, let's take a look at nine ways to lie in technical communication:

- Lying by commission
- Lying by omission
- Lying with language
- Lying with statistics
- Lying with graphics
- Lying with photographs
- Lying with multimedia
- Lying with logical fallacies
- Lying with propaganda techniques

Lying by Commission

Lying by commission refers to deliberate, outright falsification or prevarication. This form of deception has been around for a long time, as the following examples aptly show:

> *"Ye lie, ye lie, ye liar loud;*
> *So loud I hear ye lie!"*
>
> *"The Battle of Otterbourne"*
> *(Thirteenth century Medieval ballad)*

> *To advance a mission of the* Starship Enterprise *in the movie* Star Trek: The Undiscovered Country, *Mr. Spock tells Star Fleet Command the ship's warp drive is broken when it isn't:*
> *"A lie?" an ensign asks.*
> *"An error," replies Spock.*

In technical communication, lying by commission has many guises. It is altering or inventing data, as in a scientific research report. It is making false claims, such as in marketing literature for technical products. It is terming your bug-riddled software "user-friendly." It is creating phony testimonials from fictitious customers. It is claiming "millions sold" when, in fact, sales are in the thousands. It is promising a prototype exceeds standards when, in fact, it barely fulfills them. It is the outright lie.

If the technical communicator is actively altering data, making false claims, or engaging in any other form of lying, he is clearly guilty of unethical behavior. Even if he is aware that data is being altered, that claims are false, or that any other form of lying is involved, he is still ethically accountable.

What if the technical communicator doesn't *know* the material is false? Say, for example, the author has distorted data to present a theory in a more favorable light. Is the technical editor also ethically responsible? Not if he has no inkling of the deception. But if a technical communicator has reason to even suspect such dishonesty, he is duty-bound to investigate and, if the dishonesty is confirmed, to refuse to perpetuate it.

What if the author has more power than the editor—as in supervisor to employee in industry, or professor to graduate student in academe? That brings us face to face with the kind of conflicts we explored in Chapter 2. Sometimes the editor might find relief through appeal to a higher authority, such as a corporate ethics office or a college provost. Sometimes a direct exchange with the author could result in a compromise, but sometimes the editor might just have to pay a price for refusing to compromise her values—like a delayed wage increase or a lower grade in a course. Not fair, you say? We agree. But we would be unrealistic if we indicated that ethical values can always be upheld without consequence.

Lying by Omission

In the days of the Cold War, there was a joke about the Russian newspaper that ran the following news flash:

> *The Russian entry, Pride of Kiev, captured second place in yesterday's international quarterhorse sweepstakes in Moscow yesterday. The American entry, Secretariat, finished next to last.*

Every word in the story was true. Can you find the catch? The Soviet journalist neglected to report one little detail: It was a two-horse race. (Before you refer us to Chapter 10, Cultural Sensitivity, be advised that we shall even the score on the next item.)

Deceiving by omission is just as unethical as deceiving by commission. Thus, the test report that conveniently leaves out data unfavorable to a product is dishonest. The software documentation that fails to mention an annoying bug because to do so might cut into sales is dishonest. The financial report that includes record sales in January and March but ignores a disastrous slump in February is dishonest.

The technical communicator is not responsible for something that has been omitted before she gets the material—unless there's a clear indication that something is missing. But if asked to deliberately "lose" information that is unfavorable, she is faced with exactly the same conflict as in lying by commission.

A specialized form of omission is ellipsis—the intentional omission of words not necessary to meaning. For example, the statement "The books that are on the table were donated by Edith; the books that are on the shelf were donated by Agnes" loses nothing after being pruned through ellipsis: "The books . . . on the table were donated by Edith; the books on the shelf . . . by Agnes." The missing words are clearly implied.

Ellipsis is a useful technique because it makes for conciseness. Ellipsis is also helpful on research papers, since it allows you to focus on the part of a quotation you want, deleting the rest. However, when using ellipsis you have an ethical obligation to preserve the author's intent. Used dishonestly, ellipsis can distort, sometimes dramatically. Witness the movie critic who panned a bad movie by writing, "This movie is the most amazing example of pure, unadulterated trash that I have ever seen!" Imagine the critic's horror when he walked past the movie theater and saw emblazoned on the marquee the words "This movie is the most amazing . . . I have ever seen!" followed by his name.

Lying with Language

Words can be a powerful tool for deception. If our communication objective is to deceive, we are rarely at a loss for words to enable us do so. Excluding outright lies, other verbal techniques for altering or hiding the truth include the following:

- Deliberate vagueness
- Circumlocution
- Euphemisms
- False generalization
- The use of *loaded* or *slanted* language

Deliberate Vagueness

Deliberate vagueness includes the student who tries to hoodwink the professor on an exam by weaving a tapestry of verbal generalities to disguise the fact that he has virtually no specific facts to offer (this is *not* usually successful, because professors outrank students). It also includes the supervisor who denies a deserving but politically disenfranchised employee a raise by loading her performance appraisal with unsubstantiated general criticisms such as "needs improvement in interpersonal relations" or "could show greater leadership" (this *is* usually successful, because supervisors outrank employees).

Circumlocution

Circumlocution means verbally talking around an issue. It sometimes involves two parties exploiting superior vocabulary to "talk over the head" of a third party. Witness the parents who say they are considering a "family outing on the morrow to take in some sand and surf," so as not to tip their hand to their nearby six-year-old, thereby triggering a guerrilla war of badgering and if—Heaven forbid!—it should rain, sobs of anguished disappointment. (Aside to such parents: Fair warning: More often than not, the child will respond to this tactic by shouting, "Oh, goodie! We're going to the beach!" In fact, some successful technical communicators got their start this way.)

Euphemisms

A euphemism is a *nice* word used to sugar-coat an unpleasant reality. Strictly speaking, a euphemism is a form of linguistic dishonesty. Ethical judgments regarding euphemisms pivot on the motivation for using them. If the user is genuinely concerned for the feelings of the listener, the euphemism is justified. For example, referring to a person "passing away" rather than "dying" to soften the impact on a bereaved family member is an act of compassion, not deception. But resorting to niceties such as "sexually challenged" instead of "impotent" is both patronizing and potentially misleading, since the euphemism is nonspecific on the dysfunction. And deliberately using euphemisms to obfuscate crosses the line into outright dishonesty. After all, how many of us would like to be on an airplane that is "experiencing an altitude anomaly"?

Newsman Edwin Newman shared a number of his favorite deceptive euphemisms in his keynote address at the 42nd Annual Conference of the Society for Technical Communication:

- *Whether you call it "downsizing" or "rightsizing," reductions in force still mean pink slips for the workers.*
- *One forward-looking company's employees no longer "plateau," but they are "unlikely to experience further hierarchical mobility."*
- *Highway "crash barriers" have become "impact attenuators."*
- *One hospital reported a "negative patient care outcome." The patient died.*
- *In another hospital, the outcome was disguised as "the patient did not fulfill his 'wellness potential.'" Newman added dryly, "The implication is that the patient was somehow to blame."*[3]

For further discussion of "public doublespeak," refer to Chapter 11.

False Generalization

False generalization, which is also a logical fallacy (see later section) is covered here because the use of absolutes (e.g., all, none, always, never) is a common avenue for linguistic deception. Unless the absolute is verifiable (e.g., all 12 test flights *were* successful), it should not be used. Honest alternatives include qualifiers (e.g., most, few, usually, seldom), although these also can be deceptive if they are not supported by the facts. Far better to include objective data (e.g., 88%, 12%, 8 out of 10 times, only 1 out of 10 times) if it is available. And if such data is not available, it might be advisable not to make the statement at all.

Loaded and Slanted Language

Loaded language refers to terms that carry strong emotional connotations, such as "brutal baby butcher" to refer to a pro-choice advocate or "fascist Bible-thumper" to refer to a right-to-life advocate. Whatever your opinions on this controversial subject, verbal emotionalism does not contribute to an honest treatment of the underlying issues. Consider the difference between "we're going to stick Mama in a home" and "we're planning to place our mother in an ACLF (adult congregate living facility)." The latter is an honest attempt to describe an unpleasant reality; the former loses the content message in the emotional baggage.

Slanted language can distort by injecting bias. Returning to the Cold War, consider this statement, typical of a U.S. Government press release:

> *U.S. reconnaissance aircraft documented the presence of Russian spy ships.*

The planes and the ships have the same mission, but from our perspective as the "good guys," we're conducting reconnaissance. The "bad guys," of course, can't be conducting reconnaissance; they're spying. Dishonest.

Lying with Statistics

Books have been written on this subject.[4] Misuse of numbers for purposes of deception is analogous to misuse of words for deception. Actually changing numbers is commission. Conveniently leaving out numbers that don't serve one's interests is omission. Manipulating numbers to send a distorted message is like using circumlocution or slanted language. All are unethical. Of the myriad opportunities for deception with statistics, three of the most common are invalid survey techniques, misuse of percentages, and misleading use of averages.

Surveys

Surveys are tricky animals. To be valid, they must represent a sufficiently large and representative segment of the population. Questions must be worded to clearly differentiate between opinions, not to deliberately lead respondents down a primrose path to desired responses.

Consider this situation: A group of high school students is asked to survey residents of a wealthy and predominantly elderly community known to oppose tax increases on whether they would support a bond issue for a new football stadium with Astroturf. The students knock on 100 doors. In 95 cases, as soon as the homeowners learn the subject of the survey, they chase the students down the driveway wielding frying pans and broomsticks. In the other five cases, the students find one genuine supporter, one who is adamantly opposed, and three who are opposed but whom the students badger for hours until they finally give in and check "Yes" as the only way to get the students off their property. The result is "Four out of five residents surveyed recommended grassless football stadiums for those high schools that have football stadiums." The other 95? Oh, that's right—they weren't surveyed.

Percentages

Percentages are meaningful only if they are presented along with the base to which they are applied and they can be very misleading. Consider the executive who tells her employees that she will be receiving only a 3 percent raise this year while they all receive 6 percent. That sounds great until you allow for the fact that the executive's raise is 3 percent of $100,000 and the workers' raise is 6 percent of $20,000. Or consider this one: If somebody offered to give you a 50 percent raise this month if you agreed to accept a 40 percent cut the following month, would you take it? Your first impulse might be to say, "Sure." Try it out. Let's say you're making $10 an hour. With a 50 percent raise, you'd be making $15 an hour. Now apply the 40 percent cut. You drop to $11 an hour, so you're still $1 ahead, right? Think again. The 40 percent cut applies to your *new* base salary, not your original base salary and 40 percent of $15 is $6. The net result is: You'd wind up working for $9 an hour, $1 less than you started out with.

Averages

Averages come in three flavors: mean, median, and mode. When most people use the word "average," they are referring to the mean, which is computed by adding all the individual values and dividing by the total number of items. *Mean* gives an honest expression of average when is there is a fairly even distribution of values. When the values are clustered at the extremes of the range, the mean presents a distorted picture. *Median* refers to the midpoint among the values: the point at which there are an equal number of items of lesser and greater value. It is more valid than the mean for cases involving extremes; less so for even distributions. *Mode* refers to the individual value that occurs most often. It would be useful if you were trying to present the average height of a class which contained four 6'8" basketball players and one 5'2" cheerleader.

Of the three types of averages, the mean offers the most opportunity for deception. Consider the robber baron who claims that the employees in his dilapidated sweatshop of a factory live in comfort. His promotional brochure boasts: "Our employees work in a climate-controlled environment with an average temperature of 75 degrees." This deception is particularly masterful, because it demonstrates both verbal and statistical persiflage. The factory is, indeed, controlled by the climate. Since it has neither heating nor air conditioning, it is 90 degrees in the summer and 60 degrees in the winter. What is the average? You guessed it—75 degrees!

Or suppose you run into a state trooper on your way back from your bachelor's or bachelorette's party. The conversation might run something like this:

"Gee, officer, I was only averaging 50 miles an hour!"

True statement. Unfortunately, you achieved that average by going from 0 to 100.

Lying with Graphics

If one picture is worth a thousand words, then one lie with graphics is worth a thousand lies with words. Examples of deliberate deception with charts and figures abound. Pie charts can lie if the slices do not accurately reflect the percentages (which, of course, are in one-point type). Line graphs can lie by distorting the axes. Organization charts can present a misleading picture of the chain of command.

Figure 4.1a might come from the annual report of an unprincipled charitable organization seeking to raise funds. It differs from an honest depiction of the same data (Figure 4.1b) in two important respects. First, it divides the overhead costs into two subcategories to conceal the amount of expenses unrelated to patient services. Second, it visually distorts the slices to well below the actual 40 percent overhead figure. The real numbers are there to avoid fraud charges (see Chapter 5), but they're about as legible as the Surgeon General's warning on cigarette packages.

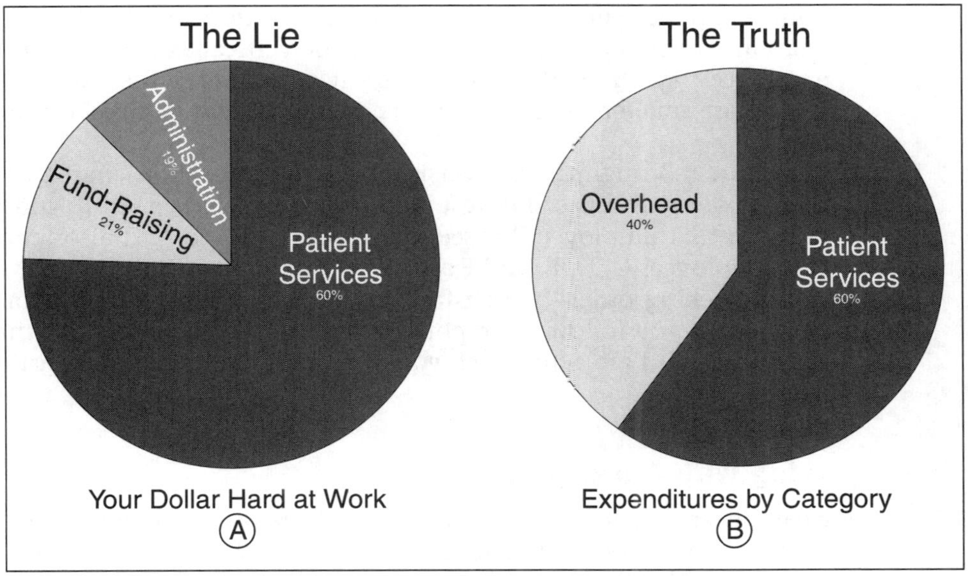

Figure 4.1 Pie charts lie when the numbers don't match the slices.

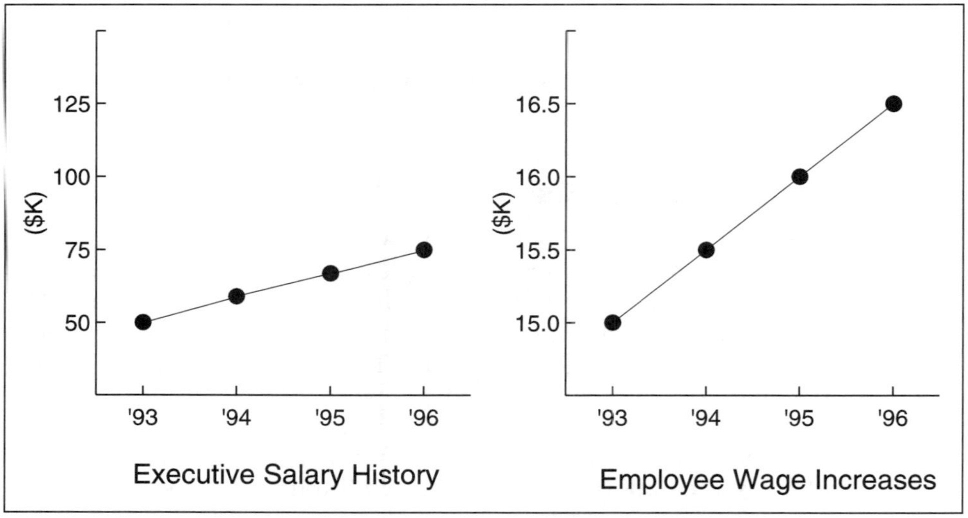

Figure 4.2 Who's been getting bigger raises? Look again.

Figure 4.2 "loads the dice" by manipulating the units of measure on the ordinate or Y-axis. At first glance, the reader might conclude that the workers' wages are climbing much more rapidly than those of the executives. Look again. The units of measure on the vertical axis of the executive pay chart are much larger than they are on the vertical axis of the similar-looking employee pay chart, thereby making it appear as if the executives' pay increases are smaller than those of the employees when, in fact, they are much larger.

Figure 4.3 might be the version of the organization chart that Employee B sends out with his or her resume. It is carefully designed to convey the impression that Employee B is second in command to A. It also tends to imply that Employees C, D, E, and F are below B but above Employees G, H, I, and J in the pecking order—which further reinforces B's "vice-emperor" message. However, if you follow the reporting lines closely, you will find that Employees B through J are actually all equal; all bear the same relationship with the boss.

Lying with Photographs

Nearly everyone in our generation recalls the use of obviously doctored photographs in tabloid newspapers back in the 1950s and 1960s: "Kitten Born with Dog's Head" screamed the headline. To document this noteworthy biological event, under the headline appeared a lurid photograph with a puppy's

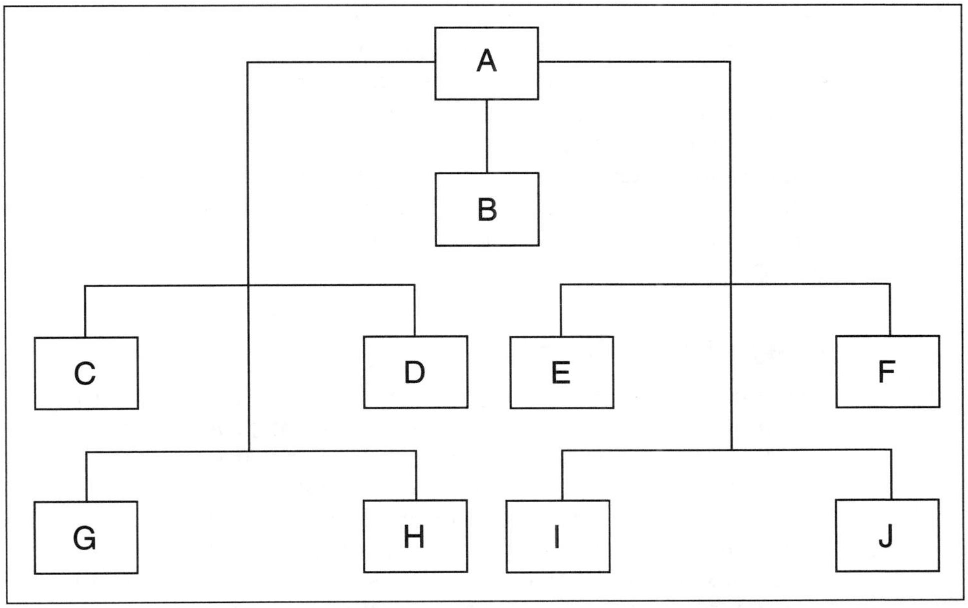

Figure 4.3 Who's more important, J or B?

head clumsily air-brushed onto a kitten's body. Sometimes if you looked closely, you could even discern the traces of the airbrush.

Well, that happens no more. No more clumsiness, that is. In today's world of dizzily onrushing technology, any entry-level illustrator with the appropriate software can, in a matter of minutes, give a kitten a dog's head. Or slap an electronic test set on a piece of equipment where it has never been used before. Or deftly whisk away any telltale plumes of pollution from the smokestacks of a factory when designing the company's environmental awareness brochure.

Given the present shift from optical to digital technology in photography, within a few years nearly all business and technical photography will be computer-based. And since any digital file can be subtly—or not so subtly—altered without a trace, it will become almost impossible to know whether a photograph is genuine or retouched.

This technological advance has upped the ethical ante for technical communicators. Increasingly, if our audiences are to trust the integrity of the photographs in the documents and visual presentations we produce, they will have to depend on our honesty.

Consider this scenario. Suppose the manufacturer of a CAT scan imaging system asks a technical illustrator in its publications department to enhance the clarity of its images for a promotional brochure. Suppose, further, that the perception of higher resolution than the product can actually provide sways a hospital purchasing agent to order one of the systems. And suppose, six months later, that a neurosurgeon depends on that equipment to pinpoint a blood clot so she can operate on a child who has been gravely injured in a car–bike accident. The neurosurgeon goes in, the clot isn't where the machine says it was, and before she can get to it, the child dies on the table. What is the ethical responsibility of the technical illustrator to the parents of the child?

Lying with Multimedia

Anyone who has seen the film *Forrest Gump* can attest to the fact that the cursor is faster than the eye. Tom Hanks was probably in diapers when Lyndon B. Johnson was in the White House, but unless you knew better, you'd swear the two were actually together in that videotape.

Videographic sleight of hand is innocent enough when it comes to entertainment, unless it reaches the point of libel or defamation (see Chapter 5), but it has serious ethical implications in technical communication. If still photographs can be altered with major consequences, consider the staggering impact of deceptive or inappropriate use of video footage, sound, and animation.

Suppose, for example, that the installation procedures for a gas water heater have been packaged in a step-by-step animation on CD-ROM or videotape rather than (or in addition to) conventional paper documentation, to increase marketability and boost sales. Successful use of animation for detailed instructions requires a high degree of proficiency on the part of both the technical writer and the graphic illustrator. Animation can be extremely effective, but if not handled by experts, it can distort—albeit unintentionally—by oversimplifying. Suppose that happens in the case of the water heater. Suppose the part that gets oversimplified involves the safety procedures for igniting the pilot light. It's not hard to finish this scenario.

Even if the situation is not this clear-cut or disastrous, just the sheer distraction resulting from excessive use of multimedia techniques can distort the message. The next few years are going to give technical communicators ever-better, ever-cheaper, and ever-faster tools for incorporating multimedia techniques in products. But just because we have these capabilities doesn't always mean we have to use them.

When it comes to safety, technical communicators have an ethical obligation to choose the simplest and clearest path. Sometimes a 3x5 index card with large red type saying "DO NOT IGNITE PILOT LIGHT UNLESS GAS GAUGE IS ON EMPTY" beats the hell out of a 500-megabyte animated presentation that lights up the screen—but blows up the house.

Lying with Logical Fallacies

The *American Heritage Electronic Dictionary*, 3d ed. offers four definitions for "fallacy":

1. A false notion.
2. A statement or an argument based on a false or an invalid inference.
3. Incorrectness of reasoning or belief; erroneousness.
4. The quality of being deceptive.

The first three relate to competency because they imply unintentional errors; the fourth is an honesty issue because it implies intentional use of fallacies to deceive. As trained linguistic specialists, technical communicators may be assumed to understand logical fallacies; it's therefore hard to imagine how they could use them without intent to deceive.

Full treatment of logical fallacies is outside the scope of this text. Most freshman composition and public speaking texts include a chapter or section on fallacies. One excellent source is *Between One and Many: The Art and Science of Public Speaking* by Steve R. Brydon and Michael D. Scott.[5] Representative fallacies with particular relevance to technical communication are as follows:

- *Post hoc ergo propter hoc*: After the fact, therefore because of the fact
- Mistaking correlation for cause: Two events may be correlated without a causal relationship
- Red herring: Diversion of attention from the real issue
- Circular reasoning: Defining something in terms of itself
- False dichotomization: "Either-or" thinking; forced choice
- *Non sequitur*: Something that "does not follow"; an "off-the-wall" conclusion

There are many more. The more gullible the audience, the more effective such techniques are at achieving deception. However, to use them for this purpose is unethical.

Lying with Propaganda Techniques

Propaganda techniques are close cousins of logical fallacies. Again, Brydon and Scott provide excellent coverage.[6] Used with intent to deceive, propaganda techniques are unethical. They are most readily observed in advertising and politics, but they can crop up in marketing communication for technical products. Representative propaganda techniques relevant to technical communication are as follows:

- *Ad hominem*: Attacking the person
- Glittering generality: As in the Army's "Be all you can be"[7]
- Testimonial: Endorsements (best *with* credentials!)
- Card Stacking: Emphasizing facts favorable to one's cause and suppressing others
- Bandwagoning: Exploiting the "herd instinct"

How to Be Honest in Technical Communication

How, specifically, does the value of honesty come into play in technical communication? This section will address that question from both an academic and a business standpoint and explore the type of conflicts that can arise.

Honesty in Research

Honesty is the core value in academic research. It applies to the manner in which you acquire, interpret, present, and attribute data.

Never Falsify Information

The following is overheard in a freshman physics lab at a prestigious university where students are industriously weighing metal fragments, heating them, dropping them into beakers of water, measuring the increase in temperature, recording the data, then plugging it into a formula to determine the specific heat of the metals:

> *[One student is reading the thermometer, the other recording the data.]*
>
> *"Whatcha got?"*
>
> *"Whatcha* need *?"*
>
> *"Well, let's see. This table says the specific heat of copper is 0.093 cal/(g)(°C). We're running too low. About 107 degrees would be nice."*
>
> *"Let me take a closer look. Wow—106.7 degrees!"*
>
> *"That's perfect."*
>
> *"Should we take any more measurements?"*
>
> *"No, let's quit while we're ahead."*

Research is a detailed and demanding process. Every researcher has experienced the disappointment of discovering that the emerging facts don't square with his expectations—or wishes. It's very tempting to assume the research results are in error for one of many reasons and simply alter the facts to what the researcher "knows they should be."

For example, suppose you and your friends are biology majors struggling through a required freshman-level course in art appreciation that you feel is superfluous for students in applied disciplines. After all, you reason, you do not intend to earn your living going to art galleries. You decide to survey all the biology students to convince the administration of the worthlessness of the art course. To your surprise, 53 percent of the students in your department say they find the course valuable. You reason that the biology students must somehow be "different" from students in the other applied disciplines. Had you included math majors, engineering majors, and the like, the results would have confirmed your belief. You therefore ignore the statistical information and include only the comments from the 47 percent who agree with your preconceived belief.

Another temptation for the beginning researcher is to develop fictitious information. Let's say the semester is drawing to a close and you have conducted only three of the five interviews required for your empirical research paper. Suppose, further, that it's the Thursday before Homecoming weekend, you have a hot date, and the paper is due on Monday. You have a choice: You can spend the weekend conducting the remaining two interviews and finishing the paper, or you can type up the paper Thursday night and enjoy the weekend. You have all the rest of your data except the last two interviews. Furthermore, the three persons you did interview have given you essentially the same information and points of view.

A solution emerges. You decide that because the other two persons would have just echoed the information you already have, you can pick two people's names at random and say those individuals agreed with everything the others said. No problem. Well, yes, *problem*—this practice is dishonest and unethical. The two persons you did not interview may be in total disagreement with the three you did.

Under no condition is it ever ethical to falsify or change information found or developed in the course of research to meet the needs of the research. It is equally unethical to present fictitious statements or concocted data as fact. The golden rule of research is that it must be able to be replicated: Someone other than the original researcher must be able to repeat the research and get approximately the same results. If you falsify or invent information in research you submit for academic credit or publish, sooner or later your dishonesty will surface upon replication and your reputation will be lost.

If Evidence Conflicts, Report It As Is

Unfortunately for those of us who must do research as part of our careers, at times we must report data that conflicts with our preconceived ideas. The ethics of research demand that we report any and all conflicting evidence—no matter how painful that may be! There are two types of conflicting evidence. The first occurs when the data or evidence you find is in direct conflict with what you expected to find. The second occurs when the evidence does not clearly point in one direction or to a specific solution.

In the first instance, even when you are certain you know what the results of your research will be, there will be times when your data conflicts with your assumption. *You must report the data, as is.* You cannot bend it to meet your needs. There is no shame in failing to support your hypothesis. The shame is in failing to report or altering data that does not support your hypothesis.

Suppose a graduate student conducting research on muscular dystrophy alters data to suggest she has discovered a potential chemical cure—then rushes the findings into print for her own glorification. It is unlikely the drug would ever be used on a human being, given the checks and balances governing the process, but what about the false hope the erroneous report might give to thousands of stricken children and their parents?

"Yes, but that's a far cry from doctoring a meaningless experiment in freshman physics!" we hear you protest. Maybe so. But the ethical issue is exactly the same. Bad habits can be hard to break.

The second type of conflicting data is very frustrating. It severely tests your powers of critical thinking and analysis. Sooner or later you will be working on a problem and discover that one piece of evidence points in one direction, a second piece points in the opposite direction, while the third piece goes off at right angles to the other two. Your research may wind up resembling the road sign shown in Figure 4.4.

Figure 4.4 Evidence that points in all directions can tempt researchers to deceive.

Professional researchers deal with this phenomenon all the time. It's their job to identify and analyze conflicting evidence and explain its possible significance—not to manipulate the evidence and force it into a pattern. An example of conflicting evidence can be seen in research on dinosaurs. Were they ectotherms or endotherms? Are they really extinct or have some of them evolved into birds? Like the paleontologists who must struggle with questions like these, you must report all conflicting evidence and analyze the influence of such evidence on the variables. This is not only an ethical requirement of research, but proves to your colleagues that you have an in-depth understanding of your subject matter. Indeed, it is precisely your ability to deal with conflicting evidence honestly that makes your research worthy of our trust.

Openly Acknowledge Flaws in Research

There is a final, difficult aspect to ethical research: the requirement that you describe in detail any and all flaws or weaknesses in your research.

What? You want me to tear down my own research? No, we want you to build it up—by presenting it honestly, with all its flaws and limitations. You must not wait for a sharp-eyed colleague to spot your errors if you already know what they are.

At times the flaws are the result of something over which you had control. Perhaps you needed to word your survey or interview questions more clearly. In retrospect, you see the need to have tested your questions more rigorously before administering the survey. Perhaps you should have limited your topic more severely. Perhaps your topic was *too* limited. Perhaps you did not include enough independent variables. Perhaps you had too many independent variables. You must be your own severest critic.

At other times the flaws in your research will be beyond your control. For example, undergraduate students may not be able to use original sources. The sources may be in a language you cannot read, like Russian, and translation might not be practical. The expert you really need to speak with may be called off to Sri Lanka just before your scheduled interview. In accordance with Murphy's First Law for Inanimate Objects, the video camera you are using to record an event may malfunction, leaving you with a garbled tape. Unfortunately, no matter how carefully you plan, errors can occur. You admit them, describe them, and show how they adversely influenced your research. Learning from our mistakes and the mistakes of others allows us and others to do new and better research in the future.

The business world has learned this. Major projects are now routinely followed by "Lessons Learned" debriefings to identify what went right and what went wrong, to avoid making the same mistakes in the future.

When we state in research that there are flaws over which we have no control, we cannot ethically include poor judgment and lack of planning on the part of the researcher. Failing to provide enough time for the surveys to come back in the mail, waiting until the last minute to schedule an interview, going to the library the week before the research is due and discovering that all the books you need are out, as well as allowing the project to fall victim to the myriad of day-to-day personal problems that beset us all are not considered flaws over which we have no control. These are errors over which you *do* have control. While evaluators are generally kind when flaws over which you have no control occur and are usually understanding when errors of inexperience cause flaws that could have been controlled, there is usually little sympathy for errors of poor judgment and lack of planning.

We frequently think of research as a test of facts to find an answer. In reality, it is a test of the ethics of the researcher to determine if we can trust the research as it is presented.

Always Provide Proper Attribution

In research writing, or in any form of communication, when we borrow material from others, it is extremely important to give credit to our sources—as we did to Michael Josephson, Steven Brydon, and Michael Scott in this chapter. This is called attribution. It's not just a matter of honesty, but of academic and professional respect. It can also be a matter of law (see Chapter 5).

Proper documentation of sources in a research paper or business report goes beyond the scope of this text. Any basic handbook for preparing a research project can provide direction for how to properly footnote borrowed material, list references, and prepare a bibliography. In a nutshell, you may include five basic types of information in a term paper or business report:

1. Your own original material, in your own words
2. Results of your own research, in numbers, survey responses, data
3. Information that may be considered "common knowledge" or public domain, which could easily be found in a basic reference book such as an encyclopedia—such as the specific heat of aluminum, the year Shakespeare was born, or the capital of Sri Lanka
4. Someone else's ideas or opinions, in his or her exact words
5. Someone else's idea or opinions, changed into your words

Items 1 through 3 do not require attribution. Items 4 and 5 do. Item 4 is called a direct quotation and requires quotation marks (or, if more than three lines in length, indentation, and single-space typing), plus a footnote, using whatever style of documentation your professor or publisher prescribes. Item 5 is called a paraphrase. It does not require quotation marks, but it does require a footnote. Items 1, 2, and 4 are clear. If it's yours, you know it. If it's lifted word for word from somebody else, you know it.

The only gray area, occasionally, may be the distinction between Items 3 and 5. For example, the fact that John Milton went to Cambridge is general knowledge. Even the fact that he had a religious upbringing is general knowledge. But a theory tying these factors in his upbringing to *Paradise Lost* crosses over into another's original ideas, which must be attributed. There can be some gray in this area. If in doubt, footnote. If you don't, you commit plagiarism, which is discussed in detail in Chapter 5.

Honesty in Professional Assessment

Honesty applies to other aspects of technical communication besides the communication products we create and the research we conduct. Another area that demands candor is professional assessment and evaluation. This section addresses that issue as it pertains to defining communication requirements, preparing cost estimates, evaluating others' communication products, and conducting personnel appraisals.

Assess Communication Requirements Accurately

A key element in a professional technical communicator's job is the ability to define communication requirements. This often involves interacting with clients or persons of higher rank in the company. Clients send checks that pay your bills. Bosses provide the paycheck that pays your bills. In either case, sometimes an honest assessment of communication requirements may lead to a message that clients or bosses don't want to hear, such as "We can't fix this problem by rewriting the documentation, because it's a basic feature of the software," or "We can't present this complex concept in terms the average user can understand unless we use animation, which exceeds the project budget."

The temptation is to remove the immediate pressure by telling those in power what they *want* to hear rather than what they *need* to hear. Here again, the old adage "honesty is the best policy" applies. In the long run, you will get into far more trouble glossing over problems than you do by confronting them honestly up front. You can also get into trouble by foisting a higher quality level upon the client or superior than her pocketbook will bear in order to boost your profit or showcase your communication prowess (see Chapter 7).

Prepare Cost Estimates Realistically

A particularly important type of professional assessment is the cost estimate. Accurate forecasting of project costs is obviously important to successful business operations; without it, profit would be in jeopardy. Cost estimating requires both skill and honesty. Let's assume the skill and focus on the honesty part. Suppose you are the manager of a technical publications department and work has been slow. You're in danger of having to lay off two of your people. A program director comes to you with a request to estimate the cost of preparing 15 technical manuals for his product area. You know through the grapevine that the director is also considering subcontracting the work. A couple phone calls to local vendors give you a pretty good idea of what price you have to come up with to keep the business in-house. (Note: In a competitive outside procurement, those calls would, in themselves, be unethical. Here, however, since the choice is between keeping work in-house and subcontracting it, getting the quotes is not unethical. What you do with them might be.)

You prepare the estimate using your best professional judgment and experience. It's 20 percent too high. That's because you have several senior people whose salaries make it hard for you to be competitive with outside sources. Your choices are: Submit an honest estimate and risk losing two people or shave 25 percent off the estimate, land the job, then try to come in as close as you can to your unrealistically low estimate. The second choice is dishonest. It may be in your department's best interest, but it is probably not in the company's best interest. We say "probably," because that suggests the only ethical option that might preserve the two jobs: You could go to the program director with an honest estimate, acknowledge that it is higher than an outside vendor's bid, but appeal for the opportunity to perform on the job anyhow with the challenge of reducing the cost as much as possible, on the basis that the layoffs would cripple your department and render it unable to respond to future work that must be done in-house.

If the program director agrees with your assessment that a valuable company resource might be eroded if she subcontracts the work, she may decide to keep it in-house even if it means a small cost penalty. If she disagrees, it may be time to consider whether you can afford to keep senior people whose productivity has not kept up with their salaries. The problem there is that your company's layoff procedures may well force you to lay off junior personnel rather than the ones who are making you uncompetitive. It's rarely easy.

Evaluate Communication Products Objectively

Both in the academic world and in the world of business, technical and business communicators are frequently called upon to evaluate the communication products of others. We are ethically bound to do so honestly.

In academe, the most common instance of product evaluation is the obvious one: professors grading student papers. Our discussion on research focused on the obligation of students to present their findings honestly. Now, the ball moves to the other side of the court. Professors must evaluate students' work just as honestly as they expect students to prepare it. The professor must grade a paper that presents an idea she personally opposes in the exact same manner she grades a paper that presents an idea she personally espouses—purely on merit. In neither case should her personal opinion have any bearing on the grade. Neither should an instructor be influenced by his personal attitude toward a student—or vice versa—when evaluating a paper or project. It shouldn't matter whether the student has been a model classroom citizen or a candidate for justifiable professorial homicide.

Believe it or not, if most professors are ever tempted to be slightly less than objective in grading a paper, it's not the temptation to downgrade a student who has been a pain; it's the inclination to cut some slack for a conscientious student who has been working hard all semester. But giving a conscientious student a higher grade than she deserves is just as unethical as giving a troublesome student a lower grade than he deserves. Note that we have talked so far only about evaluating a specific paper or project. It is not unethical for a professor to base a percentage of the semester grade on a subjective evaluation of students' participation and attitude; in this case, it would be entirely legitimate to give the cooperative student a better score than the troublemaker.

Another instance in which college professors are routinely asked to judge another's work is textbook evaluations. In this case, the publisher is paying the professor for a candid assessment of an author's work and has every right to expect exactly that. Suppose, for example, that professors in two different universities each have written textbooks within a given discipline. They give each other's names to their respective publishers for purposes of manuscript reviews. Off the record, they make an agreement with each other to write largely positive reviews, offering just enough minor criticism to be credible but avoiding any observations that could cause the manuscript to be rejected for publication or sent back for major rework—even if it needed it.

Once again, this form of dishonesty might have a short-term payoff for both parties, but in the long run it will probably hurt them—not to mention the publisher and the professors and students who try to use the books. A weak textbook will be flushed out anyhow; even if it is published, it probably won't sell many copies. Follow-up sales, in particular, will be almost nil, once word of the book's limitations gets out. That, in turn, is likely to diminish the author's reputation and discourage further sales. All of which may appear to be just deserts for a misdeed, but that doesn't compensate for the negative impact on the publisher or on the classroom users.

In business, product evaluations demand not only honesty, but timeliness and diplomacy. It's important to pinpoint product weaknesses such as poor writing or unclear graphics early, in the development and draft stages, so they can be corrected before the final product goes out. Client satisfaction surveys will provide feedback on the final product; internal reviews should help refine the product, not tear it apart after it's already done. While it's important to be honest in identifying and correcting errors, it's not necessary to run roughshod over one's colleagues in the process. This point is particularly critical for technical communicators whose job it is to edit the work of authors who are content experts but less than fluent writers.

Should the editor show the author a paper copy of the manuscript with red ink scribbled all over it? This would be honest, but probably not a good strategy. On the other hand, should the editor work online, totally rewrite the document, and hand it back to the author without a word? Perhaps this is more cowardly than dishonest, but from either perspective, it is probably not a good strategy. The author is bound to realize the material has been changed and may—perhaps with some justification—worry that the technical intent has been altered. In this case, the best strategy is probably a compromise: The editor could respectfully inform the author that she had done "a little rewriting" and ask him to revalidate the content as she hands him the manuscript. (The alert reader may call us to task for suggesting the term "a little rewriting," referring to "Lying with Language." This is almost like the old saw about the woman who has just bought an expensive but hideous new hat and asks you how you like it. Verbal dexterity in the interest of tact and sensitivity does not constitute dishonesty.)

Conduct Personnel Appraisals Candidly

Honesty is extremely important in personnel appraisals. Good performance needs to be acknowledged so the employee will know her efforts have been recognized. Poor performance needs to be clearly identified so it can be corrected. This isn't a textbook on human resource management, but the subject is relevant to technical communicators; by the very nature of our expertise, we should be expected to communicate clearly and candidly in presenting personnel evaluations if we have that responsibility.

It's not easy to look a fellow professional in the eye and tell him his performance is not up to standard. It's especially hard to do when his attitude has been good. It's tempting to gloss over the problem, bump the score up a notch, and avoid a confrontation. But to do so is not only unfair to the low performer, it is manifestly unfair to the high performers. This is especially critical if the company hits a downturn and has to lay employees off. Suppose the low performer has two more years of service than a high performer and, thanks to the supervisor's well-intended but less-than-candid evaluation, has the same score on the last appraisal. Guess who goes out the door.

Honesty in Persuasive Communication

Persuasive communication, particularly in sales, is one area where the temptation to deceive with language can be very strong. Persuasive communication covers a broad range of territory: editorials, position papers, proposals, sales presentations, advertising.

All of these carry opportunities, indeed temptations, to advance one's cause through deception. If the audience is unsophisticated, a skilled communicator can easily use fallacies and propaganda techniques to sway its opinions. This tactic is extremely common—and depressingly effective—in advertising. Does this then mean it's impossible to be honest and still be persuasive?

No. Companies win proposals without misrepresenting their capability or trashing their opposition. Reputable advertisers mount extremely effective campaigns without resorting to guile and deception. In fact, honesty can be so refreshing in a world filled with deception that it becomes extremely persuasive. Suppose you heard an advertisement for a restaurant which claimed not to have the best steak in town, but "the best steak you're going to find for $7.95." True, the steak might turn out to be like shoe leather, but we wager you might be sufficiently intrigued to give the place a try. We did. (It wasn't that bad.)

Resolving Conflicts

Most ethical conflicts involving honesty pivot on the price for being honest. In academe, this principle applies to both students and professors. In business, it applies to both supervisors and employers.

Dealing with the "Publish or Perish" Phenomenon

"Publish or perish" is unwritten policy on many campuses. Professors feel pressure to publish to win tenure, gain prestige, supplement their income, and advance within their departments. None of these goals is in itself unethical, nor is publishing articles and books as one means of attaining them. The conflict arises if an overworked professor yields to the temptation to take the same kind of shortcuts in research as the ones we discussed earlier from a student standpoint. Falsified data is falsified data no matter whether the doctoring is done for a doctorate or for a grade in freshman English.

Students may feel similar pressures (call it "Pass or Perish") and be tempted to fudge on research, skirt the edges of plagiarism on papers, and read Cliff notes instead of assigned novels—you name it. They may conceivably also experience peer pressure, as in the case of lab partners where one student suggests the data might be "reconfigured" to reach the desired result.

There is no easy answer here. It boils down to how much a person is willing to sacrifice to uphold a core value.

Toeing the Bottom Line without Toppling Over!

The business world's answer to "Publish or Perish" is the famous "bottom line." Every day, barracuda-like businessmen and women sacrifice their moral values on the altar of profit and personal advancement. It can be tough to advance—or even survive—in such a ruthless world without adapting the unethical tactics of some of its inhabitants.

If a computer-illiterate supervisor nearing retirement honestly attributes an innovative idea to an aggressive and computer-savvy subordinate when presenting the concept to upper management, might not she run the risk of being nudged out the door during the next round of overhead cutbacks? Wouldn't she be wiser to claim the idea herself or at least to downplay the role of her subordinate?

If you know that an unprincipled colleague is looking for every opportunity to stab you in the back, is it unethical to stab him first? Don't you have the right to defend yourself? After all, what's wrong with a little preemptive backstabbing? When in Rome, do as the Romans do, right? As Josephson wryly asks, is it wrong to lie to a liar? Those are tough questions, with no easy answers. It all depends on your basic values.

Summary

Technical communicators strive for honesty, clarity, and accuracy in their communication products. They do not falsify, omit, or slant information to deceive their audience. They are honest in their professional evaluations, cost estimates, and research. Because we as technical communicators have particular expertise with language and communication, we carry a commensurately higher ethical responsibility to apply our talents honestly.

Honesty is sometimes not an easy choice; it often comes with a price. On the other hand, the long-term price of dishonesty often exceeds its short-term benefits.

Communicating honestly means avoiding deception, intentional or unintentional. One may lie through commission or omission; one may deceive with language, statistics, graphics, photographs, or multimedia; one may mislead through fallacies or propaganda. The ethical communicator does not resort to such tactics.

Honesty in technical communication applies to research and scientific writing, to professional assessments, and to persuasive communication. Ethical researchers do not falsify data, nor do they suppress conflicting data.

They openly acknowledge flaws in their research, whether or not they had control over the problems. They properly attribute material or ideas they quote or paraphrase from other sources. Failure to do so, whether in academic research or in business, is known as plagiarism—a serious form of dishonesty.

Ethical communicators are candid in their professional assessments, including definition of communication requirements, cost estimates, product evaluations (be they student papers, textbook manuscripts, or business documents), and personnel appraisals.

In persuasive communication, it can be tempting to use techniques of deception, especially with unsophisticated audiences. The professional challenge is to be honest and persuasive, too.

Most ethical conflicts concerning honesty revolve around its consequences. In academe, professors may feel pressure to "publish or perish"; the student equivalent is "pass or perish." In business, the pressure is to serve the bottom line. Internal competition for personal advancement is as intense as external competition for business. The price for honesty in such an environment can be high.

Application Exercises

1. Identify the following as deception by commission or omission.
   a. Changing the temperature readings in a chemistry experiment to make the results support the hypothesis.
   b. Presenting someone's comments out of context (e.g., lifting a quote without the surrounding material).
   c. Attributing a quote to someone who didn't say it.
   d. Conducting an interview, then presenting only the comments favorable to your position.
2. Use ellipsis to shorten this sentence without changing its meaning:

   The minister cautioned members of the youth group about the dangers of premarital sex, illegal drugs, and gratuitous violence in films.
3. Use circumlocution to inform your spouse what you intend to buy your six-year-old daughter for her birthday, without giving away the secret to the child, who is avidly listening.
4. Translate the following euphemisms:
   a. Little Johnny has a "vivid imagination."
   b. Bertha is "big-boned."

c. Paula has "imbibed rather liberally."

d. The submarine is experiencing an "unplanned descent."

e. Bubba is "cognitively challenged." Engaging him in a battle of wits is therefore unethical, because he is unarmed.

5. Revise these statements to eliminate false generalization.

   a. All seniors oppose the imposition of a school dress code.

   b. Sherolyn never gets to class on time.

   c. The Canadian Mounted Police "always get their man [person?]."

6. Rewrite these statements into one statement, without the slant.

   Truncheon-wielding cops brutally attacked the innocent students who had peaceably assembled in the intersection of Main Street and First Avenue.

   Dedicated officers of the peace overcame fierce resistance from wild-eyed student rebels to reestablish vital traffic flow at Main Street and First Avenue.

7. Rewrite these statements without loaded language.

   a. David's father is a war-mongerer.

   b. Melvin is a religious fanatic.

   c. Mary is a slut.*

   *As a corollary to Item C, come up with an equally loaded word for the same concept in a male. Give up? That should tell you something about the tenacity of the "Double Standard" in our culture.

8. What's wrong with Figure 4.5[8]? How does it deceive?

9. Suppose you are asked to present the average value of a home in a community which includes 2 multimillion-dollar mansions, 10 middle-income homes with slightly varying values, and 3 low-income dwellings. Identify and consider each of the three types of average. Which is the most valid measure in this case? Why?

10. You're a defense attorney. The prosecuting attorney puts in evidence a photograph that shows your client in the act of the crime of which he is accused. How might you use a technical communicator as a witness to discredit this ostensibly damning piece of evidence?

11. Create two original euphemisms.

12. Below is a brief paragraph that might appear in a term paper on literature. The sentences are numbered a through e. For each sentence, indicate whether it should be footnoted and explain why.

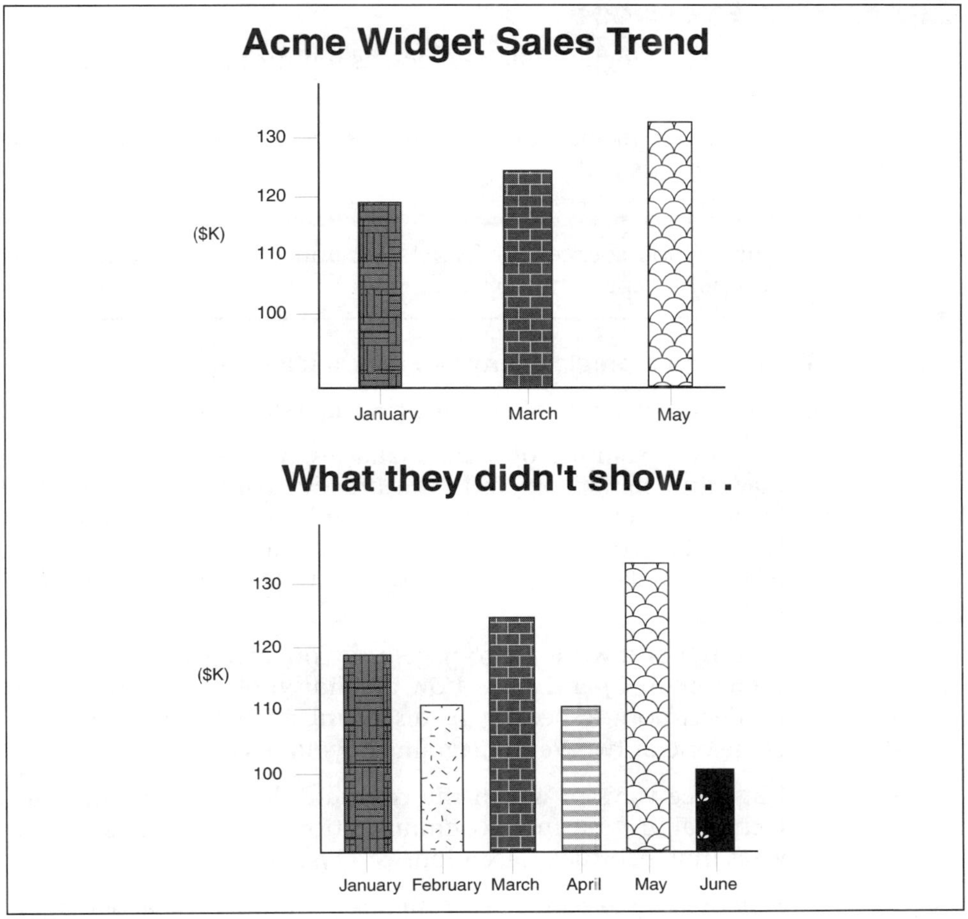

Figure 4.5 What's wrong with this picture?

(a) Aldous Huxley was born in 1894. (b) His grandfather was the famous biologist Thomas Henry Huxley and his great-uncle was the poet Matthew Arnold. (c) Perhaps as a result of this heritage, throughout his life, Huxley was interested in both science and literature, and his most famous novel, *Brave New World*, reflects that blend. (d) Huxley majored in biology at Eton University, where he developed eye trouble which left him nearly blind—a disability which he courageously overcame. (e) His early novels are cynical and satirical, reflecting his conviction that civilization was headed for self-destruction and that the chief architect of its demise would be science.

Exploration Exercises

1. Define five logical fallacies such as (but not limited to) the ones listed in this text.
2. Define five propaganda techniques such as (but not limited to) the ones listed in this text.

Note: Good sources for Exercises 1 and 2 include freshman composition and public speaking texts.

3. Create an example of any two logical fallacies.
4. Create an example of any two propaganda techniques.
5. Find five examples of logical fallacies and/or propaganda techniques in magazine, newspaper, or television advertising. In each case, identify the fallacy and explain why your example meets the definition of that fallacy. Attach the magazine or newspaper ads to your paper. In the case of television ads, briefly describe the ad in its entirety; then focus on the fallacy.
6. Research and write a two-page essay on syllogistic reasoning (A=B, B=C, therefore A=C) and show how invalid syllogism can be used deliberately for deception—especially by disguising a fallacy in the initial premise. Include one or two Venn diagrams in your essay.
7. List three ways in which the computer has made it easier to deceive in technical and business communication. Then for each example suggest ways in which such deceptions can be detected.
8. Suppose you are a first-year high school English teacher fresh out of college. A student with a C average in your course submits an essay which, based on the sophistication of the vocabulary and syntax, appears to have been copied verbatim or paraphrased from a source, perhaps not in its entirety but at least in several places. You check in the library, but you are unable to locate the source. You don't want to risk falsely accusing the student of plagiarism, but neither do you want to reward such behavior. The paper's good; it's either an A (if it's original) or an F (if it's plagiarized). How would you handle the situation?
9. Suppose you are an advertising account executive with a reputation for creating hard-hitting but honest ads. However, your largest client, a cosmetics company, has been taking a beating the past several months on sales of its traditional bestseller, hair coloring. One of its competitors has come up with a rinse which matches most of the features of your client's

product and sells for 30 percent less. Pressure is mounting for you to shift strategies and attempt to "trash" the opposition. To serve that goal, your boss has somehow managed to lay her hands on confidential correspondence to the customer service division of the other company. She wants you to take the juiciest complaints and weave them into an ad that implies the competition's product is cheaper because it is grossly inferior—when, in fact, investigation of your client's customer service records would, in all probability, reveal the same sort of complaints. Oh, by the way, your spouse just got laid off and you have two kids in college. How do you proceed?

10. Create a scenario involving honesty and technical communication. Analyze the underlying issues and suggest possible courses of action, as well as probable consequences for each.

11. Pretend you're a talk show host interviewing Diogenes, the Greek philosopher who ran around with a lantern searching for an honest man. With the help of a time machine, Diogenes has carried his quest into the twentieth century. Write a two-page script for how such an interview might go, using one of the following scenarios:

    a. Diogenes has just returned from a week in Washington, D.C.
    b. Diogenes has just completed a two-week internship in your department at work.
    c. Diogenes is fresh off a week-long visit to your college campus.
    d. Your choice.

References

1. Michael Josephson, videotape *On Ethics*, Josephson Institute of Ethics, Washington, D.C., c. 1992. Presented within inservice training program on ethics at Martin Marietta Corporation, Orlando, FL (now Lockheed Martin Corporation).
2. Norman R. Augustine, President and Chief Executive Officer of Lockheed Martin Corporation, "The Cost of Being Ethical," *Executive Excellence*, (February 1994): 5.
3. Edwin Newman, keynote address, 42nd Annual Conference of the Society for Technical Communication, Washington, D.C., April 1995.
4. Darrell Huff, *How to Lie with Statistics*, W.W. Norton & Co., Inc., 1954.
5. Steve R. Brydon and Michael D. Scott, *Between One and Many: The Art and Science of Public Speaking*, (Mountain View, CA: Mayfield Publishing, 1994), 410-416.

6. *Ibid.*, pp. 417-419.
7. *Ibid.*
8. Jules Harcourt, A.C. "Buddy" Krizan, and Patricia Merrier, *Business Communication*, 3d ed. (Cincinnati: Southwestern Educational Publishing, 1996), 516.

Chapter 5 Overview

Legality . . .

The Mission: Legality in technical communication means abiding by our duty to follow the laws and regulations that govern our profession, including meeting all terms and obligations of legal contracts we undertake.

Chapter Roadmap

Definition and Focus. This section provides a philosophical framework for legality as a core value.

Laws Protect Intellectual Property. Intellectual property is the tangible expression of information. This section familiarizes you with three specific issues regarding its creation, protection, and use:

- ❑ *Understand Copyright*
- ❑ *Practice Fair Use*
- ❑ *Beware Plagiarism*

Legality on the Information Highway. This section acquaints you with the tangle of legal and ethical problems regarding information that is available via the Internet and the World Wide Web.

Safeguarding Electronic Images. This section addresses how the manipulation of electronic impressions blurs the legal control over the use and integrity of information.

What Laws Govern Collaboration? This section identifies the responsibilities and pitfalls involved in using intellectual property provided to us by someone else.

Regulating Multimedia. Because multimedia combines audio, video, animation, narration, and text, the legalities can be confusing. This section introduces some of the concerns.

Securing Company Information. Corporations possess a great deal of intellectual property that is crucial to their success. Because communicators have access to that information, this section introduces the laws that control its use:

- *Trademarks and Service Marks*
- *Patents*
- *Trade Secrets and Confidentiality*
- *Nondisclosure Agreements*

Specifications Control Document Content. This section notes some regulations and specifications that dictate the content of communication products.

The Law and Contract Issues. This section presents the concept of honoring the terms of all contracts we enter into.

Summary, Practical Exercises, and References

. . . Our Duty to Obey the Law

CHAPTER 5

Legality: When Laws Don't Suffice, We Must Be Guided by Ethics

To put one's theft into print is to have it forever on the library shelves,
guiltily stacked just an aisle away from the volume it victimized,
a stain that doesn't wash but forever circulates. —Thomas Mallon[1]

When you steal from one author, it's plagiarism;
if you steal from many, it's research. —Wilson Mizner[2]

Definition and Focus

You are driving on a straight, flat stretch of interstate highway on a clear, summer day. You see no cars ahead of or behind you and only an occasional car going in the opposite direction on the other side of a wide, grassy median. The posted speed limit is 65 miles per hour. The speedometer is pushing 80. Is your action illegal? Is it unethical?

Clearly, if that particular stretch of highway is patrolled by an aircraft that you do not see, you will likely receive a speeding ticket in the mail, a gentle reminder that your action was indeed illegal. Was it also unethical for you to drive above the posted speed limit—weather, road conditions, and other safety factors permitting? That is your decision.

The point is that some actions are illegal, some are unethical, and some are both. Some legal actions are unethical; some ethical actions are illegal. Others are gray. Take, for example, the case of Rosa Parks. It was quite legal in the 1950s for the bus company in Montgomery, Alabama, to require Ms. Parks to sit at the back of the bus. But was it ethical? Was it ethical for Ms. Parks to refuse to follow that law and, instead, to sit at the front of the bus? In the culture of the 1990s, you will likely find general agreement that the bus company's action, while legal, was unethical and Ms. Parks' action, while illegal, was ethically justified.

These few examples from everyday life illustrate the complexities of ethics and the law. In the context of this book, *legality is our duty to follow the laws and regulations that govern our profession, including meeting all terms and obligations of legal contracts we undertake.* To do so, we must be aware of the laws that govern both the processes by which we create our communications and the content of those communications. There are many communication situations in which behavior most of us would consider unethical is also illegal. Our legal responsibilities are well documented.

Laws represent the highest shared values of the land. They are critical shared values in the hierarchy suggested by Payne and others (Chapter 2). These values reflect the highest consensus in society and, consequently, are backed by law. In point of fact, many laws were originally codified because occasions arose in which people chose to behave unethically, resulting in injury to others. Extreme and harmful instances of dishonesty, for example, led to the implementation of truth in lending laws.

Consider again the movie *Forest Gump*. Tom Hanks' character appears to meet and shake hands with presidents Kennedy, Johnson, and Nixon. Thanks to masterful manipulation of media, the film's editors create the impression that these meetings did, in fact, take place. As we said in Chapter 4, such videographic manipulation is good entertainment. Carried to an extreme, however, it and similar deceptions can be criminal.

Let's say, for instance, that instead of splicing film of an actor into newsreel footage of a presidential press conference, a campaign manager splices film of the opposing candidate into video of a meeting between prominent crime figures. Of course, the "meeting" never actually took place, but he releases the video to the press. It leads the 6:00 o'clock newscast. An extreme and unlikely example, to be sure, but one that illustrates how unscrupulous dishonesty can cross the line to blatant illegality—in this case, defamation. (You defame someone when you publish, or otherwise make public, lies that masquerade as fact—opinions are allowed—and injure his character.) Although the two forms of defamation, libel in written communication and slander in oral communication, have traditionally been associated with yellow journalism rather than technical communication, you could potentially defame a competitor by implying untrue criticism in a marketing brochure. More to the point here, though, is the fact that many laws govern our profession, protecting communicators and audiences alike.

Unfortunately, these laws do not cover every conceivable circumstance. So, let's go back to those everyday scenarios with which we can readily identify. Let's say, for example, your professor in an introductory technical communication course reads your most recent assignment aloud to the class as an example of a report that is badly written. He stops frequently to emphasize the poor choice of words, the weak sentence structure, and the lack of adequate transitions. Although he does not identify the paper as yours, he places it on top of the stack of assignments and immediately returns the papers. Clearly, it is impossible for your classmates not to realize that the offending report is yours.

Did your professor break any laws? No. He did not reveal your grade, thereby violating your right to privacy. He did not slander or libel you, or defame your character.

Did he behave unethically? We think so. Publicly pointing out the embarrassing shortcomings of your work, even under the guise of anonymity, was, at the very least, inconsiderate, boorish, and unprofessional. Most of us would agree that his behavior was also unethical. Obviously, similar thoughtless behavior occurs thousands of times daily, in professional as well as academic settings. Bosses ridicule employees in like fashion in corporations throughout the land. However, the scope of such behavior is too broad to effectively regulate and govern. The sphere of what legislation can address is quite narrow, leaving many gaps that can be filled only with human decency.

Legislation governing communication has such gaps. For example, technical communication is very much shaped by technology; courts and legislators can't possibly keep pace with the flood of advances in technology, so laws always lag behind.

If we can't be guided by laws, we must be guided by ethics. When facing situations that fall between the cracks of the law, then, we must make ethical decisions; but ethical responsibilities often fall within the gray zone. Indeed, life would be easier if we had prescriptive rules to follow in every situation. The legal perspective, however, is clearly narrower than the ethical perspective.

Obviously, we can't present an exhaustive examination of all of the laws—international, federal, state, and local—that bear on our profession. But we *can* identify areas with which you should become familiar; address some of the more frequently encountered issues; and examine how they can affect you, your company, and our profession. We aren't engaged in providing professional, or even off-the-cuff, legal advice. We do, however, advise you to be familiar enough with the legal implications of your actions to know when it is advisable to contact your personal or corporate attorney. And when the law doesn't address a particular situation in which you must decide on a course of action, be keenly attuned to the ethical ramifications of your behavior.

Laws Protect Intellectual Property

For the most part, the laws with which technical communicators deal most often govern the products of people's minds, known as intellectual property. To look at it another way, intellectual property is the tangible expression of information, which can be a significant corporate asset. As technical communicators, our primary responsibility is to uncover, develop, and disseminate information—this intellectual property—so its legal and ethical use is of primary concern to us. Copyrights, patents, trademarks, and other regulations protect original works such as documents, art, designs, inventions, and the like, that are the expressions of our thoughts.

To qualify for legal protection, intellectual property must meet certain criteria; clearly, not all products of someone's thought can meet the criteria. Situations in which intellectual property is not protected by law become the object of ethical decisions for the communicator.

Understand Copyright

We use and create copyright-protected material on the job every day. The manuals we write, the videos we produce, and the illustrations we draw are all protected by copyright law. Much of the source material we uncover in our research is also copyrighted. Clearly, then, we need to understand both the legal and ethical implications of creating and using such material.

United States copyright laws protect works of art, scholarship, and intellectual inquiry. Whether published, displayed, or performed, a work need not be made public to be protected by copyright. In essence, the law protects the products of intellectual property immediately upon their creation and for the life of the creator plus 50 years. If, however, you create a work while in someone else's employ, as part of that employment, the copyright may belong to your employer rather than to you. The duration of copyright protection for such works for hire is 100 years from creation or 75 years from publication, whichever period is shorter.

Specifically, the law addresses the rights of both the author and the users of protected material, establishes requirements for use of the © notice, and specifies procedures for registering intellectual property for protection. Numerous texts, including many listed in our bibliography, provide more thorough discussion of these rights. We recommend that you consult them to familiarize yourself with the regulations, as we provide only the broadest overview. We also recommend that you consult your personal or corporate attorney for direction when you create works that may be eligible for protection.

Practice Fair Use

Although copyright laws are intended to protect intellectual property, no one wants to hinder the dissemination of knowledge. Therefore, in deliberating copyright cases, courts have long upheld the doctrine of *fair use*. Fair use holds that limited use can be made of copyright-protected material for personal or scholarly purposes. Fair use is defined by four factors:

- Nature of the copyrighted work
- Purpose and character of the use of the copy
- Proportion of material copied
- Effect of the copy on the commercial value of the protected work

Each of these criteria is somewhat vague, however. For instance, to qualify for fair use, the nature of the protected material must be *critical* to its use in the copy. (By the way, the use of the term *copy* in this context refers both to photocopies and to using someone else's words in your own writing.) *Critical* is a broad term, one that is not clearly defined. When a critic reviews a new novel, she may need to use quotes to support the claim that the novelist's descriptions are fresh and carefully crafted. Certainly in that case, the quote is critical so it falls within the guidelines of fair use. Quoting the lyrics of a song on a newly released compact disc is very likely critical for a reviewer to illustrate that the words are timeless and moving. Putting those lyrics to

"Stop me when I hit the limit for fair use."

another melody and performing them on stage, however, hardly fits the definition of critical.

The definitions of "purpose and character" and "proportion" are equally murky. Libraries, for instance, receive special dispensation to make restricted numbers of copies for use within the library. But in other circumstances, just how many copies are too many to be considered fair use?

Examining some court decisions concerning fair use may provide some clarification. In the 1989 case of Norell versus Freeman, the Ninth Circuit Court ruled that a romance novelist's use of historical facts and "ordinary" phrases copied from a history book constituted fair use. The court based its decision on two fair use principles: Historic facts and everyday expressions are not original creations, so they are not protected by copyright; use of the material did not adversely affect the value of the history book.

In an earlier case, however, Marcus versus Rowley in 1983, the same court found that a home economics teacher's classroom use of recipes from a cookbook without attribution constituted copyright infringement rather than fair use. This decision was due, in part, to the fact that the teacher violated the proportion of use criteria by using fully one-third of the cookbook in her class.

Clearly, you can use copyright-protected material in a limited manner in your scholarship, research, and teaching, in ways that are beneficial to the public and fair to the material's creator. However, you would be wise to be much more wary of using copyrighted material in commercial ventures. In short, only specific instances comply with fair-use doctrine. You may freely quote from this book without fear of infringement, for example, if you are reviewing it for a professional journal or writing a news story about the authors for the public media. You may also quote us in your next term paper, assuming, of course, that you give us proper credit for our words. (Giving credit to the creator of intellectual property is called attribution, which this chapter discusses later.) In either case, we won't prosecute you for infringing on copyright. But when you write your own book about ethics for publication, you had better be careful to acquire our permission before lifting entire chapters!

Beware Plagiarism

Some works that meet all of the criteria for protection can't be copyrighted. United States government documents do not qualify for copyright protection. It is assumed that all work published by the federal government belongs to the citizens who are served by that government. Therefore, the law does not protect such publications. Does that mean it is ethical to make unlimited use of such works? Of course not. The unethical use of others' works is *plagiarism.* Plagiarism is what happens when attribution doesn't. This is true whether the original document is copyrighted.

Whenever you wish to use material or borrow an idea from a work that is not copyright-protected, you must give proper credit to the creator of that work, just as you would if that work were copyrighted. If you do not properly document your use of someone else's intellectual property, you are engaging in plagiarism. Quite simply, plagiarism is the use of someone else's work or ideas without crediting that person as the creator of that intellectual property. Keep in mind that the original work need not be published. Copyright laws protect intellectual property immediately upon its creation, not after its dissemination. The same principle applies to plagiarism.

Perhaps some of your earliest memories of writing—those term papers you produced in middle school—are rife with teachers' warnings against plagiarism. You probably spent many hours in English class learning the sundry rules and formats for footnotes and bibliographies so that you could properly attribute information, quotations, and ideas to those who created them. In those days, you were expected to complete your own work as an exercise in collecting and critically evaluating information. In teachers' lectures and in school honor codes, therefore, the focus was on misrepresentation, the dishonor of fraudulently claiming someone else's work as your own.

Plagiarism is not restricted to the halls of academe. If, for instance, a business colleague creates a particularly clever turn-of-phrase during a brainstorming session, you are plagiarizing if you use it in your next document without attributing it to your colleague. Your response to such conceptual pickpocketing may be, "But, it's all in-house." Yet, the ethics of dissemination of information require that the person who creates the intellectual property get the credit, even if it cannot be protected under copyright law. There are no *if*s, *and*s, or *but*s. You would not take a colleague's computer from her desk for your own use, would you? No, you answer. Well, that's "in-house," too. Likewise, you should not take her ideas. Laws governing copyright, patents, and trade secrets are not concerned with the fraudulent presentation of material; they are concerned with the theft of intellectual property.

Plagiarism is a serious matter. In academic circles, it can lead to failure on an assignment, failure in a course, or expulsion from an institution. In business, where profit is involved, plagiarism is closely allied to copyright infringement, which can lead to lawsuits and sizable financial settlements, among other consequences. In addition to compensation for damages, someone who is found guilty of copyright infringement may have to pay the copyright's owner all profits made as a result of the infringements as well as all legal fees involved in the owner's pursuance of the case. Moreover, numerous court decisions have held that an employer is liable when an employee infringes on copyright in the course of pursuing the duties of employment. So if an ambitious writer whom you supervise includes a dynamic slogan from someone else's work in your company's next brochure, both you and your employer share liability.

Legality on the Information Highway

The further proliferation of information available electronically—via the Internet for instance—complicates the ethical matter. The speed and immediacy with which we can share and access information electronically makes acts of plagiarism, even copyright infringement, easy to accomplish but difficult to detect and prosecute.

The information highway also blurs geographic and legal boundaries. Many foreign countries have copyright laws similar to those in the United States. Many even grant copyright protection to U.S. citizens. The © symbol is recognized by a Universal Copyright Convention. Similarly, the unpublished works of foreign citizens in the United States are protected, as are their works when first published in this country. Finally, works published abroad by non-U.S. citizens may also be protected by U.S. copyright when the author's country of citizenship has formalized copyright agreements with the United States. The point is, when copyright-protected information has international distribution and availability, a plethora of additional regulations and all their permutations applies—further expanding the zones of gray.

Safeguarding Electronic Images

A related ethical and legal dilemma, and perhaps one of the grayest areas of copyright protection and fair use, is the manipulation of electronic images. Clearly, technology advances faster than courts and legislators can keep up. Scanning, desktop publishing, and image manipulation software give communicators the means to modify images quickly and easily, whether the images are copyright protected. As a result, creators of such images have less control over the use and integrity of their creations than do communicators.

As a technical communicator, you will often find yourself in a position to manipulate images or to be a party to such manipulation. You may, for instance, find a photograph that is almost, but not quite, perfect for the cover of your latest manual. Just a few changes—delete a person here, add a piece of equipment there—and it will portray your company and its products beautifully. In just a few minutes and half a dozen mouse strokes, your designer can accomplish the changes. But is it right? If you have any doubt, get permission from the owner of the copyright. Use a stock photo that better fits your needs instead of manipulating one that doesn't. And be sure you understand the agency's rules for appropriate use of its photo. (For a discussion about the honest use of photographs, refer to Chapter 4.)

What Laws Govern Collaboration?

The predominance of collaboration in technical communication can cloud these issues even further. Often, we depend on subject matter experts—the engineers or software developers with whom we work—for the technical information we include in our publications. In other instances, project responsibilities are divided in a way that gives another team member the research tasks. As John G. Bryan, professor at the University of Cincinnati and manager of STC's ethics committee, points out, the result of the collaborative nature of our profession is that "opportunity and rationale for deflecting responsibility are myriad" and defy regulation.[3] Because we rely on others for much of the specialized knowledge we incorporate in our work, we must be particularly careful of the pitfalls involved in using intellectual property provided to us by someone else.

Regulating Multimedia

Accelerating advances in technology afford us extraordinary and diverse media for expressing our message. Unfortunately, these same advances deepen the quagmire of legal and ethical difficulties. Multimedia capabilities enable us to combine video, sound, animation, narration, and other media with our text. Each of these media is governed by its own set of regulations and procedures for licensing, granting permission for use, and other legal rights. Moreover, as technology improves and the legal profession becomes more familiar with multimedia, issues will continue to evolve. When developing multimedia projects, you need to carefully identify the origin of every element you use and seek legal advice to navigate the regulatory mire.

It is in your own best interest, then, to be knowledgeable about ownership of intellectual property and about copyright law. It is equally important to be aware of the ethical implications of the using what other people create.

Securing Company Information

Every corporation develops and maintains a substantial amount of intellectual property that can spell the difference between success and failure, profit and loss. Frequently communicators have access to this material as they develop publications, so they must be aware of the laws that control use of corporate intellectual property.

Trademarks and Service Marks

Technical communicators frequently deal with trademarks and service marks. It's likely that your own company's products or services are protected in this way, as are those of other companies you may need to mention or acknowledge in your documentation.

A trademark allows a company to distinguish its products from those of its competition; service marks function the same for services. Distinguishing features that can be protected include words, designs, slogans, names, sounds, and so on. Of course, thanks to advertising, you're familiar with trademarks such as Coca-Cola®, Frito-Lay® or the checkerboard design of Ralston Purina® and with service marks such as Federal Express. Note that the use of the symbol ® identifies a trademark or service mark that is registered with the federal government. Use of the symbols ™ and ℠ for trademark and service mark, respectively, serves notice that federal registration has been or is likely to be sought. Only by registering a trademark or service mark can you protect it, however.

Owners of these marks must vigorously guard them to maintain legal protection. In fact, the owner can actually lose protection by failing to protest or prosecute the use of similar marks or by allowing the marks to become descriptive of a group of similar goods. Have you ever used the word *Kleenex* to refer to a tissue made by another manufacturer or said you were going to *Xerox* a paper rather than photocopy it? Did you know that the word *aspirin* began as a trademark but lost its protection under the law when the word became commonly applied to an entire class of pain relievers? That's why you'll often see advertisements and notices in trade and professional journals aimed at protecting trademarks and service marks, such as those shown in Figure 5.1.

Although protecting these marks against infringement is the responsibility of the owner, communicators have an ethical responsibility to respect the rights of those owners. Infringing on these marks means expensive legal battles and, if you're found guilty, even more costly penalties. So you must be careful that if, for instance, the software you are documenting runs on a particular hardware platform, you identify all hardware trademarks, as we have identified marks in this chapter. But that is not the only caution. You must also beware that when you use another company's trademark or service mark, you do not suggest or imply that the company endorses your products.

Patents

What about the protection of intellectual property that has practical, utilitarian application and, therefore, cannot be secure under copyright law? United States law doesn't overlook such property. Patent laws protect inventions and the applications of those inventions, including processes, compositions of matter, new life forms, and designs. Although patent laws don't provide for exclusive ownership of the invention itself, they do grant a monopoly on the development, use, and manufacture of the invention for 17 years.

Although technical communicators have traditionally had less occasion to be concerned with patent regulations than with copyright and trademark law, technological advances are having as significant an impact in this legal arena as in others we have examined. Patent infringement is often difficult to

Figure 5.1 Clorox ran this ad in a publication directed at writers. Such efforts are imperative to maintain trademarks on familiar products.

avoid. Because patents aren't as obviously marked as the © and ® notices, you may be unaware that a patent exists. And, unlike copyright infringement in which you clearly copy something tangible, you can violate patent protection even if you aren't deliberately copying someone's product.

What are the repercussions of patent law violation? A 1986 ruling against Eastman Kodak for infringing against a patent held by Polaroid resulted in the recall of (and reimbursement for) millions of cameras. That, in turn, led to the closure of a Kodak manufacturing facility and the layoff of hundreds of employees—all before damages were even awarded. Rather chilling.

The aspect of patent law most likely to affect our profession is the recent increase in software-related patents. Both the processes incorporated in software as well as combinations of software and hardware are patentable. In other words, patents can protect such software features as the particular way in which data is presented or the way programs manipulate data.

These regulations are currently of more concern to software developers than to software documentation specialists. Whether you are documenting products that may contain patented processes or gathering material for a multimedia presentation, however, you need to be vigilant in identifying ownership of all forms of intellectual property to avoid possible infringement.

Trade Secrets and Confidentiality

As business practices become more complex, copyright and patent laws no longer afford full protection in all circumstances. For example, when a company develops a product formula that gives it a competitive advantage, protection under patent laws is valid for only 17 years. If the formula is likely to be valuable for longer than 17 years, it may not be in the company's best competitive interest to patent it. In other instances, competitively valuable ideas may not meet the criteria for protection under patent law. Instead, they may meet three criteria to qualify as a trade secret:

Economic value. The information must be used for some business purpose; in other words, an idea that pops up in a brainstorm session does not count as a trade secret until it is actually put into use by your company.

Limited knowledge. The protected material must result from some intellectual activity. It cannot be common knowledge to the public or to others in the same or a related industry.

Secrecy. The secrecy of the material must be aggressively protected by the company that claims it as a trade secret.

Technical communicators frequently have access to trade secrets. After all, to adequately document our company's goods and services we must fully understand both the product and the market. While writing the marketing brochure for the mountain bicycles your company produces, you might interview a design engineer who chatters endlessly and in specific technical detail about the composition of the bike's frame. In fact, he brags, it is that very feature that is predicted to boost the company's sales 53 percent over the next year. Obviously, then, this is a key selling point worthy of coverage in your sales materials. But before you rush to the keyboard, think a moment.

Companies must make every effort to protect trade secrets. Safeguards include restricted access, controlled distribution, contractual obligations, even special procedures for the disposal of trash. Why such elaborate precautions? Even if the trade secret is disclosed accidentally or stolen and used with malicious intent, its disclosure invalidates the protection. Companies can sue those who disclose or steal information, but the competitive value that is lost can't be recovered.

This brings us back to the bicycle frame. If, in fact, its composition gives your company a demonstrable advantage in the market, it is a trade secret. Publishing that information, then, negates its protected status. In fact, *any* unrestricted disclosure of the bike's composition will end its status as trade secret. The proud engineer violated the secrecy provision by sharing the information with you, someone without a need to know. By including it in a brochure, you would certainly finish the deed. That is why it is important for communicators to understand the nature of trade secrets or confidential information and the ramifications of disclosing such information. Keep in mind, that as an employee you are legally responsible for the protection of that information. So we, as technical communicators, must have both the knowledge to recognize when we come into possession of this kind of information and the judgment to know not to use it.

Some valuable or competitive information that cannot be safeguarded under any of the laws we've examined may be protected by laws of confidentiality. A company may copyright its marketing documents, for example, but if a former employee-turned-entrepreneur remembers and uses the information contained in those documents without copying it, copyright law provides no protection. Laws governing the confidentiality of information may, however, protect the company against the entrepreneur's use of that information. Typically, companies establish policies and procedures to identify and safeguard confidential material similar to those for trade secrets. Be sure that you understand and follow these policies, too.

One more caution about dealing with these types of protected information. Often, you may supervise the work of freelance, contract, or other temporary employees. Be especially careful to oversee the nature of the information to which they are given access.

Nondisclosure Agreements

Aggressive maintenance of the sensitive nature of a trade secret or confidential information depends on the willingness of employees to protect that information. So it is typical for companies to require that employees, consultants, contractors, and others who have access to valuable information sign agreements stating that they will not disclose that information. In the past, such nondisclosure statements asked personnel to agree not to divulge information of a confidential or competitive nature. In fact, many employers are now taking even more stringent steps to safeguard intellectual property. Recent nondisclosure agreements stipulate specifically which pieces of information are considered trade secrets; personnel may be asked to sign new agreements as they gain access to new or additional protected information. In this way, both the company and the employee acknowledge specifically which pieces of information are protected, giving the company greater leverage in court.

Don't treat nondisclosure agreements lightly. Before signing one, know what you're agreeing to abide by. If you agree never to use or disclose specific information, you can be held to that promise even if what once was secret becomes public. And remember that such agreements are enforceable; violations carry strict penalties like injunctions against using the material and payment of damages.

Specifications Control Document Content

Our discussion so far has centered on those laws and regulations that protect our own intellectual property and govern our use of the intellectual property of others. These are not the only laws that technical communicators need to know about. A significant number of laws govern what we produce and what that work contains. Often the content of our documents is also dictated by law, regulation, or government guideline. Those laws vary depending on our fields of specialization. It is, therefore, impossible to identify and discuss every such law in this text. Suffice it to say, you must become familiar with all of the regulations governing your particular area of expertise. It is, of course, in the best interest of our continued employment to be certain that we fulfill all laws and regulations governing our respective industries. A few examples will demonstrate the nature of the regulations that communicators must deal with routinely.

One field that is heavily regulated is the environmental industry. Several federal environmental laws govern the use of communication for companies and agencies that deal with hazardous materials; consequently, this arena serves as a good example of the types of laws we are examining.

A number of states have enacted corresponding laws, and many federal, state, and local regulations provide communication guidelines. For now, however, let's look at the scope of the federal laws as representative of the types of legal situations we must be aware of as technical communicators.

Two laws, in particular, govern communications relating to hazardous materials. The Emergency Planning and Community Right-to-Know Act mandates that the public be provided with information about the existence of hazardous chemicals in the community. The National Environmental Policy Act requires the creation of environmental impact assessments and statements, listing a variety of publications an agency must produce for public access depending on the potential of the action to have an impact on the community and its environment. Clearly, if you are the communications specialist for an agency or company working in the realm of environmental risk, assessment, or remediation, it is your job to know what documents are required and to be familiar with the guidelines for producing those documents.

Other laws are not directed solely at identifying mandatory documentation; they do, however, have such requirements embedded in them. The Comprehensive Environmental Response, Compensation, and Liability Act, better know as the Superfund Act, is primarily intended to outline the procedures that must be implemented in assessing the release of hazardous substances at inactive waste sites. Buried within the legalese are requirements for communication pieces intended to support the required involvement of the public in the cleanup of such sites. A community relations plan, for instance, figures prominently in the required community relations process. Again, if your employer must abide by the mandates of the Superfund Act, you had better be familiar with the communications pieces also mandated by that law.

Clearly, the repercussions of not following the regulations can be severe. The most obvious penalty is the lawsuit, from which you obviously want to protect yourself and your employer. Many citizens and environmental protection groups have an astute understanding of the laws governing that industry. It is not unlikely, then, that failure to comply with the regulations can result in a lawsuit alleging failure to follow due process. Such a lawsuit has additional repercussions. Pending outcome of the suit, for instance, your company can be enjoined from continuing operations, resulting in lost revenue, fines, and irate clients. Funding can be suspended or cut, again with unpleasant results. It certainly does behoove us, then, to be familiar with all of the laws and regulations that govern communications in our areas of specialization.

Legal regulations are not limited to the environmental field. Many other specialized fields are governed by similar laws, regulations, and requirements. Department of Defense specifications dictate the content and organization of maintenance manuals and other publications. Statements of work impose deliverable publications as part of a contract. Consumer protection laws compel us to provide documentation that ensures the safe use of our company's products. Regulations governing procurement outline rules for estimating the costs of our publications. The list is extensive and growing. Check with federal, state, and local agencies as well as your corporate or personal attorney to ensure that you understand the communication efforts required by law and make certain that your publications comply with those laws.

The Law and Contract Issues

Part of our legal responsibility is to abide by all terms of contracts we enter into. Whether you are a full-time permanent employee, a consultant, or a contractor, you may be asked to sign an employment contract. Of course, major contract violations can land both parties in court. But minor breaches may never go that far. They call for ethical behavior—honoring our commitments.

Contracts protect both the employer and the employee. For the company, a contract can spell out policies, such as nondisclosure agreements, identify security requirements, and stipulate length of employment. On the employee's part, contracts can specify salary and benefits, define responsibilities, and designate performance bonuses. Contracts can also be crucial in avoiding potential ethical conflicts: They can define ownership of work made for hire, designate licensing agreements for patentable inventions, delineate limitations on subsequent employment, and clarify conflict-of-interest issues. A typical scenario, for instance, might stipulate the specific circumstances under which you would be considered to be involved in conflict of interest by agreeing to write software documentation as a freelancer while you are employed full time by a software development company.

Contracts take many shapes, from the proverbial handshake to extensive forms laced with legalese. Bear in mind that verbal contracts are upheld in court, although proof of any stipulations is nearly impossible without witnesses. Later chapters look at other aspects of contractual responsibility, such as conflict of interest (Chapter 9) and quality (Chapter 7).

Summary

Many ethical issues cross the line to become legal issues. Our duty as technical communicators is to obey the laws governing our profession.

Many of these laws pertain to intellectual property, the products of our minds. Copyright, trademarks and service marks, and patents all provide legal protection for individual creations. Copyright protects the tangible expression of ideas in words, pictures, music, and so on. Trademarks and service marks protect the quality of our products. Patents protect utilitarian inventions.

Some intellectual property does not meet the criteria for these types of protection. In other instances, it is in our commercial best interest not to register for protection. In such cases, valuable competitive intellectual property can be safeguarded under restrictions concerning trade secrets and confidentiality.

Owners of material that is protected under all of these legal devices bear the responsibility for protecting their intellectual property and detecting infringements. With the vast wealth of information that is readily available today, particularly in electronic format, it may be an easy task to use protected information improperly—and to get away with it! It is our duty, then, to abide by the laws and consider the ethical ramifications of our actions in using the creations of someone else's intellect. Finally, we have both a legal and ethical obligation to abide by all terms of contracts to which we agree.

Application Exercises

1. Match the description below with the laws that protect the intellectual property.

    a. A company invents a new method of securing information transmitted via Internet through encoding.

    b. A company develops a list of businesses that are likely prospects for purchasing an innovative product.

    c. A professor develops extensive lecture notes to use in class.

2. Would it qualify as fair use to use an exact replica of Hobbes, the cartoon tiger of *Calvin and Hobbes*, in the masthead of a student-produced newspaper for a high school whose mascot was a tiger? Would it qualify as fair use to use a modified sketch of Hobbes? Justify your answers.
3. You and a colleague collaborate to write a book, which you publish and copyright yourselves. Each of you writes the draft of half of the chapters, after which you edit and rewrite each others' material. Several years later, you want to write another book by yourself, using some of the material from the book you wrote with your colleague. Are there any legal issues involving copyright? Any ethical issues?

Exploration Exercises

1. You are a student in a senior-level technical communications course. The major assignment is to write a paper of publishable quality. You submit a proposal to present your paper at a professional conference and it is accepted. You are finalizing the paper, which you will submit both to the professor for a grade and to the professional organization for publication in the conference proceedings. Subtly, the professor lets it be known that she expects to be listed as coauthor. After all, she has guided your research throughout the semester, even though she did not write a word of the final paper. What would you do? What should you do?
2. Spend some time browsing the Internet. What material do you find that you could easily misuse—either unethically or illegally?
3. Create your own scenario involving copyright law and technical communication. Analyze the underlying issues and suggest possible courses of action, as well as probable consequences of each.
4. Several court decisions within the last decade have had a major impact on the way in which businesses handle copyright issues. Look up one of the following decisions at a law library and write a brief paper analyzing how the decision could influence business: Sony Corp. of America v. Universal Studios, Inc., 1984; Harper & Row Publishers v. Nation Enterprises, 1985; Basic Books, Inc. v. Kinkos Graphics, 1991.
5. Although laws are slow to change, court decisions can more quickly alter the spirit of the law. Visit the local law library, and write a brief paper explaining the impact and ramifications of decisions regarding copyright made by the United States Supreme Court within the last five years.

References

1. Our thanks to a colleague for providing this quotation, attributed to Thomas Mallon, although we have been unable to track down the precise source.
2. According to *Bartlett's Familiar Quotations*, this version of a popular saying is most often attributed to Mizner.
3. John Bryan, University of Cincinnati, copyright discussion on TECHWR-L, October 1995.

Other Sources Consulted

Amernick, Barton A. *Patent Law for the Nonlawyer: A Guide for the Engineer, Scientist, Technologist, and Manager*, 2d ed. New York: Van Nostrand Reinhold, 1991.

Branscomb, Anne W. *Who Owns Information?: From Privacy to Public Access*. New York: Basic Books, 1994.

Don, Robert. *Protecting Trade Secrets, Patents, Copyright and Trademarks*. New York: John Wiley, 1995.

Elias, Stephen. *Patent, Copyright, and Trademarks: A Desk Reference to Intellectual Property*. Berkeley, CA: Nolo Press, 1996.

Foltz, Raymond. *Handbook for Protecting Ideas and Inventions*. Cleveland: Penn Institute, 1992.

Rose, Mark. *Authors and Owners: The Invention of Copyright*. Cambridge, MA: Harvard University Press, 1993.

Samuels, Jeffrey M., ed. *Patent, Trademark and Copyright Laws*. Washington, D.C.: Bureau of National Affairs, 1993.

Chapter 6 Overview

Privacy . . .

The Mission: **Privacy in technical communication means making every attempt to respect the rights of others. It means protecting the rights of both individuals and our employers in gathering and disseminating confidential information.**

Chapter Roadmap

Definition and Focus. This section provides a philosophical framework for privacy as a core value.

The Emergence of Surveillance. This historical overview explains how privacy emerged as a treasured American value.

Communicators Face a Multitude of Privacy Concerns. This section familiarizes you with several ethical issues communicators must deal with regarding the products we develop and the way we gather and share information. Two particularly thorny issues are:

- *Whom Must We Protect?*
- *Communicating via E-Mail*

Is Corporate Big Brother Watching? A great deal of information about us resides in company files. This section points out some ethical issues you must confront on the job.

Summary, Practical Exercises, and References

. . . Our Duty to Respect the Rights of Others

CHAPTER 6

Privacy: It's a Delicate Balance —the Need for Protection and the Desire for Privacy

Information is sacred. Information is power.
And society must protect it at all costs.

—*The Net,*
Columbia Pictures, 1995

Scene: *A bookstore in New York City*

Marie: *So, I just happened to see his American Express bill . . .*

Sally: *What do you mean you* ***just happened*** *to see his American Express bill?*

Marie: *He was shaving and there it was in his briefcase . . .*

Sally: *What if he came out and saw you looking through his briefcase?*

Marie: *You're missing the point. I'm telling you what I found.*
He just spent $120 on a new nightgown for his wife.
I don't think he's ever going to leave her.

—*When Harry Met Sally,*
Castle Rock Entertainment, 1989

Definition and Focus

Most of us have never had occasion to furtively search through the briefcase of a married lover as did Marie, played by Carrie Fisher, in the scene previously quoted, or through his pockets as she does in an earlier scene. And even if we found ourselves in Marie's precarious romantic situation, we surely would never so blatantly invade a lover's privacy, would we?

But as a parent, have you rifled through your child's sock drawer or rummaged through the piles of dirty jeans and tee-shirts in the bottom of her closet? In this day of peer pressure, widespread drug abuse, and other grave dangers, many parents feel justified in covertly keeping tabs. In their eyes, it is a form of protection. In their children's eyes, it is an invasion of privacy.

These familiar scenarios illustrate the same question—what are the acceptable bounds and limits of an individual's right to privacy? Of course, the puerile foolishness we engaged in as kids has few, if any, consequences, whereas the repercussions of parents' actions can be (and often are) more severe. Let's look at a case where the stakes may be even higher.

A felon convicted of a series of particularly brutal rapes of teenage girls is about to be released from prison on parole. He hopes to begin anew, quietly residing incognito in a nearby neighborhood. But a citizens' group, led by

the parents of his victims, insists that the rapist's new address be made public. It isn't fair, they repeatedly tell the media and the parole board, for upstanding taxpayers to be kept in the dark about a heinous threat lurking in the very suburbs where they moved to keep their daughters safe.

This situation, which you realistically might hear on this evening's news or read in tomorrow morning's headlines, is typical of what is certainly one of the most hotly debated ethical issues of our time. How do we balance the need to protect the public with the individual's right to privacy? Nor is the dilemma limited to the public need to know—corporations find themselves in a similar quandary. How does a company square its need to acquire the information necessary to protect the interests of stakeholders with its desire to protect employee privacy?

In her book *Mind Your Own Business*, Dr. Gini Scott, consultant on privacy issues, identifies this conflict over personal privacy as an historic watershed because one's sense of privacy:

> *. . . goes to the core of one's very persona, since it is central in defining the self and distinguishing the boundaries between self and society. . . . However, as more and more of the private self becomes open to public inspection, that boundary between public and private becomes swept away.*[1]

Individual privacy is, indeed, being assailed from all sides. Every documentary news show from *Geraldo* to *60 Minutes* has enlightened us about how security guards watch shoppers in dressing rooms through two-way mirrors, how employers spy on employees with tiny cameras hidden in pencil sharpeners, and how parents observe babysitters with video equipment concealed in the coat closet. And that's just the tip of the iceberg. All sorts of companies, government agencies, and individuals gather, maintain, and disseminate all sorts of information about us.

So how do we, individuals trying to maintain the boundaries between self and society, deal with this conflict ethically? And more to the point of this book, what are our ethical obligations in regard to privacy as professional communicators?

Privacy in technical communication means making every attempt to respect the rights of others. It means protecting the rights of both individuals and our employers in gathering and disseminating confidential information. You'll also note that some of the ethical dilemmas we discuss are not directly related to our jobs. They are, however, related to our individual privacy and that of our friends and families. As experts in communication, we have a unique ability and opportunity to confront such issues; therefore, we have an ethical responsibility to do so.

In particular, we'll look at such areas of privacy as intrusion into private affairs, disclosure of private facts, or collection and dissemination of personal information. We will not, in this context, examine another commonly debated aspect of privacy—the individual's freedom to make personal lifestyle choices.

Before we examine our ethical responsibilities as communicators, let's begin our foray into privacy by briefly looking at its history as a cherished American value and the laws and court decisions that govern it, so we have an historical framework on which to hang our discussion.

The Emergence of Surveillance

Privacy is fundamental to our society. Drive around any Denver suburb and you'll be greeted by miles of "privacy fencing." We post signs identifying our yards as "private property." We join "private clubs." We send our kids to "private school."

Interestingly, this was not always the case. Unlike values such as honesty, quality, and legality that have been held dear through the ages, privacy is a relatively new concept. Privacy was difficult when multiple generations of a family lived in the same small house, for instance, and undesirable when people relied heavily on their family, friends, neighbors, and community for their very survival. But as the population increased, cities grew, and society moved west, people increasingly became strangers and individualism became a treasured concept. In other words, the value of privacy strengthened as community and family bonds weakened.

One side effect of this growing level of estrangement within communities was a requisite decrease in the trust that derives from familiarity. Without close ties throughout the community, people were no longer under the direct scrutiny and influence of neighbors, family, and friends. (Think about it—how often have you seen a reporter interview the neighbor of a felon and heard, "he was so quiet, always kept to himself"?) Without familiarity, then, we had to find other means of verifying people's integrity. We can no longer feel even relatively secure that people will conform to our expectations; therefore, we feel compelled to gather credentials and watch their activities in an attempt to protect our interests.

That gathering and watching, in turn, affects each of our lives. Your social security card, driver's license, credit cards, and telephone orders all "put you in the computer"—well within the reach of anyone, scrupulous or otherwise, who has the capability and desire to track you down.

The 1995 movie *The Net* captured the public's attention—and attendance—by dramatizing our worst fears. In that film, a computer hacker, played by Sandra Bullock, accidentally uncovers a secret corporate conspiracy. The perpetrators retaliate by electronically removing all traces of her existence. As the plot progresses, she realizes that the bad guys learned the details of her life by tracking her activities on the Internet. They knew what she ate, drank, and smoked; the movies she watched; and where she vacationed. As Sandra Bullock's character explains:

> *Our whole world is sitting there on a computer . . . your DMV [Department of Motor Vehicles] records, your social security number, your credit cards, your medical history . . . everyone is stored in there. It's like this little electronic shadow on each and every one of us just begging for somebody to screw with.*

As numerous and varied forms of surveillance replaced mutual trust, people began to clamor for ways to maintain personal autonomy and a sense of self. One way in which individuals waged this battle was via legal suits. We also expected legislators to respond to the public concern over increased surveillance. Certainly, strides have been and continue to be made in both arenas to protect personal privacy. However, as we saw in the last chapter, it is impossible for legislators to pass laws at a pace which keeps abreast of the technological advances that continuously amplify the means of gathering, maintaining, disseminating, and misusing personal data. Moreover, as political pressures and climates fluctuate, so does the protection afforded individual privacy. Nor are court cases broad enough to protect all rights. Without adequate legal or legislative coverage, then, many privacy issues are ethical concerns.

Communicators Face a Multitude of Privacy Concerns

In the previous chapter we examined the legal issues we face as communicators, noting that some of our products that are covered by copyright law are also subject to additional ethical values such as privacy. Let's examine some of those privacy issues in greater depth.

Whom Must We Protect?

For the most part, concerns about the right to privacy in relation to communication have traditionally been associated with the media. Many of you, and much of the public, would agree that individuals must be protected from an intrusive press corps. Recall your reaction on an occasion when a particularly aggressive reporter stuck a microphone in the face of bereaved parents whose children were suddenly killed in a horrible traffic accident and badgered them with the question, "How do you feel?" You probably thought, "Just leave them alone. They deserve some privacy now." But that's a journalist, not a technical communicator, you say. We don't deal with such situations.

Well, laws and ethical issues of privacy can also play a role in the life of the technical communicator. For example, you may want to videotape a product in use at a client site for a sales brochure. You may want to photograph your company's facility in its bustling downtown location with the teeming lunch crowds for a college recruiting poster. Or you may want to tape an interview with a customer for use in a testimonial. All of these are privacy-related issues. So we do need to be aware of the ethical issues involved in the privacy of those who have a role in our communication products.

In general, the right to privacy ensures that we can be left alone to live our lives free from invasion except as justifiable to maintain the law. So we assume that unless we are breaking the law (or in the middle of a major news event), we will not be photographed, quoted, or otherwise intruded upon in our day-to-day activities.

In terms of photography, privacy means that photographers must obtain permission to publish anyone's picture. In fact, secondary subjects in a photo, even those included by accident, must also give their permission. A 1996 lawsuit brings the point home. A Merrill Lynch advertisement in *USA Today* depicts a man working on an airplane while a couple sleeps in the adjoining seats. Two people who claim to be that couple and who did not give permission for use of the picture, sued Merrill Lynch for nearly a quarter of a million dollars for invasion of privacy.[2]

Legally, whenever we use photographs of people, it is necessary to have them sign a release, as shown in Figure 6.1. Ethically, this is an issue, as well. For instance, we may assume that because we're photographing our company's facility for use in a corporate brochure, we need not seek legal permission from friends and fellow employees who wander into the shots. After all, they are employed by the company being touted in the sales brochure, so their participation should be a given. Right? Not right. Not ethical. Ethically, we owe our colleagues the same right to privacy as our clients and the public.

We need to be equally careful with the privacy of people's words. When developing technical marketing materials, we drool over those letters from customers extolling the fine performance of our company's widgets and the quick response of our support personnel. Before rushing to include those testimonials in proposals and brochures, though, we owe the correspondents the same ethical courtesy we give to those we photograph—asking permission to use the testimonial via a release form. Some companies prohibit the use of their names or those of their employees in this way, so we must respect their right to privacy.

Our ethical responsibility extends to finished communications as well as the materials that go into them. We often recycle portions of previously published materials in later projects. That testing procedure written for a proposal may be lifted for the bid on a similar project a year later. Photos from a

Privacy Release

I, ____________________, do give ____________________ [his/her assignees, licensees, and legal representatives] permission to use my [picture, photograph, name] in the form and media identified below for advertising, trade, or other legal reasons:

I hereby waive any right to approve final versions of the above. I have read this release and am fully aware of its content.

Signed: ____________________________ Date ________

If subject is a minor, release must be given by legal parent or guardian. By signing above, you certify that you are the legal parent or guardian, that you have legal authority to execute the release, and that you approve the release and waive all rights.

Witnessed: ____________________________ Date ________

Figure 6.1 Always request a privacy release when using photographs of people in your publications.

sales brochure may be useful in the annual report. Clips from an informational video may be useful in a training tape. Generally, such reuse is no problem within a company; however, ethical conflicts can surface when teaming agreements or multiple clients are involved.

If the proposal containing the original testing procedure was a collaboration between Company A and Company B, with Company B's engineers providing the bulk of the plan, good ethics demand that Company A get appropriate permission from them before using that material in a subsequent bid that does not involve them. Similarly, if several of the photographs you used in a sales brochure for Southern Gadget Corporation are ideally suited for the annual report you are now developing for Rocky Mountain Novelties, you should procure a release from Southern Gadget before using them. (This assumes, of course, that there are no legal or financial restrictions on using the photos. Refer to Chapter 5.)

There are other circumstances in which we communicators like to "recycle" the products we develop. Often, when we write, edit, or produce a particularly effective manual, newsletter, or multimedia show, we have good reason to use the material on our own behalf rather than the company's. We might enter it in a competition or include it in our personal professional portfolios. Bear in mind that each of these products is, in fact, the property of the employer or client for whom we produce it, so once again, we are ethically bound to acquire permission to use it for our own purposes.

Lately, we have begun to indulge in another type of recycling of information that is communicated to us—the reuse of data we receive and uncover via electronic mail. Now there is a hotbed of ethical *shades of gray*!

Communicating via E-Mail

Electronic mail and the Internet have been a boon to technical communicators, greatly expanding our capabilities to research and maintain contact with our peers worldwide. When we have a question on the job, the Internet enables us to quickly and easily find an answer. Simply send out a query before leaving the office tonight and tomorrow morning you're likely to have a number of responses from colleagues who have solutions. Unfortunately, the privacy concerns surrounding the use of the Internet are as extensive as the opportunities it opens for us.

For the most part, the speeding technology of the Information Super Highway far outstrips the government's ability to legislate about it. Some early legislation, not directed specifically at e-mail, does apply. The 1968 Omnibus Crime Control and Safe Streets Act prohibits wiretapping, but it was not sufficiently broad to cover new technology; therefore, the Electronic Communications Privacy Act of 1986 was passed to extend the protection of the 1968 regulation to all electronic communication. The recent statute, however, does not explicitly address all of the privacy concerns. Over the last few years, therefore, there has been a rash of articles, papers, and books addressing this ever-changing landscape. We recommend that you read a variety of such publications to acquaint yourself with all sides of the controversy. Our purpose here is simply to familiarize you with the ethical dilemmas that can, and likely will, arise and to suggest how you may apply value analysis to make appropriate decisions.

The overriding question about e-mail is this: *Is e-mail private*? Certainly, federal law prohibits us from opening and reading mail sent to someone else via the U. S. Postal Service. Unfortunately, we apparently lose that protection if we elect to correspond via the much faster option, e-mail. Of course, once we receive mail, it is ours to do with as we will. Generally, we assume that a personal letter on flowered stationary sent to a friend or loved one will be kept private; we probably also make the same assumption about electronic mail. After all, why would a trusted friend post a personal note to a public bulletin board?

One accused murderer in New Jersey learned a lesson about privacy (or lack of it) on the Internet the hard way. (Granted, this isn't a workplace example, but it makes our point so dramatically we couldn't resist!) The suspect allegedly communicated explicit details about the murder to an acquaintance via a chat session on America Online, fully believing his electronic correspondence was secure—particularly because the recipient deleted the message after reading it. However, the police confiscated America Online's master files and records, which were still intact, using them as evidence to warrant the suspect's arrest.

In an interview on *Today*, Sherry Steele of the Electronic Frontier Foundation said that consumers have a "misconception that [electronic] messages are totally private."[3] Not true. Statements in fine print found in the service agreements of most companies like America Online give service providers permission to read customers' e-mail. Appearing on the same show, David Phillips of America Online assured viewers that the company does "take client's privacy seriously," but admitted that company representatives "cooperate with law enforcement as appropriate,"[4] which they did in this murder case. The lesson for us in the workplace is clear—he who *owns* the system owns the e-mail on it and can read it if he chooses to do so. Indeed, most recent court decisions have upheld this principle.

By the way, this case was still pending at press time. It will be interesting to track its progress, and the arguments pertaining to this evidence, through the courts.

Most often, the privacy issues that arise concerning electronic mail lie not with the people with whom we correspond but with their employers. Employer monitoring of employee e-mail has sparked a raging debate about privacy. Courts and legislators are just beginning to grapple with all of the ramifications of this issue. In general, however, court rulings have favored system owners. In keeping with this philosophy, some companies are openly monitoring employee e-mail. According to a *USA Today* report posted to Internet, "Texaco tells employees they're being watched via the computer activity logs and no funny business will be tolerated—or they could be out of a job."[5]

So it is possible that your own employer, or the employer of the people to whom you send e-mail, can and will legally read your correspondence. Given that framework, what are the ethical implications for us as communicators?

First, be careful that your electronic communications are truly related to the duties you perform for your employer. That's not to say that you should never indulge in personal electronic exchanges. For instance, professional development is as valid a reason for corresponding with colleagues electronically as it was for writing to them on company letterhead in the days of "snail mail." In the case of a valid personal exchange such as that, however, the possibility of having your e-mail monitored is a matter of principle rather than fear of punishment for misusing company tools. (This issue is discussed more fully in Chapter 9.)

Uncertainty about whether electronic communications will be monitored can, much to our dismay, affect the quality of our communications. For instance, e-mail can be an excellent medium for brainstorming and sharing early drafts in the development stages of a project. As we toss around ideas and approaches, we can run them past our coworkers in-house and our peers elsewhere for constructive criticism and suggestions. If we were working on paper, we would mail copies to everyone whose input we seek without fear that a *rough* draft, far from ready for management consumption, would fall into the wrong hands, so to speak.

But we cannot be so certain if we send our brainstorming ideas or drafts electronically. What if a manager somewhere sees what is not yet intended for his eyes and judges our performance on that early work? That nightmarish possibility can severely inhibit the creative process! Ethically, then, our second responsibility, to ourselves and to those with whom we collaborate in this way, is to determine just what our company policy is on monitoring e-mail, to ensure that we pass along that information as appropriate, and to address any concerns we have about those policies within our companies.

Third, you should respect the privacy of all electronic communications you receive. Never forward them without permission from their originators, and certainly never post them to a public electronic forum such as a bulletin board.

If you do make use of the treasure trove of information available electronically, be sure to give credit where it is due, just as you would if you used material from a hardbound book. In Chapter 8 you'll see we did just that when we made use of some excellent information that a colleague posted to TECHWR-L.

Finally, we are obligated to ensure that the electronic access provided to us free of charge by our employers is not misused. Needless to say, it's unethical to use corporate e-mail access to complain about your employer, carry on an illicit love affair, or use inappropriate language or objectionable material. Using your current employer's system to correspond with a prospective employer or interview an expert for a freelance article also comes to mind. But there are more significant issues. For instance, our employers maintain certain private information, such as proprietary data, that we are ethically bound not to transmit over electronic communications. So before you press the Send key, pause for a moment to consider this ethical question: Are you treating others' e-mail as you would have them treat yours?

Is Corporate Big Brother Watching?

Much of what occurs in our lives is related to our jobs. We spend 40 hours or more each week in the office conducting business, writing correspondence, conversing with colleagues, placing or receiving phone calls, and communicating electronically. And much of what we do on our jobs is documented. Personnel files, benefit and payroll records, travel authorizations and trip reports, voice mail and e-mail, even performance appraisals, document what we do and how well we do it 50 weeks each year. Clearly, then, the corporate version of Big Brother is an issue in any discussion of privacy.

Some companies are aware of such concerns and address them in their codes of ethics or business conduct. In his compilation of model business conduct codes, Walter Manley excerpts samples of statements regarding employee privacy from several corporations. Two examples typify the way in which employers "sit on the fence," acknowledging employees' right to privacy while safeguarding their own freedom to gather and use personal information about those same employees. The St. Paul Companies address the issue in this way:

> *The company recognizes the paramount importance of preserving the dignity and privacy of each individual. Of course, in today's complex society, certain information must be collected and maintained for current employees as well as applicants for employment.*[6]

This statement is typical of all of those Manley presents. Most are general; few speak to specific privacy issues that raise ethical controversy. In particular, we are concerned here about employee privacy rights in matters related to surveillance by employers and to electronic communication, which we discussed in the previous pages.

Whatever the corporate policies say, surveillance of employees is growing by the proverbial leaps and bounds. Many companies believe that such surveillance is both useful and necessary. Unfortunate realities such as increased substance abuse, growing employee theft, the intensifying need to ensure high quality and productivity, soaring numbers of fraudulent workman's compensation claims, and the rising concern about corporate waste, fraud, and abuse are all factors that entice executives to begin monitoring the activities of workers. Certainly, if an employee who is under the influence of drugs or alcohol makes a mistake on the job that harms or injures other employees or clients, for example, the employer could be held legally liable. If an aerospace company employee misuses equipment funded by the Department of Defense, the company could face charges of waste, fraud, or abuse. It's understandable, then, that corporations feel the need to protect themselves.

Surveillance has many forms. Video surveillance is becoming an extremely high-tech and profitable business. Other kinds of devices can be useful in tracking employees' locations. Electronic badges provide security; they also detect where workers go and for how long during the work day. Telephone monitoring devices record when employees make calls and for how long they talk—a practice that can be useful in improving customer service or detecting the amount of time spent on personal business. Everything employees do on a computer is very easily observed—who logs on, what files they access, how many keystrokes they type, and how long they are away from the workstation. In fact, a *MacWorld* survey in 1993 furnished some amazing (or frightening, depending on your perspective) statistics. Through a survey of 300 companies, the publication estimated that 20 million workers are monitored on the job and that 30 percent of all companies do, in fact, participate in such practices.[7] Another survey suggests that sales of monitoring software will increase 50 percent per year.[8]

The explosive proliferation of the World Wide Web has added a new dimension to electronic surveillance. Although your employer may not have installed miniature cameras in pencil sharpeners to snoop on employees who "borrow" a few minutes of company time to read a favorite magazine, the ability to track exactly what Web sites an employee visits, and for how long, is inherent in the electronic network through which the Web is accessed. In other words, if you have Netscape at work, Big Brother knows where you've ventured electronically!

Monitoring need not be high-tech, however. A questionnaire distributed to TWA managers in 1992 is a prime example. Concerned about perceived abuse of sick time, TWA executives questioned managers about employees who called in sick. The questions were so personal and offensive—seeking information about family members, friends, and physicians; automobile make, model, and license plate numbers; eating, smoking, and drinking habits; identifying scars and marks—that managers called the press, which caused TWA to discontinue the practice.[9]

The TWA incident demonstrates more than the extent to which some companies will go to protect the bottom line; it demonstrates the reaction of most employees to being "spied on." Obviously, we don't routinely see headlines about situations like that at TWA, but a number of lawsuits have been filed over this issue in the last 15 years. For the most part, decisions have favored employers, as long as the surveillance is reasonable and doesn't invade employees' private space.

Of course, that begs the question of just what "space" is private. As discussed earlier, the hot issue in that regard is electronic mail. Most of us would be aghast to learn that our supervisor had opened a letter addressed to us at the office. Can we say the same for e-mail? Without a doubt, much of the controversy over e-mail is of a personal nature; however, because e-mail is, in fact, a form of corporate communication, the ethical framework surrounding it is complex. There are abundant opportunities for gray to creep in.

So we have to deal with plenty of privacy worries in the workplace. How do we cope with these ethical dilemmas? Short of filing civil suits against our employers or becoming CEO so we can institute changes, our best course of action is once again to draw on our unique expertise—our communication skills. We can take action. If our employers collect, use, or disseminate personal information, we can use our communication skills to begin debates and open discussions in the workplace about the ethical concerns and ramifications of their practices. We can research policies, procedures, standards, laws, and regulations. We can disseminate our findings through hallway conversations, memos, newsletter articles; and spark debate—bring concerns to the attention of our corporate public. We can discuss with our colleagues and managers the ethical issues concerning workplace privacy, and help everyone work toward consensus. If it's not our job, then whose is it? Will it get done at all?

And one more point—remember that privacy works both ways. Like you, your employer has information that must be kept private—trade secrets, proprietary information, marketing plans, client lists, and the like. You are ethically obligated to respect and maintain this privacy. For a discussion of the legal and ethical issues in this arena, see Chapter 5.

Summary

Individual privacy has become a significant value only in recent history. As close community ties weakened, people felt compelled to find means of "keeping an eye on" others in society. As a result, the individual privacy of many is being assailed more and more often in an attempt to protect the public, corporations, and other institutions from the few who pose a legitimate threat. Although a few laws protecting the individual's right to privacy have been enacted, neither legislation nor litigation can keep pace with the technology intended to maintain surveillance, intruding on privacy.

In relation to technical communication, privacy means making every attempt to respect the rights of others, both individuals and the companies for which we work. In particular, technical communicators are concerned with privacy in regard to collecting and disseminating confidential information.

Several privacy issues are particularly thorny in communication. For instance, we must be careful to seek permission, via a privacy release, when we use the pictures or words of individuals in our communication products.

Privacy in the workplace poses several dilemmas. In what ways can employers ethically conduct surveillance of employees? And to what uses can they ethically put that information?

As we note so often throughout this book, we have an ethical obligation to avoid abusing our unique language skills. We have an equal obligation to put them to use in instances such as those identified in this chapter in which individual privacy is being assaulted. We have the gift, the education, and the ability to research, write, debate, and generally spark public discussion of these issues anywhere we can raise our voices—and we are ethically obligated to do so.

Application Exercises

1. Has anyone invaded your privacy? What were the circumstances? How did you feel? Compare your reaction to those of your classmates in similar situations. How was the situation resolved?
2. Write a letter that you might send to clients asking permission to use their comments to your company as testimonials in sales literature.
3. Do you believe a parent is justified in searching a child's personal effects on a specific suspicion of drug use or other misdoings? How about on a preventive basis, without specific grounds?
4. Apply question #3 to high school officials (e.g., random searches of lockers for drugs).
5. If you happened to find what appeared to be an intimate letter to your spouse or significant other, would you read it? If not, how would you deal with the situation? If yes, how would you proceed: (1) if your fears were confirmed and (2) if your fears were dispelled?
6. Under what circumstances, if any, do you feel an employer is justified in reading employees' personal e-mail, monitoring their phone calls, or searching their desks? Explain.

Exploration Exercises

1. Research the regulations and penalties for reading U.S. mail intended for someone else. What parallels could be drawn with electronic mail? Should similar regulations apply? Why or why not?
2. Research the latest in surveillance technology available for use by corporations. Write a brief paper or prepare a brief presentation that informs your classmates about the variety and quantity of surveillance equipment in use.
3. Select several heroes from American folklore or literature and discuss how they illustrate the ways we value privacy.
4. Contact the e-mail administrator at your place of employment or school. Find out how it is possible for someone to read the e-mail of another student or employee. Report your findings to the class.

References

1. Gini Graham Scott, Ph.D., *Mind Your Own Business: The Battle for Personal Privacy* (New York: Plenum Books, 1995), 2.
2. "Pair sues, claims ad defamatory," *The Denver Post*, (20 January 1996): 3A.
3. "Today," NBC Broadcasting, February 7, 1996.
4. *Ibid.*
5. *USA Today*, via Internet, February 6, 1996.
6. Walter Manley, *Executive's Handbook of Model Business Conduct Codes* (Englewood Cliffs, NJ: Prentice Hall, 1991),70-71.
7. Charles Piller, "Bosses with X-Ray Eyes," *MacWorld* (July 1993): 118-123.
8. Laurie Flynn, "They're Watching You: Electronic Surveillance of Workers Raises Privacy Concerns," *San Jose Mercury News* (13 June 1993): 1F. Cited in Scott, *Mind Your Own Business,* p. 82.
9. Scott, *Mind Your Own Business,* p. 82.

Other Sources Consulted

Bacard, Andre. *The Computer Privacy Handbook*. Berkeley, CA: Peachpit Press, 1993.

Dolan, Edward. *Your Privacy: Protecting It in a Nosy World*. New York: Cobblehill Books, 1995.

Gilliam, John. *Surveillance, Privacy, and the Law: Employee Drug Testing and the Politics of Social Control*. Ann Arbor, MI: University of Michigan Press, 1994.

McLean, Deckle. *Privacy and Its Invasion*. Westport, Connecticut: Praeger, 1995.

Smith, H. Jeff. *Managing Privacy: Information Technology and Corporate America*. Chapel Hill, NC: University of North Carolina Press, 1994.

Chapter 7 Overview

Quality . . .

The Mission: **In technical communication we define quality as defining and performing to standards that best serve the user's unique communication requirements.**

Chapter Roadmap

Definition and Focus. This section opens with a brief historical perspective on quality, then explains how industry has redefined quality in terms not of perfection but "meeting requirements." It discusses Total Quality Management, or TQM, as it relates to technical communication.

- ❑ *In the Arts, Quality Means Striving for Perfection*
- ❑ *In Business, Quality Means Meeting Requirements*
- ❑ *Quality Is Inseparable from Cost and Schedule*

Quality Demands Competency: AOC^7. This section surveys nine core skills that add up to competency in technical communication and helps you gauge your mastery.

- ❑ A*ccuracy. Is it right?*
- ❑ O*bjectivity. Is it fair?*
- ❑ C*ompleteness. Is it thorough?*
- ❑ C*orrectness. Is it mechanically sound?*
- ❑ C*onsistency. Is it stylistically uniform?*
- ❑ C*larity. Is it simple and easy to understand?*
- ❑ C*onciseness. Is it brief and to the point?*

- ❑ *Coherence. Is it easy to follow?*
- ❑ *Creativity. Is it memorable?*

How to Weigh Budget versus Quality. This section explores the classic dilemma of how to provide appropriate quality when the budget is tight. It helps you understand when to hold the line on standards and when to make compromises to save costs.

Summary, Practical Exercises, and References

. . . Our Duty to Provide Products and Services That Will Best Serve Our Users

CHAPTER 7

Quality Is Not Perfection, But Striving for the Perfect Match between Products and Requirements

Put quality first—and
cost and schedule will follow.

—Traditional Motto at Lockheed Martin Electronics and Missiles Co.

The Four Absolutes of Quality:

1. *The definition of quality is conformance to requirements.*
2. *The system of quality is prevention.*
3. *The performance standard is zero defects.*
4. *The measurement of quality is the price of nonconformance.*

—Philip B. Crosby, *Quality without Tears: The Art of Hassle-free Management*

Definition and Focus

Quality—long the holy grail of both the arts and business. Exactly what *is* it? And what *isn't* it? And how does it relate to ethics and technical communication?

In the Arts, Quality Means Striving for Perfection

Simple yet elegant in design, perfect in symmetry, grand beyond expression even in ruins—the Parthenon is the embodiment of quality in human achievement. Few would argue with us if we cited the famous Greek edifice as an example of *quality* in architecture.

Humans have always strived for perfection. From Platonic idealism in ancient Greece, to the birth of scientific inquiry and flowering of the arts in the Renaissance, to the urgent imperatives of today's information explosion, we have pursued the holy grail of quality. Until very recently, the very word "quality" carried connotations of an absolute, an ideal, something that in this life, at least, is never fully attainable, which is to be sought with a passion nonetheless.

That definition works well when we are talking in terms of the Parthenon, the Sistine Chapel, the *Mona Lisa*, or *The Old Man and the Sea*. Hemingway was said to have revised the last page of the latter more than a hundred times, polishing and repolishing every phrase, every word, before he was satisfied. Edgar Allan Poe does not miss a rhyme or a beat in his consummately crafted poetic masterpiece, "The Raven." That didn't happen in one pass, or two, or a dozen.

Most of us nod our heads approvingly when we hear of such dedication to perfection in the arts. After all, we would hardly expect works that have become benchmarks of human artistic expression to have been created in slapdash fashion. They are the *best* humankind has to offer; we would *expect* blood, sweat, and tears behind their creation.

That's it, then—quality means being the best. And when it comes to winning a Nobel Prize for Literature, capturing first place in the Cannes Film Festival, or being First Violin in the New York Philharmonic, the concept is congruent with our perception of the arts and our expectations of the artists.

In Business, Quality Means Meeting Requirements

But when we attempt to apply the same definition to a software manual—well, that's when the fun starts! Hardly anything has received more attention in the business world over the past decade than quality—from Quality Circles to Total Quality Management. Books on the subject abound; full treatment is obviously outside the scope of this text. However, we would be remiss if did not at least establish an operational framework to serve as a point of reference for resolving ethical conflicts that involve the value of quality.

The key issue is the difference between an *absolute* definition of quality ("We will settle for nothing less than the best!") and a *relative* definition ("We will determine the appropriate level of quality to meet the user's requirements and perform to that level").

The relative definition of quality currently occupies center stage in the business theater. Internationally renowned quality guru Philip Crosby pioneered this concept for American industry; it is embodied in the First Absolute previously cited: *The definition of quality is conformance to requirements.* Whereas we as technical communicators once strove for perfection as a measure of quality, editing and preparing art to meet stringent and unyielding standards, we now try to adjust the level of effort to match the client's communication requirements. This does not mean producing trash (see Chapter 13), but it also rules out producing a "Cadillac" when the client needs a "Volkswagen." Thus, if we insist on creating a four-color flyer for a simple order list, we are not serving our client well because we are wasting her resources.

In their definitive article in the STC technical journal, Smart, Seawright, and DeTienne stress the complex nature of quality, urging a holistic approach:

> *Quality in technical communication has different components, with the importance of a component varying depending on the context. A holistic approach to quality acknowledges the contextual situations that affect its application.*[1]

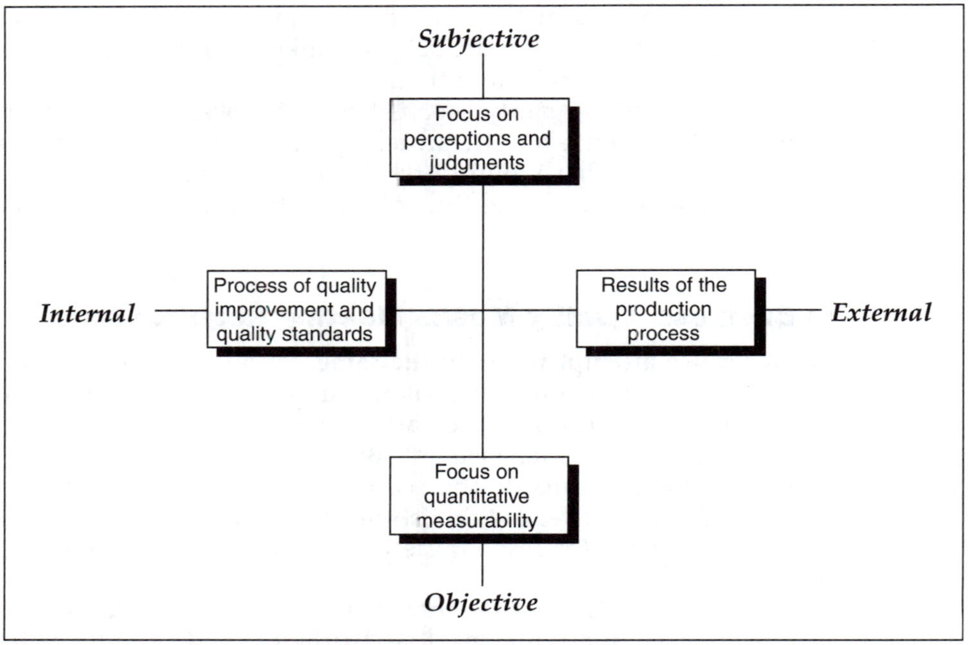

Figure 7.1 Two-dimensional model for comparing definitions of quality shows different perspectives.

They go on to explain how "definitions of quality fall in one of four quadrants on a grid (Figure 7.1)[2], depending on the emphasis of the definition."

> *An* internal focus *concentrates on the processes of improving quality and establishing internal quality standards. In contrast, an* external focus *directs attention to the results of design, writing, and production processes—emphasizing the ends rather than the means. In addition, quality standards depend on frames of reference external to the producer. A* subjective focus *relies on individual perceptions and assessments, whereas an* objective focus *uses tools that facilitate measurement and replication of results.*[3]

Clearly, definitions of quality abound. By and large, however, we feel that the best approach is to define quality in terms of user need. *In technical communication, we define quality as defining and performing to standards that best serve the user's unique communication requirements.*

The challenge that leaves us with is knowing when to hold the line on quality standards, and when to "flex" them to meet user requirements.

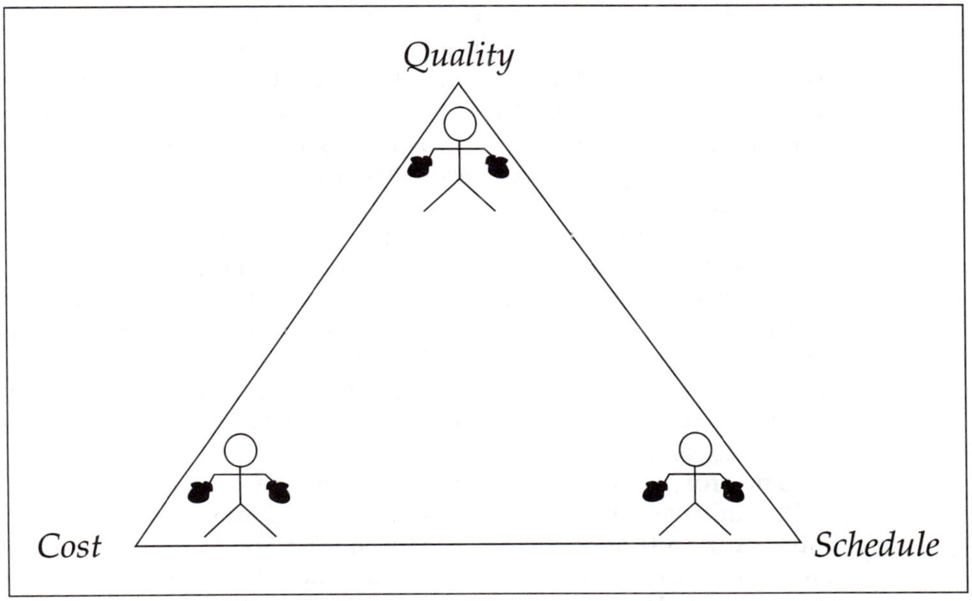

Figure 7.2 If you want perfection overnight and on the cheap, you don't understand quality in technical communication—or anywhere else!

Quality Is Inseparable from Cost and Schedule

That brings us back to the quotations on cost, schedule, and quality at the beginning of the chapter. The three parameters form the classic triangle shown in Figure 7.2. They are not mutually exclusive, but they are definitely related in terms of how much can be achieved on a given job. Tradeoffs must be made; that is, if budget is low or turnaround is fast, certain concessions may have to be made on quality. One jaundiced veteran of corporate cost-schedule-quality tradeoffs commented: "You can have it fast, you can have it cheap, or you can have it good. Pick one."

While the comment is cynical, it is not without wisdom. It stands to reason that a client with a severely limited budget does not have the same range of communication options as one with ample resources; that the client who must have a document in a weekend cannot expect the level of editorial treatment that would be possible over a 30-day period. *The key is to provide the best quality that can be provided within the constraints; this, in turn, pivots on strategic selection of communication media and realistic definition of treatments* (for example, using black-and-white printing instead of four-color, or playing silent video clips with the speaker providing the explanation rather than a fully narrated video).

Again, this does *not* mean producing a defective product. We still do a good job; we just do less than we *might* have done with a larger budget and a friendlier schedule. We stress *might* because the mere fact that the client has the time or the funding to allow a full-blown, four-color brochure or a narrated video does not mean this is necessarily the best communication solution for the situation, nor does it necessarily represent the wisest use of the client's resources.

Indeed, there is a point of diminishing returns when it comes to editing the "nth degree" of stylistic quality into a document. Consider a software documentation manual for a new release of a software application in a highly competitive market niche. Yes, the documentation should read clearly and be free of distracting mechanical errors. But it doesn't necessarily have to read like a Pulitzer prize-winning novel; getting the documentation on the street before three more versions of the software have been released by a competitor is important too!

The story of the Russian tank commander comes to mind. During the Cold War, there was a time when the United States had a far superior main battle tank to that of the Soviet Union. The Soviet tank was so confining to the crew that the gunner position had to be filled by a soldier 5' 6" or less and there was no decent air conditioning, resulting in sweltering temperatures inside during warm-weather combat. The U.S. tank, on the other hand, accommodated a full range of troop sizes and was equipped with all the latest creature comforts. The only problem was that the superior U.S. tank was still in design and development while the inferior Soviet tank was rolling off the production line by the thousands. This led to the story, possibly apocryphal (since it came from the U.S. side!) about the motto of the Russian tank commander:

Better is worst enemy of good enough!

The simple pragmatism is appealing. However, to suggest that we enter every communication situation automatically prepared to compromise quality is not the right message either. So, how do we balance our drive for top quality with the practical realities that militate against perfection?

TQM Defines Quality as a Process

One good solution is found in the concept of Total Quality Management, or TQM, which redefines quality as an all-encompassing management *philosophy* and product development *process* rather than pursuit of an abstract icon of perfection. Philip Crosby captures this concept when he says:

Quality improvement is a process, not a program, and it takes a long time for it to become a normal part of the scene.[4]

Before TQM, within most companies quality assurance was the special province of a discipline called "Quality" or "Product Assurance," in which trained quality engineers ferreted out design flaws and trained quality inspectors held the finished product up against stringent quality standards, demanding it be reworked until it was up to snuff. This approach is what the proponents of TQM call "*inspecting* quality in." As long as management gives quality inspectors the authority to *stop the line*, this technique works, but it is costly and can be demoralizing.

TQM operates instead on the concept of "*designing* quality in." Members of all disciplines (design engineers, manufacturing engineers, product support specialists, as well as quality engineers) seek to establish progressive quality checkpoints at all points in the product development cycle, rather than pinning everything upon key inspections at the end. Final inspection still has a role, but under TQM much of the battle on quality has been waged long before the approval tag is placed on the product. Furthermore, *any* employee—not just a quality inspector—is empowered to stop the line if she feels quality is being compromised.

This actually happened on the Patriot missile production line at Lockheed Martin in Orlando, Florida—which is one reason why when Saddam Hussein began raining Scud missiles upon Saudi Arabia and Israel during Desert Storm, Improved Patriot was there in time to intercept them. For those suspenseful days in Riyadh, the empowered quality commitment of assembly line workers in Orlando provided vital protection for our troops as well as the civilian population.

TQM is what Crosby is driving at with his Second Absolute: *"The system of quality is prevention."* It is far more effective to prevent problems in the first place than to correct them after a finished product is rejected in final inspection: *It's called getting it right the first time.*

Applied to technical communication, *the principles of TQM dictate early involvement by editors, producers, illustrators, and multimedia specialists in the planning and development of communication products.* Even under short schedules and limited budgets, communication specialists can develop amazingly creative communication solutions if given the parameters up front. Of course, we must also be aggressive in involving ourselves in such situations whenever we sense they are imminent. If we are truly blind-sided by a budget- and schedule-constrained job, then all we can do is react—coming up with a suitable and achievable product at the best quality level possible under the circumstances. If we have even a modicum of forewarning, we can often do much more.

The Goal Is Zero Defects; the Benefit Is Reduced Cost

How good must we be? Crosby's Third Absolute seems to demand perfection: *"The performance standard is zero defects."* At first glance, such a philosophy

"The performance standard is zero defects."

would appear to be unrealistic, doomed to cause frustration and demoralization on the first failure or problem. After all, nobody's perfect, right? That's true. But if we enter a project *expecting* to have a certain amount of failures or rework, then the failures or rework are going to happen. We get less because we expect less.

The key is to define requirements realistically. *Zero defects* does not mean imposing an inappropriate level of quality upon a job (e.g., insisting on an animated multimedia project when a simple visual aid will do the job). It does mean setting a goal of error-free performance within the defined requirements (e.g., this brochure will have no typos in it).

So what if a typo creeps through? Under zero defects, that means we have failed, right? Yes and no. If we just shrug off the typo as inevitable, then we *have* failed—philosophically, at least. But if we plan to run our next brochure past an extra pair of eyes just before press to improve our chances of picking up that last, perfidious typo, then we are operating properly within the framework of zero defects.

Zero defects does not mean flagellating ourselves—or a fellow professional—after a job is complete for the peccadillo that slips through all our quality safety nets. It *does* mean refusing to accept the inevitability of error up front—to set the standard of error-free performance, then come as close to that standard as we humanly can.

How can we measure quality? In his Fourth Absolute, Crosby tells us: *"The measurement of quality is the cost of nonconformance."* In other words, how much are our mistakes costing us? Poor documentation means a higher volume of service calls to the technical assistance department, which means longer hours on the phone, more personnel, higher cost. Inadequate usability testing means client frustration, reduced follow-on sales, lost business. These parameters can be quantified, at least within a reasonable tolerance of estimating accuracy. If it adds X dollars to the cost of a project to ensure thorough, clear, accurate documentation, but the improved documentation drives down service calls enough to save 2.5X dollars, it becomes clear that *good quality is good business*.

If, on the other hand, it costs 3X dollars to print the documentation in a laminated, five-color, "scratch-'n-sniff" musical loose-leaf binder . . . well, you get the idea!

As technical communicators, we contribute greatly to our clients by helping them accurately define their communication requirements, then performing flawlessly—or as close to flawlessly as we can—to those requirements. That's what commitment to quality is all about.

Quality Demands Competency: AOC^7

We're quick to recognize *incompetency* in technical communication. Incompetency is the enumerated list with faulty grammatical parallelism. Incompetency is the technical illustration with inconsistent line weights. Incompetency is the instructional videotape that leaves one image on the screen unchanged for a minute and a half while the narrator drones on endlessly.

But what characterizes *competency* in technical communication? That's a taller order. There are so many job descriptions within technical communication, so many specific skills within each of those descriptions, that it is well-nigh impossible to devise an all-inclusive formula for competency. That should not, however, prevent us from developing some broad guidelines, some general standards against which to gauge our performance as professional communication specialists.

We give you, therefore, what we call AOC^7—not because it is the *only* standard for performance in our profession, but because it provides a useful quality template that can readily be applied to the communication products we produce for our clients and employers.

So what, then, is AOC^7? Here's the formula:

1. Accuracy. Is it right?
2. Objectivity. Is it fair?
3. Completeness. Is it thorough?
4. Correctness. Is it mechanically sound?

5. Consistency. Is it stylistically uniform?
6. Clarity. Is it simple and easy to understand?
7. Conciseness. Is it brief and to the point?
8. Coherence. Is it easy to follow?
9. Creativity. Is it memorable?

Accuracy . . . Is It Right?

According to legend, the *Wall Street Journal* sets a demanding quality standard for its reporters: Three errors in fact, and you're gone.

Accuracy is the first law of good journalism. After all, a simple reporting error such as announcing an important public hearing for 9:00 P.M. when in fact the hearing is at 7:00 P.M. could not only decimate attendance, but also inconvenience hundreds of people—all subscribers to the newspaper!

The imperative for accurate reporting leads to an adage that is drilled into the heads of neophytes in many scholastic journalism programs:

> *If in doubt, check it out.*

And, if that fails, the corollary applies:

> *If in doubt,* leave *it out.*

This is not a journalism textbook. However, the above principle constitutes excellent advice for technical communicators as well. Technical communication, by definition, includes a specialized vocabulary; words often have a precise meaning. We often face tough technical questions as we edit text and prepare graphics for documents and other communication products. One of the authors recalls with chagrin an early career embarrassment resulting from correcting a "typo" in a wind tunnel report, so the line would read "the model was attached to the *string*," instead of the obviously incorrect *sting*. The only problem is that the technical term for the device to which models are attached in a wind tunnel happens to be *sting*. Oh, the pain of true confessions! But we wanted you to know that nobody is exempt from the tenet of accuracy; and there is no inoculation for the virus of inaccuracy. So if you want to avoid the "sting" of inaccuracy, unless you have been gifted with omniscience, we humbly submit that there is only one safe passage: *Ask questions*!

- First, we should make every effort to get the question answered. This can be extremely frustrating when the author is out of pocket and the schedule is short. However, if we yield to the temptation of glossing over the unanswered question, we are doing our users a gross disservice. Inaccurate information is worse than no information at all.

- Second, if we fail utterly in getting the question resolved, and schedule demands we move on, then we must omit the material in question—unless to do so is to fundamentally flaw the document (such as leaving out a key safety instruction for a piece of potentially dangerous equipment).
- Third, if we can't get the information in time to meet the deadline, and omission *will* create a fundamental flaw, then we must delay the schedule. Won't that get us some negative attention? You bet. But better to use that attention to shine a spotlight on the problem and get it fixed than to take the expedient path of omission and become an accomplice in the release of an inferior and potentially dangerous product.

Good editors are aggressive, but they are also careful. They respect the sanctity of information and are ethically committed to presenting it accurately. Figure 7.3 depicts three types of editing.[5] The line in the middle symbolizes *technical meaning*—that sacred icon so zealously defended by authors. Crossing the line is inaccurate editing—editing that changes the author's in-

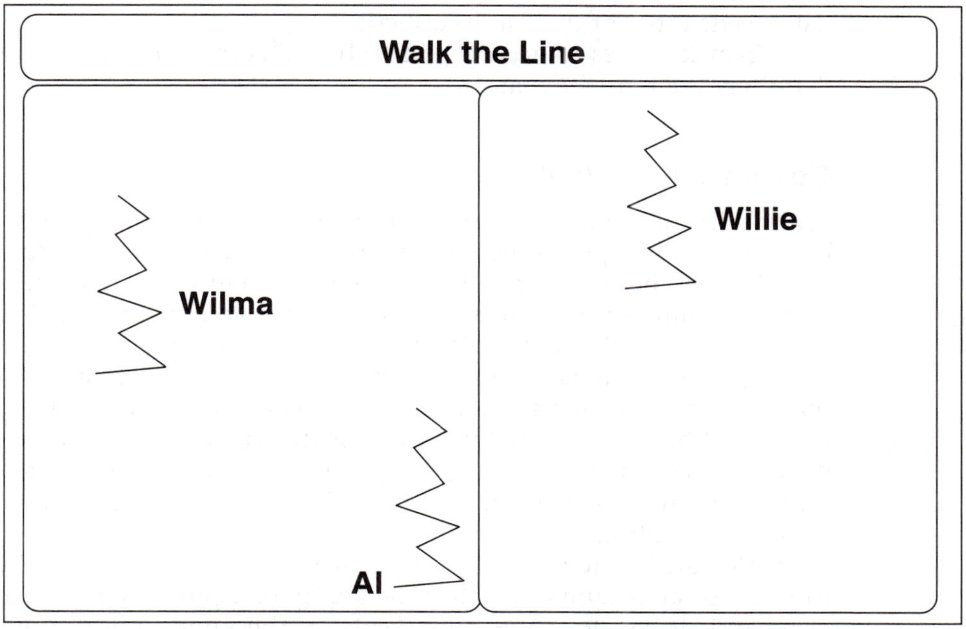

Figure 7.3 The skillful editor tempers aggressiveness with care.

tent ("you changed my *meaning*!"). Crossing the line is editing from right to wrong. Crossing the line is the cardinal sin.

The line on the right represents the marauding linguistic lunacy of Willie Wildman, an editor who needs to be sent to an island in the North Atlantic where he can do no further harm to innocent copy. The line on the left, on the other hand, represents the timid, nonsubstantive, "comma-punching" edit of Wilma the Wimp. Wilma will be more popular than Willie with most authors, especially those afflicted with "ego involvement" (for a full discussion of this dread malady, which applies to editors as well as authors, see Chapter 13), but discerning authors will realize that the safe but impactless edit is not serving them because it adds no value.

The competent, professional editor, Al, edits along the middle line, aggressively adding value to information but taking great care not to alter meaning in the process. When you are editing highly technical material, the only way to "walk the line" successfully is to ask questions, questions, and more questions; to brush nothing under the rug; to thoroughly grasp the content of the document—after all, how can you help somebody else understand what you don't understand yourself?

The same principle holds true for accuracy in technical illustration. If two key data points are missing on a plot, it is not okay to just "smooth the curve over." Accuracy demands the illustrator take the time to find the missing data needed to plot the curve correctly.

When it comes to accuracy in technical communication, zero defects is definitely the right standard!

Objectivity . . . Is It Fair?

Science, like journalism, is assumed to be an objective discipline. Facts are to be gathered and presented impartially, untinged with personal bias. That sounds great, but it can be more challenging than you might think. It would be an oversimplification to say scientists are totally neutral and objective or even that science itself is totally neutral and objective.

For example, an environmental scientist may *care* intensely about the importance of preserving the environment. This, indeed, could be the avowed purpose of her research. But if she allows this value to cause her to exaggerate data on environmental damage to galvanize public opinion against a paper mill, then she has lost objectivity and is not being fair to the company, which is unethical.

Let's start in the general arena, drawing again from the field of journalism. A typical assignment to test the ability of aspiring reporters to write objectively is to ask for three news leads on the same subject: one slanted in each direction, and one objective (down the middle). For example:

- *The campus police report*: An unruly mob of radical hippies threatened violence.
- *The student newspaper*: Courageous supporters of student rights assembled peaceably to express their views.
- *The local press report*: Student protesters staged a demonstration.

Slanted writing is a form of deception. For further discussion from that perspective, refer to the discussion on slanted language in Chapter 4.

The ultimate test of objectivity is the ability to express an opposing position so fairly that a proponent would agree that you understand her position and have expressed it correctly. Test yourself. Pick an issue about which you feel strongly, find somebody whose views are diametrically opposed to your own, write down a one- or two-sentence statement summarizing the opposing position, and have your "opponent" critique it for objectivity. Be prepared to do some rewriting!

Now let's move into technical communication. In scientific research and writing, objectivity means gathering and presenting data in a logical, consistent fashion, not skewing the sample or twisting the data to support a favorite hypothesis (see the discussion on honesty in research in Chapter 4).

In business communication, objectivity means addressing potentially divisive subjects in a quiet, rational tone that signals task-oriented, analytical problem-solving rather than egocentric, personal ax-grinding.

In marketing communication, objectivity means not distorting the facts to give one's product an undeserved edge over the competition. For example, it would be unfair to compare Acme's 1996 widget with Ace's 1995 model to exploit the absence of features that Ace added in 1996. On the other hand, as long as the playing field is level (e.g., 1996 versus 1996), it is entirely appropriate to aggressively seek out legitimate performance advantages and drive them home forcefully. In a competitive business marketplace, fair play doesn't mean lowering one's gloves and thrusting one's chin out. It means not putting brass knuckles inside the gloves.

For a discussion on how to avoid distortion in graphics, see Chapter 4.

Completeness . . . Is It Thorough?

Our commitment to provide quality products and services that best serve our end users means providing *all* the data they need to make well-informed, intelligent business decisions. A product fact sheet that lists several benefits of the product but offers no price will be of limited value to users because they will be unable to weigh the benefits against the impact on their budget. A lawn-mower maintenance manual that explains how to sharpen the blade but not how to change the oil short-changes the buyer. Assembly instructions that provide a detailed parts list and intricate directions on how to put the parts together but provide no pictures of the parts and no diagram of the stages of assembly are woefully incomplete. A recipe that neglects to mention the need to grease the pan before baking could reduce a novice cook's first attempt at biscuits to charred rubble.

To test the completeness of a news lead, journalists use the tried-and-true formula of *5W&H*: who, what, when, where, why, and how. The newspaper reader is not served if he learns when the meeting is, but not where. Similarly, the buyer of a home computer is not served if all the user documentation is online and there are no printed instructions on how to physically set up the computer. In this case, the instructions may actually be complete, but they are not accessible—so they may as well be nonexistent.

Usability testing experts will vouch for how difficult it is to be truly thorough in writing procedures. Sometimes the obvious step is the one that is omitted. The step may be obvious to the writer; but unless it is equally obvious to the reader, it must be spelled out if the procedures are to be used successfully.

For this reason, procedures should routinely be debugged by thorough usability testing before they are released; that is, users unfamiliar with the

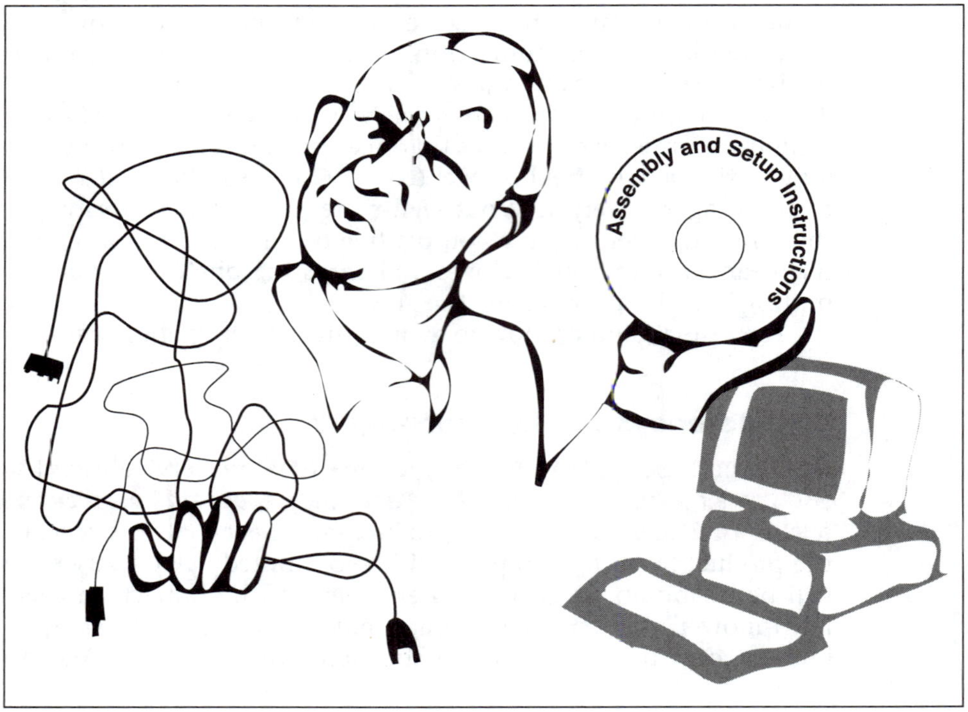

product should implement the procedures step-by-step to see if anything "intuitive" has been left out. To do less is to deliver an incomplete product.

In scientific writing, completeness demands explicit presentation of *all* the steps in an equation or proof. The authors recall a perfidious freshman calculus textbook (not the one cited!) that was famed for its one-step leaps from something like:

$$(a+b)^k = a^k + ka^{k-1}b + \frac{k(k-1)}{2!}a^{k-2}b^2 + \frac{k(k-1)(k-2)}{3!}a^{k-3}b^3$$

$$+ \text{---} + \frac{k(k-1)(k-2),,,(k-n+1)}{3!}a^{k-n}b^n$$

$$+ \text{---} + kab^{k-1} + b^k$$

to: *It is intuitively clear that a = -1.*[6]

Well, maybe Einstein could "intuit" it; the rest of us poor blokes in Calculus 101 didn't have a clue. That wasn't a good way to write a freshman calculus text and it sure isn't a good way to produce quality scientific writing.

We'll sum up completeness with yet another spin on the old adage:

If in doubt, spell it out!

Correctness . . . Is It Mechanically Sound?

Good, clean grammar and mechanics cannot be separated from quality content, in technical writing or any other form of writing—both on paper and online. As a minimum, errors in grammar and mechanics are distracting to the reader, taking attention away from the message. Worse, they can affect meaning—distorting the message itself.

Consider the effect of a missing comma and a subject-verb disagreement on the following statement:

> *The flow of the coolant to the environmental control unit located near the aft stabilizing fins were blocked.*

A comma is needed after "unit" to signal that the clause "located near the aft stabilizing fins" is nonrestrictive; otherwise, it is unclear whether there is one environmental control unit or more than one. The use of "were" can cause unwary readers to conclude that the blockage refers to the fins, not the coolant flow. True, if they took the time, most knowledgeable readers would be able to get past the grammatical errors and discern the writer's true intent, but the editor's job is to make the user's life easier, not harder. We should save them time, not waste it.

Then there's the ubiquitous misplaced modifier, with a capitalization error tossed in for good measure:

An adapter was developed by engineering that will allow simultaneous operation of the fan and the defroster.

Nifty, huh? Failure to capitalize "Engineering" is a serious error here, because the proper name is needed to distinguish the group that did the design from the design itself. The long modifier "that will . . . defroster" needs to come right after the word it modifies: "adapter." Putting it anywhere else is at best awkward and, at worst, outright misleading. A much better way to write this sentence would be as follows:

The Engineering Department developed an adapter that will allow simultaneous operation of the fan and the defroster.

Now, all chance of ambiguity is eliminated. And that is our job.

Sometimes even the most innocuous-looking mechanical error can have a potentially catastrophic impact on meaning. Consider this properly written instruction in a gun-owner's manual:

To correct a hang-fire, the gun must be carefully opened by a trained expert.

Now let's put a comma where the hyphen used to be and lose the existing comma:

To correct a hang, fire the gun must be carefully . . .

Finally, let's say we're at the end of a page, and the next page happens to be missing:

To correct a hang, fire the gun . . .

A little far-fetched? Perhaps. But stranger things have happened. Anything that converts "fire" from a noun to a verb in a gun-owner's manual is clearly suspect!

Consistency . . . Is It Stylistically Uniform?

Was it Shaw who once said, "Consistency is the hobgoblin of foolish minds"?

Beleaguered technical editors laboring late into the night over their fifth cup of coffee may often feel as if they are the only ones in the entire world who give a hoot about stylistic consistency. After all, how often do we hear the technical-minded engineer, the sales-minded marketeer, the cost-conscious manager question: "What difference does it make if we have bold type on this item and italic type on that item?"

Reducing stylistic inconsistency to the micro level, taking it out of context of the overall document, it is fairly easy to question the need to correct it. After all, what editor wants to take the rap for insisting upon a nonsubstantive, "trivial" correction—especially in the middle of the night when tempers are worn thin and everybody just wants to go home?

And, realistically, there *are* times when the cost and political factors involved in a situation like this do make it advisable for the editor to yield—to sacrifice the last full measure of stylistic uniformity in the interest of schedule, cost, or even morale.

But having a pragmatic sense of flexibility when it comes to the rigorousness of stylistic editing does not mean tossing the concept of uniformity to the wind. Quite to the contrary. Considering stylistic consistency at the macro level, within the context of the entire document, its importance is much easier to defend.

Consider consistency in headings, for example. For readability, this book uses four levels of headings instead of a complex decimal system. That means the only way to distinguish which section is subordinate to which is to follow the sequence:

Level 1 Head

Level 2 Head

Level 3 Head

Level 4 Head (run-in with text)

By adhering consistently to this pattern, hopefully we have made things a little easier for you, our user. Actually, we wouldn't be at all surprised if you had never taken notice of the pattern of headings one way or the other—and had we not suddenly called them to your attention, you might have gone merrily on your way without ever having given them a second thought. That's fine; the headings still served their purpose—to organize the material into manageable "chunks" of cohesively arranged information. Rest assured, if we *hadn't* stuck to any pattern, you definitely *would* have noticed them!

The point, of course, is not how we arranged the headings in this book. The point is that *a consistent, uniform format makes a document much more us-*

able than it would be without such a format. So while it may not be earth-shattering that the United Kingdom is spelled out on page 47 and abbreviated as U.K. on page 447, it *is* disruptive for a document to be inconsistent in its basic format from page to page or chapter to chapter.

Inconsistency in graphics definitely can affect meaning. For example, consider the classic colors used for quality audits:

- Red = major problem, "out of control"
- Yellow = problem being worked, "in control"
- Green = no significant problems, "on track"
- Blue = outstanding in all respects, quality "above and beyond" expectations

If Figures 1 through 4 use these colors in the above fashion to depict audit results, and Figures 14 through 17 later use the colors entirely differently on a related subject, much confusion could result. And no, it is not perfectionism for an editor to dig in her heels and insist on consistency in a situation like this.

Nowhere, perhaps, is consistency more critical than it is in online documentation—particularly in helping users navigate HyperCard™ stacks and World Wide Web sites. Without the familiar linear progression and the historical cues of the printed page, the user depends on the designer's consistency in the use of icons and navigation aids to maintain a sense of location and equilibrium. If the Home button on one screen is a chartreuse trapezoid and on the next screen it is a pomegranate octagon, the user is likely to wind up with that "lost in hyperspace" feeling.

Consistent format is especially important for World Wide Web sites, because many browsers enter a site not through the "front door" but by means of a direct Uniform Resource Locator (URL) link into a field three layers deep in the stack. Without a consistent logo icon identifying each screen with the Web site, the casual user might not even realize where he is. And without a consistent panel of navigational buttons leading back to the main menu, many browsers can easily be "dead-ended," never realizing that the little chamber at which they're marveling is attached to a vast cavern that is infinitely richer in information.

To answer Shaw, then: "Inconsistency is the hobgoblin of incompetent communicators."

Clarity . . . Is It Simple and Easy to Understand?

Close your eyes. Picture yourself cuddled up in front of a crackling fire in a ski lodge with your significant other, about to take your first sip of Grand Marnier from an exquisite crystalline liqueur glass which reflects the darting flames. But first, the toast. Your glasses meet. Clink.

Freeze frame.

Capture that sound.

Hold that clear, crystalline chime that rings out like a carillon through the lodge, then recedes like the expanding concentric circles after a pebble shatters the sapphire mirror of a placid mountain lake.

That's clear writing. Just like the moment previously captured, it's very hard to come by.

As any writer can tell you, writing clearly is one of the most deceptively difficult arts known to man (or woman). Words are elusive. They have a perfidious way of getting tangled—all on their own, of course!—somewhere between their conception and their grudging capture with pen or keyboard.

We have been practicing and/or teaching the exasperating craft of writing for more years than we care to admit. Many techniques exist for honing one's ability to write with crystalline clarity—we provide a sampling below—but when all is said and done, we've found that clarity in writing boils down to an elementary tautology:

The best way to write clearly
is not to write unclearly.

—Allen and Voss, 1996.

Profound, yes? We thought so.

Seriously, however, a kernel of truth lurks behind our tautology. While some clarity problems in writing can be traced to specific causes such as misplaced modifiers or unclear antecedents, others defy simple diagnosis. Somehow the words just don't work. They stumble over themselves and meaning is garbled. When the diagnosis is specific, so is the cure (put the modifier where it belongs; make the antecedent unambiguous). When there is no single cause for the confusion, there is no simple cure. The sentence just staggers along like a syntactical drunkard, thumbing its nose at anyone who tries to decipher it. Sentences like this are terminal. If they occur in a classroom assignment, you will recognize them from the word "Awkward" scribbled in the margin, which amounts to a professorial judgment that the Mystery Malady of Mangled Meaning (the dread 4M Syndrome) has struck again. The best thing to do when you create a 4M sentence—and face it, we *all* do from time to time—is to put it out of its misery quickly (before it spreads!) and start over.

Working with students as they struggle to recast a particularly resistant awkward sentence, we have found it useful to ask them to express the idea verbally: "Put your pen down. Just *tell* me what you mean." Nine times out of ten, the simple, lucid explanation that escapes the pen flows like liquid platinum off the lips. When it does, we say, "Write that down, quickly!" Sometimes the new sentence works exactly the way it is stated; sometimes it has to be tweaked to make it more appropriate for the document. If you're laboring alone, outside the bounds of professorial catalysis, worry not; you can still use this technique. Just utter the offending phrase aloud; force it to emerge from the murky shadows of obfuscation and stand exposed before the light of clarity, where it will rapidly wither and die. Then replace it with a simpler, more direct expression—and voice the new sentence *aloud,* as if you were explaining the concept to someone. Try this technique; it really works!

If you get caught doing this and your friends think you're strange, chin up—think of it as the price of dedication to your chosen craft. Or you can simply blame it on us; we'll be glad to take the blame, as long as you rewrite the sentence!

And take heart! You are not alone in struggling for clarity in writing. Indeed, so prevalent is linguistic "gobbledygook" that even the government, itself a primary practitioner, has finally responded to the public's clarion call for clarity in communication: Obscure legalese is slowly but surely falling before the pruning shears of "plain English" laws.

The definitive work on this subject is a marvelous little—its very brevity (64 pages) underscoring its mission—handbook called "Fog," by Robert Gunning, inventor of the famed Gunning Fox Index, a formula for computing the reading level of a passage.[7]

Edwin Newman applauded a federal judge who ruled that a nearly undecipherable form letter from the Welfare Department denying benefits be rewritten in plain English. Newman cited the judge's order, which specified that "doublespeak, bureaucratese, officialese, double-talk, and gobbledygook do not constitute English" and therefore must be translated into something readers could understand.[8] Here, here.

Newman went on to observe that clear, succinct writing is more than a philosophical or literary crusade: It is a business imperative. He warned that "wasted and wayward words waste money," contending that "productivity demands verbal economy," that "simplification improves communication," and that "*writing clearly pays off.*"

The facts bear Mr. Newman out. In 1993, the Society for Technical Communication researched and published an impressive cavalcade of documented cost savings directly attributable to quality document design—in particular, clarity and conciseness—in technical and business communication. We excerpt three compelling examples from that selection, which appeared in the Second Quarter edition of STC's journal, *Technical Communication*:

> *Southern California Gas Company simplified its billing statement and is saving an estimated $252,000 a year from reduced customer inquiries.*[9]
>
> *Since the British Government began its review of forms in 1982, it has scrapped 27,000 forms, redesigned 41,000 forms, and saved over $28,000,000.*[10]

Between 1980 and 1983, the U.S. government reduced paperwork government-wide by 32%, far surpassing its 25% goal . . . over 400 million hours of paperwork have been eliminated since 1981.[11]

We rest our case.

Here are 10 common impediments to clarity in writing ("maladies"), along with recommended cures:

1. Vagueness: Be specific.
2. Misplaced modifiers: Put them where they belong.
3. Dangling participles: Undangle them. Provide a referent for the modifier.
4. Unclear antecedents: Eliminate ambiguity.
5. Runaway sentences: Break them down into shorter sentences.
6. Misuse of restrictive and nonrestrictive: Brush up on comma usage rules.
7. Misreading caused by missing commas: Insert commas where they are needed.
8. Long and confusing series: Use semicolons to separate items with internal commas.
9. One thought too many for a sentence: Use ( ), —, or *. Or leave it out.
10. Fog: Defog. Easier said than done, but important!

For full treatment of stylistic clarity in technical writing, we refer you to *Technical Writing Style* by Dan Jones at the University of Central Florida, due off the press from Allyn & Bacon in Fall 1997.

Conciseness . . . Is It Brief and to the Point?

Brevity is the soul of wit. —*Hamlet,* Act II Sc ii

Pretend you're a busy executive. The In basket on your desk is bulging with paper; the Incoming Mail folder on your e-mail is overflowing with messages, half of them flagged URGENT. Paper first. You seize the first handful of memos.

"Some paperless society!" you mutter, as you spread the memos out on your desk. "What have we accomplished? Now we get it online and on paper, too."

Three of the memos are multiple-page monstrosities. The fourth is a full page of gruesome single-spaced type. The fifth is about 10 lines, double-spaced.

Which will you read first? Which might you not read at all?

As business communicators, we should never lose sight of the close relationship between wasted words and wasted time. Even in the days of just paper, conciseness provided a critical communication advantage. E-mail has upped the ante. The ease with which information can be disseminated has inundated the business world with material that must be read—or, as a minimum, screened to see which messages really *do* need to be read.

As message *receivers* bombarded with information, we have all learned to put both the circular file and the Delete button to regular use. As message *senders*, we need to place ourselves in the shoes of our receivers and realize that our message will be locked in deathly competition with scores, perhaps hundreds, of other messages.

The moral of the story is simple: Get to the point, or get deleted.

The person who can write straightforward, crisp, to-the-point business messages has a decided advantage in today's information-clogged marketplace. Here are 11 tips on how to communicate more concisely:

1. Don't use expletives. Constructions using *there is*, *there are*, and *it is* are word wasters.
2. Avoid *deadwood*. Don't use many words where one will do. Don't use "in the event that"; use "if." Don't write "at the present time." What's wrong with "now"?

3. Use active voice—unless you have a specific reason for using the passive (for instance, the doer is unknown or insignificant). Use "firemen extinguished the blaze," not "the blaze was extinguished by firemen."
4. Eliminate ~~needless~~ redundancy. Like that one. Consider "round circles." As opposed to what, trapezoidal circles? And how about "dead corpses"? Most are.
5. Drop "that" in indirect quotations. He said ~~that~~ it was raining.
6. Use hyphenated modifiers. They compress action and description, as in the "drought-parched Midwest" or the "injury-riddled Dallas Cowboys."
7. Avoid "which was" and "who was" constructions wherever possible.
8. Simplify grammatical structures as much as possible. Use a clause instead of a sentence, a phrase instead of a clause, a word instead of a phrase. Don't refer to "people who ride on the airplane." Try "passengers."
9. Avoid needless repetition. Deliberate repetition for dramatic effect is okay.
10. Use strong verbs. And don't smother them. Why "effect a reconciliation with" when you can "reconcile"; "perform an analysis of" when you can simply "analyze"?
11. Avoid unnecessary complexity in graphics. Don't clutter the axes with excessively detailed units of measure.

The list is by no means exhaustive, but it should get you moving in the right direction. For full treatment of the subject, we recommend taking a journalism course that focuses on newswriting. Compressing information into news leads or headlines is a particularly valuable exercise for honing your ability to write succinctly.

Coherence . . . Is It Easy to Follow?

The word "coherence" derives from the Latin "cohere," to stick together. Coherence in a document means the reader can move easily from one part to another, without getting lost. Four factors combine to establish coherence:

1. Underlying organization
2. A clearly delineated introduction and conclusion
3. Clear signposts throughout
4. Internal transitional expressions

The key to organization in any document—and we use the term "document" broadly, to include electronic media as well as traditional paper documents—boils down to one word: OUTLINE!

Outlining is critical to achieving coherence both in academic writing and in business communication. Electronic media such as hypertext (for example, the World Wide Web) have carried outlining into a whole new dimension, moving away from traditional linear progression into linked files with multidirectional reader access. Wherever possible, technical communicators should aggressively involve themselves in the planning and outlining stages of communication products. Logical flaws in structure are far easier to correct in a 3-page outline than they are in a 150-page draft or a storyboard treatment rather than in a finished Web site or narrated video.

Signposts, always a hallmark of coherence in paper documentation, also take on new significance in online documentation and electronic communication products. Can you imagine embarking on an auto tour of a foreign nation without a roadmap?

Asking a reader to take on a 30-page thesis or an Internet surfer to ride a 150-screen World Wide Web site with no signposts amounts to the same thing. In a written document, signposts consist of:

1. A clear preview of main points, in the introductory paragraph
2. Clear transitional statements at the beginning of each main point
3. A brief summary, in the concluding paragraph

In an interactive online presentation, signposts are called navigation aids ("navaids" for short, if you want to sound Web-savvy) and include:

1. A main menu, accessible from every element in the stack
2. Submenus, accessible from all subordinate elements
3. Consistent icons for returning to the main menu and submenus
4. A "Help" element, accessible from every element in the stack, explaining basic navigation
5. A "Find" feature, with brief instructions on how to plug in descriptors to search the stack for specific information

In graphics, signposts include:

1. A clear legend for all symbology, including color, line variations, and the like
2. A title that embodies the main message
3. In visual aids, the words in a *takeaway* box at the bottom drive home the key message the presenter wants the viewer to "take away" from the pitch

Creativity . . . Is It Memorable?

Oh, come on, we hear you say—this isn't a book on *creative* writing; it's a book on ethics in *technical* communication. That all depends on your definition of creativity.

If you define creativity with the written word as spinning an exquisite sonnet, crafting a pithy haiku, or penning the Great American Novel, well then, yes, you're outside the boundaries of technical communication.

But we disagree strenuously with such a narrow definition of creativity. We define it less in terms of blinding inspiration and more in terms of flexibility and originality in exercising communication options. *Creativity is willingness to seek new communication solutions.* In fact, we believe original approaches to presenting technical material often differentiate quality from mediocrity in our field.

On the other hand, creativity does not mean being different just for its own sake, when an established pattern has proven effective:

> *If it ain't broke, don't fix it!*

By the way, one school of thought on quality demands we do precisely that: "reengineer" everything, whether or not it's "broke," in the interest of making it better. That catches us on the horns of a dilemma: Without creative reengineering, many potential improvements would be lost, in technical communication and elsewhere. On the other hand, is it wise to always be tinkering with something that works fine as it is? Remember the motto of the Russian tank commander earlier in the chapter.

We can't make this decision for you—all we can do is present both sides. Personally, we are leery of reworking something that meets all requirements just for the sake of reworking it—especially if there's something else around that needs attention, which there usually is. It might be okay to redesign the perfectly good starter of an automobile to make it better, but not in lieu of fixing brakes that are "metal-to-metal"!

What do we mean by flexibility and originality? Obviously, we can offer but a glimmer of an answer to this far-reaching question. Six possible paths to creativity in developing solutions in technical communication are as follows:

1. Use originality in choice of communication media. Sometimes selection of the right medium is the most important part of the communication process.
2. Use titles and captions effectively. A short, catchy tag on an e-mail message, for example, may spell the difference on whether a busy recipient with a full In box reads the message or deletes it unopened.

3. Use figurative language (judiciously). Techniques such as simile and metaphor can add life to technical writing, but the tone must still be appropriate to the communication situation. Which communicates more effectively: "The requirements for this document are extremely hard to define" or "Defining the requirements for this document is like trying to nail jello to a tree"?
4. Use attractive packaging. By this we mean breaking up deadly expanses of text—be they on paper or, even worse, online!—by using subheads, enumerations, bullet lists, pull quotes, white space, an accent color, or a simple graphic.
5. Use a picture. Showing is often better than telling.
6. Use humor where appropriate—but never at the expense of clarity or safety.

Let's talk about humor for a moment. We believe humor does have a place in technical communication; nowhere is it written that everything we touch in technical communication has to turn into sawdust.

Indeed, judicious use of humor can lower barriers of intimidation. Humor can be the lubricant that allows highly technical material to flow comfortably to the user. Thus, the little cartoon character who walks timid users through an online tutorial for a complicated new software package makes a substantive contribution to communication by increasing their confidence with the product.

Another avenue where humor can serve us well in technical communication is to relieve tension in a high-stress work situation. Deftly applied, the styptic of humor can ease painful clashes on the perilous pathways of interpersonal relations. Rare is the technical communicator who does not have to navigate these pathways in the routine performance of her job responsibilities. The rules of thumb here are as follows:

- Never laugh at somebody else. Laugh at the *situation*. Find irony, incongruity, or—if you feel your audience will be receptive—sheer zaniness.
- If you laugh at anybody, laugh at yourself. Self-deprecating humor can be contagious, defusing a potentially volatile situation before it explodes into communication meltdown.

Yes, humor can serve on occasion. However, heed this important warning: *Never allow humor to oversimplify or trivialize content, especially when safety is involved*!

We recall a horror story shared by a colleague at an ethics workshop: A naive, non-native-English-speaking user assumed that because the cartoon character who told her not to plug in the appliance while standing in a puddle of water was humorous, the advice was not to be taken seriously. The results were nearly tragic and the ensuing litigation was substantial. Comedy and tragedy don't mix. And humor simply does not translate (see Chapter 10). So before you use humor to leaven technical communication, be sure there is no danger of misinterpretation.

How to Weigh Budget versus Quality

In the perennial tradeoffs among cost, schedule, and quality, the most common dilemma faced by technical communicators is the imperative to deliver top quality with inadequate budget. Too often, our clients have "champagne tastes and beer pocketbooks."

What to do?

The first step is to analyze the communication situation and determine the appropriate level of quality that will best meet the communication requirement. Sometimes, if we are honestly committed to our customer's best interests, such an analysis will reveal that the budget actually *is* adequate after all; to do more may represent overkill for the situation. And if we are serving our *own* interests rather than our client's by creating an overly elaborate and expensive product, then we are behaving unethically.

If, however, the requirements analysis shows budget is indeed inadequate to get the job done properly, the next step is to communicate that fact, directly and honestly, to the client. The objective is to get sufficient budget to achieve the quality standard needed to meet the communication requirement.

And if that fails?

Here, the choices get tougher. If we are an independent contractor, it may be time to hit the road. Producing a product we know will not meet the client's requirement because he will not (or cannot) provide adequate funding is not a wise professional decision (see Chapter 13). The long-term damage a shoddy job inflicts upon our reputation will more than erase the benefits of the short-term paycheck.

If we are in the employ of another who insists we go ahead and generate the product on inadequate budget (or an unreasonable schedule, or both), then about all we can do is communicate clearly, candidly, and *specifically* exactly what the requester can expect:

- "You understand, then, that if we scan all these photos in overnight, there will be no time to retouch or enhance them?"
- "You realize, then, that if we produce this video in a weekend, there will be no soundtrack? You will have to explain the clips as you present."

This lessens—though it does not eliminate—the possibility of disappointment after the fact. All we can say then, but not in so many words, is "I told you so." And, to draw from aerospace lingo, always remember:

> *The bitter taste of poor quality remains long after the sweetness of cost savings or rapid delivery is forgotten.*[12]

Summary

Quality in technical communication is defined not in absolute but in relative terms: It means performing to standards that best meet the end user's unique communication requirements.

The three major parameters by which a communication product is measured are cost, schedule, and quality. The technical communicator often must conduct difficult tradeoffs among these three, in which case clear definition of the product with the client in the planning stages is essential.

Total Quality Management, or TQM, operates on the principle that quality is "designed in," not "inspected in." Applied to technical communication, the principles of TQM dictate early involvement by editors, producers, illustrators, and multimedia specialists in the planning and development of communication products.

The *zero defects* standard reflects not blind perfectionism but refusal to acknowledge error up front. If we expect less, we get less.

Quality is measured by the cost of nonconformance to requirements: That's how much our mistakes cost us.

Competency in technical communication may be measured in terms of AOC[7]: accuracy, objectivity, completeness, correctness, consistency, clarity, conciseness, coherence, and creativity.

Budget constraints can play havoc with quality. When this happens, the first recourse for the communicator is to get more budget. Failing at that, the options are to decline to work the project (if independently employed) or, if leveraged into quality compromises by an employer, to communicate clearly and specifically exactly what kind of communication product the available funds will buy.

Application Exercises

1. What is the difference between an absolute and a relative definition of quality? Which is better suited to technical communication? Why?

2. Explain the rationale behind the motto of the Russian tank commander. Do you agree? Why or why not?

3. How does Total Quality Management, or TQM, differ from traditional approaches to quality? What are the implications of TQM for technical communication?

4. How can you edit aggressively and still avoid the "cardinal sin" of changing meaning?

5. Why is outlining important? Identify three written communication products to which outlining may be applied.

6. Explain how signposts can improve coherence:

   a. In a written document
   b. In an interactive online document
   c. In graphics

7. Find an example of humor appropriately (or inappropriately) applied in technical communication.

8. Take a memo, e-mail message, or information sheet that is dull from a layout and packaging standpoint, and red-pencil it for improvement. Incorporate headlines also, if appropriate.

Exploration Exercises

1. Interview a quality engineer or inspector. Ask him or her how quality procedures have changed over the past several years.
2. Read *Quality is Free: The Art of Making Quality Certain* by Philip B. Crosby. Write a short (2- to 3-page) reaction paper or prepare a brief (3- to 4-minute) oral presentation.
3. How do you feel about "Zero Defects"? Is this philosophy realistic? Is it a useful approach to achieving quality? Would you sign up to it? Explain.

Directions for 4 through 12: For each parameter governing quality in technical communication (AOC^7), come up with an example of *competency* in that parameter or an example of *incompetency* in that parameter. For the examples of incompetency, suggest or make revisions. Clip magazine and newspaper articles; collect literature from government, businesses, civic organizations, and the like.

4. Accuracy
5. Objectivity
6. Completeness
7. Correctness
8. Consistency
9. Clarity
10. Conciseness
11. Coherence
12. Creativity
13. Analyze the following communication scenario, derive an effective communication strategy, and select the most appropriate medium (media) through which to implement that strategy.

    Situation: Mandatory weekend overtime is going to force you to miss an out-of-town family reunion your mother has been working on and looking forward to for two years. Too many other family members are involved to reschedule the event and defying your boss on the overtime could cost you your job.

    Communication Objective: Deliver the message to Mother as gently as possible, yet firmly enough to make it clear there is no appeal.

    Audience Analysis: Your mother, while warm and loving, also has a tendency to use guilt manipulation to get children and family members to do what she wants. She is sensitive, her feelings are easily hurt, and when crossed, she's extremely persistent, using every tool at her command to try to get what she wants.

Communication Strategy/Media Selection: As a communication expert, how would you approach this dilemma?

14. Use the index in a basic grammar text to check out the first nine maladies that impede clarity. Find and complete one or two exercises on each to confirm your mastery. Stress comma usage.
15. For clarity malady #10, acquire a copy of Gunning's "Fog," read it, and complete the five exercises on defogging.
16. To pursue conciseness, chase down lists of *deadwood* and redundancy (good sources are writing handbooks and journalism texts), and share them with your classmates or coworkers.
17. Read the chapter on news leads in a journalism text and complete a few exercises.
18. Read the chapter on headlines in a journalism text and complete a few exercises.
19. Consider the following three philosophies on quality:
    - "If it ain't broke, don't fix it."
    - "Anything can be improved."
    - "Better is worst enemy of good enough."

    What's your personal position on each? Explain.

References

1. Karl L. Smart, Kristie K. Seawright, and Kristen Bell DeTienne, "Defining Quality in Technical Communication: A Holistic Approach," *Technical Communication* (Third Quarter 1995): 474-481.
2. *Ibid.*
3. *Ibid.*
4. Philip B. Crosby, *Quality without Tears: The Art of Hassle-Free Management* (New York: McGraw-Hill, 1984), p. 50.
5. Daniel W. Voss, "Editorial Training," Martin Marietta Corporation, Orlando, FL, 1988 (now Lockheed Martin Corporation). Presentation at STC's 37th Annual International Technical Communication Conference, Santa Clara, CA, May 1990; in *Proceedings*, ET-58-61.

6. Stewart James, *Calculus*, 2d ed. (Arlington, VA: Brookes/Cole Publishing Co., 1991), 664.
7. Robert Gunning, "Fog," (Chicago: Dartnell Press, 1964).
8. Edwin Newman, Keynote Address, 42nd Annual Conference of the Society for Technical Communication, Washington, D.C., April 1995.
9. *Simply Stated*, "Gas Utilities Switch to Plain Language," American Institutes for Research, Washington, D.C., March 1986, p. 2; cited by Karen A. Shriver, "Quality in Technical Communication," STC's *Technical Communication* (Second Quarter 1993): 250-51.
10. Robert D. Eagleson, *Writing in Plain English*, (Canberra, Australia: Australian Government Publishing Service), p. 6; cited by Shriver, *"Quality in Technical Communication."*
11. I. A. Etzkorn, "Why Government Has Difficulty Communicating," in *Plain Language: Principles and Practice*, ed. E. R. Steinberg, (Detroit: Wayne State University Press, 1991), 224; cited by Shriver, *"Quality in Technical Communication."*
12. Traditional motto at Lockheed Martin Electronics and Missiles Co., Orlando, FL.

Chapter 8 Overview

Teamwork . . .

The Mission: In technical communication, we define teamwork as the ability to work together effectively with clients, employers, and coworkers to develop and produce quality communication products that meet user requirements.

Chapter Roadmap

Definition and Focus. This section provides a brief overview of teamwork. Teaming is "in." This chapter explores in detail how to collaborate effectively in technical communication projects, whether in the classroom, on the job, or within other organizations, and how to understand and fulfill our roles not only as communication specialists but also as respected team players.

Collaboration Demands Sacrifice but Reaps Rewards. This section can help you apply tried-and-proven techniques to increase your effectiveness in collaborative projects.

- *Collaborative Classroom Projects Prepare Students for the Workplace.* This section provides a systematic review of the steps in a successful group project and ties in the underlying principles of group dynamics.

  Accomplishing the Task Means Following a Sequence of Steps

  Working Together Involves Group Dynamics

  Collaboration Is a Two-Edged Sword: It Has Benefits and Risks

  Helping Students with Writing Assignments: What Are the Ethical Boundaries?

- *Collaboration in the Workplace Yields Better Projects.* This section explores the role of teamwork in developing software documentation and in preparing winning proposals.

Early Involvement of Technical Communicators Means Better Documentation

Competitive Proposals Impose the Acid Test of Teamwork

- *Professional Association Initiatives Require "Writing by* Committee"
- *Writing for Publication Often Entails Coauthorship*
- *Communication Specialists Can Be Community Leaders*

Teaming in the Workplace is the Trend of the 1990s. This section provides tips on how to survive and flourish in changing industry environments and how, specifically, to deal with the ethical dilemmas these environments can create.

- *Teaming in a Traditional Vertical Hierarchy Brings Ethical Conflicts*
- *Integrated Product Development Teams Exercise Concurrent Engineering*
- *Self-Directed Work Teams Empower Employees—IF Management Is Enlightened*

Summary, Practical Exercises, and References

. . . Our Duty to Work Together to Meet Mutual Objectives

CHAPTER 8

Teamwork in Technical Communication Means Putting Clients, Coworkers, and Products Ahead of Self

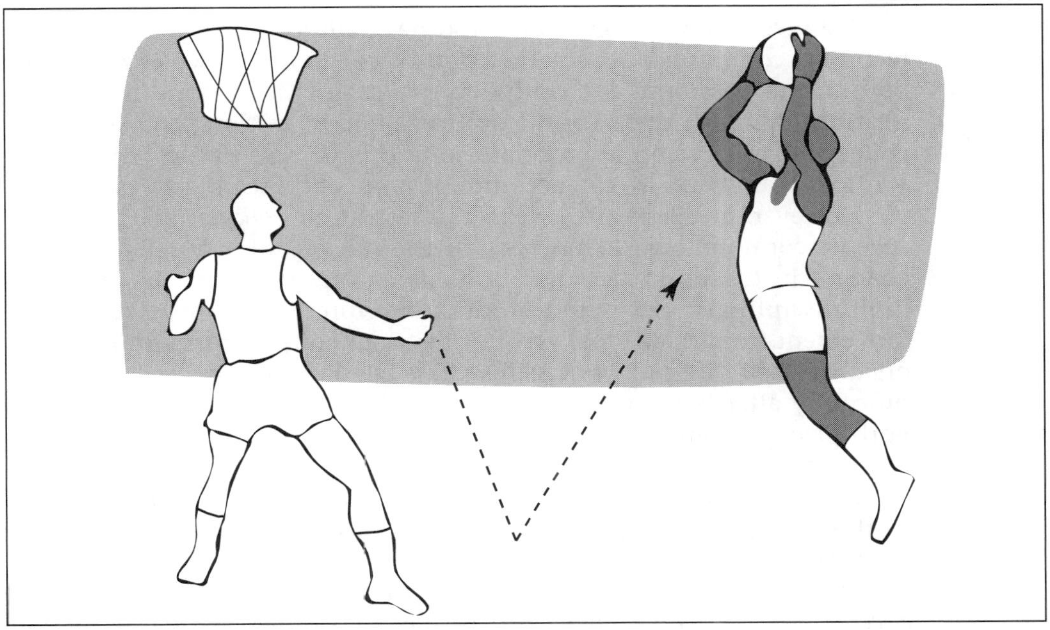

If we don't hang together,
we shall most assuredly hang separately.

—The character of Ben Franklin
in the Broadway musical *1776*

Definition and Focus

When a professional basketball team charges down the court on a fast break, teamwork means physically and mentally acting as one unit, subjugating all personal goals and interests to the accomplishment of one overriding mutual objective—putting the ball in the basket.

When the 13 colonies declared their independence from the mightiest nation on the face of the earth, teamwork meant overcoming serious regional and philosophical differences to present a united front and sustain what appeared to be an infinitesimal chance of success. For the signers of the Declaration, much more than political independence was riding on the outcome. After all, when they affixed their "John Hancocks" to that parchment, they were, in the eyes of King George at least, committing treason.

That's where the hangman came in.

While technical communicators are not likely to encounter quite that severe a penalty for failing to exercise good teamwork, in today's changing business environment the ability to work together effectively is more important than ever—both to successful accomplishment of a job and personal success in one's career.

Teamwork is not a new concept. It has traditionally applied to the field of technical communication in the form of collaborative writing assignments, both in the classroom and on the job; of maintaining smooth interpersonal relationships with clients and employers; and in working cooperatively with colleagues to develop appropriate solutions to communication challenges. Without teamwork, we can accomplish none of these objectives.

However, in the past few years the word *teamwork* has taken on new and specific connotations. It has been at the nucleus of a number of profound changes in the business world. Notable among these are the emergence of multidisciplinary integrated product-development teams (IPDTs) and the movement from traditional vertical management structures into horizontally organized self-directed work teams (SDWTs). Both of these concepts have significantly altered the role of the technical communicator in the workplace. Both place a premium on the ability to work together with a wide range of professionals.

Understanding of and commitment to the principles of effective teamwork in collaborative projects are a major asset in your academic training, job performance, and career development.

In technical communication, we define teamwork as the ability to work together effectively with clients, employers, and coworkers to develop and produce quality communication products that meet user requirements.

Collaboration Demands Sacrifice but Reaps Rewards

One of the common situations demanding teamwork in technical writing is the collaborative writing project, whether it be a classroom assignment, a science research report, or a business report or proposal. *Experts in business and technical writing universally acknowledge the importance of collaborative writing*:

> *Technical writers benefit in important ways from interaction with one another, with representative readers . . . and with managerial and technical personnel, whose feedback can be valuable for job-specific writing projects.*[1]

> *The ability to work effectively on writing teams is an indispensable skill . . . groups can generate a larger and more creative pool of ideas than can one person working alone.*[2]

> *Collaborative writing gives you the advantage of being able to share your knowledge and hear what others know. "Two heads are better than one" often proves true.*[3]

> *Sometimes . . . a writing project is too large for one person. Or it may require special knowledge that would only be available within a group of persons. Typical messages written by a team of writers are long reports or complex proposals.*[4]

> *Both aerospace engineering professionals and students found collaborative writing to be more productive than writing alone.*[5]

Let's take a look at how collaborative writing works best. We'll start in the classroom, then branch out to the workplace.

Collaborative Classroom Projects Prepare Students for the Workplace

Most technical communication programs include one or more courses that require a collaborative writing project. A 1992 survey of 200 college graduates showed that 73.5 percent had been involved in at least one collaborative writing project during their undergraduate training.[6] At first, students tend to be uncomfortable with this type of assignment, because it introduces requirements different from those involved in traditional solitary assignments. Yet collaborative writing assignments are an important part of the curriculum, because they prepare students for similar assignments in the workplace.

The key to success in collaborative academic projects is twofold:

1. Follow a clear *sequence of steps* (task-oriented)
2. Understand the *group dynamics* involved in the process (relationship-oriented)

Accomplishing the Task Means Following a Sequence of Steps

1. **Forming the Team.** The composition of a team is obviously a key element in the determining the success of its project. Classroom work teams form in two ways: The professor appoints them or the students form their own groups. In either event, the two key criteria to consider in forming the alignments are personal competencies and technical knowledge or expertise.
2. **Orienting the Team.** Clearly, the first order of business for a newly formed team is for its members to get to know each other and to select a leader.
3. **Defining the Task and Requirements.** This step, or at least a skeleton guideline, is generally provided by the instructor. It is extremely important for the work group to understand exactly what it is expected to produce—a written report, an oral presentation, a demonstration, a videotape, some combination thereof—as well as the specific criteria by which the product will be evaluated. If you wish to work together on your own initiative and the instructor has not specified collaboration, be sure to check whether such an arrangement is acceptable; otherwise, you run the risk of appearing to have cheated or plagiarized on the assignment (see Chapter 5).
4. **Brainstorming.** This step is a free-wheeling, loosely structured exchange of ideas designed to encourage creativity. The golden rule of brainstorming is that nothing is "too stupid" to toss up for discussion. Often it is precisely the wackiest idea that spins off a creative solution which might otherwise never have been conceived.
5. **Product Planning.** Precision is critical on product planning. Unclear directions mean wasted effort, which breeds frustration for individuals, discord for the team, and diminished quality for the product. Whatever the project is, to be successful it must be planned and *outlined* or, in the case of a videotape, *storyboarded* (see the discussion on coherence in Chapter 7). An important subset of product planning is to develop a *consistent format.* For written reports, it's important to come up with a short *style guide* governing the usage of common terms.
6. **Dividing Responsibility.** "Divide and conquer" is the name of the game here. The key is to divide up assignments so as to distribute the workload as evenly as possible and exploit specific skills of individual team members—play to strengths and work around weaknesses. The leader must do her best to ensure an equitable workshare; unequal levels of effort are one of the most common pitfalls of collaborative projects.

7. **Scheduling.** Classroom projects have due dates, just as jobs in the workplace have deadlines. To ensure the project is ready on time, the team must draw up a detailed schedule at the very beginning, identifying the major milestones in the project and establishing when each must be completed in order for the project to be finished on time. In other words, *plan the work and work to the plan.*
8. **Establishing Working Procedures.** As obvious as it may seem, we should not neglect to mention the importance of establishing team working procedures during project planning. Basic questions such as times and places for meetings and avenues for communication can be crucial. An error as simple as forgetting to exchange phone numbers and e-mail addresses can cause a serious schedule impact if a team member with a critical segment of the project due on Monday cannot get a work-stopping issue resolved on Saturday. To be sure you've covered all bases, we recommend using the journalist's *5W&H* model: who, what, when, where, why, and how.[7]
9. **Running Efficient Meetings.** More time is wasted in inefficiently run business meetings than perhaps any other activity in the workplace. Key precepts for running efficient meetings are:

    - ❑ Start on time.
    - ❑ Have an agenda.
    - ❑ Stick to the agenda.
    - ❑ Take minutes.
    - ❑ Keep the tone friendly.
    - ❑ Keep the meeting short (an hour or less).
    - ❑ Finish on time.
    - ❑ Meet regularly, but not excessively.
    - ❑ Use e-mail to limit meetings.

10. **Drafting.** Whatever the project, it should come in three stages: outline, draft, and final. Drafts are usually done by individuals, then reviewed by the group to ensure consistency and coherence. In all cases, we stress that the word *draft* does *not* mean shoddy work. In fact, the opposite is true. The draft should represent its creator's "best shot," ready for group review and criticism.
11. **Reviewing and Revising.** Just like individual writing assignments, collaborative projects are rarely, if ever, perfected on the first pass. In fact, revision is even more critical in collaborative projects because of the natural tendency of individuals to diverge from the overall plan.

The purpose of the group review is to prevent divergence; rather, *to converge in both content and style.* For the review-and-revision process to be harmonious and effective, it's important to establish specific criteria and procedures for evaluation during project planning, and make sure that all team members "sign up" to them.

12. **Finalizing the Product.** After the substantive revisions are made to the draft, the product is finalized; text is typed, illustrations are made "pretty," raw video footage is edited into final form and narration—if applicable—is added. The final step in production is a quality check: proofreading text, checking illustrations for accuracy, ensuring visuals are legible when projected, viewing video for continuity and maintenance of viewer interest.
13. **Debriefing.** An important epilogue in the collaborative project is the instructor's evaluation. This should be more than a grade; it should include feedback on team performance. The team should meet with the instructor to go over the evaluation and also to form its own conclusions about how well the project went—and why. Debriefings should accentuate the positive, but should also include "lessons learned"—things that could be done better the next time around.

Working Together Involves Group Dynamics

Let's return to our basketball team. What happens to teams whose players are more concerned with individual accomplishments than team success? They generally lose—especially to teams whose players are willing to put winning ahead of personal glory.

The same holds true in collaborative work projects. No matter how much talent a team has, no matter how well planned the project is, if team members cannot work together effectively the project will not be successful. That's the special challenge of collaborative writing. As you are well aware, writing is hard enough by itself. Superimposing group dynamics makes it harder still.

But just as teamwork on the basketball court can bring a championship, teamwork in the classroom can produce outstanding projects (Figure 8.1).

What, then, is the secret of effective collaboration? It comes down to two words: *respect* and *communication.* Of course, those two words are far easier said than achieved! Here are a few guidelines for achieving them:

1. **Choose the Leader Wisely.** The leader has responsibilities critical to team success:

   - Providing overall structure and direction
   - Ensuring that work is divided equitably
   - Involving *all* team members in discussions

Figure 8.1 Respect and communication are the keys to effective teamwork.

- ❑ Making sure assignments are completed on schedule
- ❑ Resolving conflicts
- ❑ Running efficient meetings

2. **Allow Time for Social Interaction.** The relational aspects of collaboration cannot be divorced from the task-oriented aspects. This is especially important to remember in the group's formative stages. One classic pattern for teaming describes four stages in the formation of an effective work team: "forming, storming, norming, and performing."[8] In other words, after the "hellos" are finished, some conflict is inevitable before procedures are ironed out and a "modus operandi" developed.
3. **Be Sure Everyone Understands the Goal and Is Committed to Achieving It.** Before getting down to the detailed assignments, it's important to make sure everyone is playing to the same sheet of music. If half the team is creating opera and the other half is generating ragtime, it's hard to imagine successful project integration!

4. **Agree Upon a Method for Handling Conflict—Before It Happens!** If there are differences of opinion or style, it's far better to resolve them up front than to allow them to smolder throughout project development, poisoning the atmosphere and eroding productivity. Even if such differences are not immediately apparent, they are likely to surface under the pressure of approaching deadlines.
5. **Be Open-Minded.** The more receptive individual members of the team are to the ideas of others, the more successful the team will be. Giving another's ideas a fair hearing does not mean abandoning your own ideas—it's simply showing respect. Sometimes you may change your position, sometimes you may not, but in all cases you should at least be willing to listen.
6. **Listen.** Again, l-i-s-t-e-n! We all like to think we *listen*, but we rarely do—not really. True listening means setting aside our preconceived notions about the subject, our attitudes toward the speaker—yes, even our preoccupation with our date that evening!—and *focusing* on what our teammates have to say. It means asking genuine questions for clarification, not trap questions designed to expose the drawbacks of the speaker's plan. It means *showing interest*. The acid test of listening is the ability to repeat what has just been said in a manner that shows the speaker you understand it. Try it. True listening is difficult to accomplish. When it *is* accomplished, it is also invaluable to the collaborative writing process.
7. **Be Considerate of Others' Feelings.** This advice has many applications. It means:

   - ❑ **Give credit where credit is due.** Acknowledging others' ideas not only makes *them* feel good, it also shows *you* to be an ethical communicator.
   - ❑ **Share.** Share the workload, share the workstation, share the floor in discussions, share the instructor's attention.
   - ❑ **Maintain trust and confidence.** A breach of confidence spells discord—and is very tough to mend.
   - ❑ **Criticize diplomatically.** This is especially critical during draft review and evaluation. How can you suggest improvements without stepping on a teammate's toes? The answer is you can't, not completely. The key is to step gently. Be positive. Be specific. Don't attack; suggest. You can nudge a few toes without crushing a foot! Fred D. White provides excellent direction for critiquing drafts diplomatically, differentiating between three common types of responses:[9]

Type of Response	*Function*	*Example*
Directional	Tells writer specifically what to do	"Give more concrete examples here."
Judgmental	Draws an evaluative conclusion	"These assertions are too general."
Facilitative	Suggests possible revision strategy	"Can we provide examples that will convince our readers?"

The best kind of criticism is facilitative; the worst kind is judgmental. Thus, instead of saying "You have failed to provide any examples here," try saying "That's a good point. Did you run across any juicy examples that we could use to back it up?"

8. **Be *Task*-Oriented, not Ego-Driven.** Being task-oriented means avoiding what we call *ego involvement*—investing your self-esteem in the project to the point where a constructive suggestion on your work becomes a personal attack at you. Ego involvement is extremely contraproductive in collaborative work projects. For a thorough discussion, see Chapter 13. Being task-oriented means accepting criticism graciously—putting product improvement (task) ahead of personal competition (ego). Protecting *others'* toes means stepping gently by criticizing diplomatically. Protecting your *own* toes means wearing the toeplate of task orientation. *Criticism is an essential part of collaboration: Don't take it personally*!

Collaboration Is a Two-Edged Sword: It Has Benefits and Risks

The advantages and disadvantages of collaborative writing, in the classroom as well as elsewhere, can be summarized as follows:

Aspect	*Advantages*	*Disadvantages*
Workshare	A team can accomplish much more than an individual. One person can't do everything.	Sometimes the workshare is inequitable. One or two team members *may* wind up doing nearly everything.

(*continues*)

Benefits and Risks of Collaborative Writing (*Continued*)

Aspect	*Advantages*	*Disadvantages*
Deadlines	Sometimes a team can get the job done much more quickly than an individual. That is particularly true when the task is noncontroversial and can easily be segmented (e.g., finding examples of artwork from four different countries).	Sometimes it can take *longer* for a team to finish the job than one individual. This is particularly true when the task calls for consensus among members with sharply opposing positions.
Stress	A team can reduce the stress of individual responsibility.	A team can also *increase* stress through conflict and competition.
Knowledge base	The collective knowledge of the team exceeds that of any individual.	Those who know less may ride the coattails of those who know more.
Skills diversification	A team can bring a wide range of skills to bear on the project.	Those who can do less may ride the coattails of those who can do more.
Perspectives	A team brings more world views, provides greater opportunity for depth and breadth.	With differences comes the possibility of disagreement and the need to establish consensus.
Feedback	The team can provide valuable feedback on individual performance.	Candid feedback can be threatening.

Clearly, then, collaborative writing is a valuable technique, but it is not a panacea for every technical communication situation. It offers many advantages, but it entails a number of risks as well. Many of the risks can be mitigated by implementing the guidelines for effective group dynamics previously outlined.

Helping Students with Writing Assignments: What Are the Ethical Boundaries?

Any discussion of collaborative writing in the context of professional ethics would be incomplete without addressing the hotly debated question: *Can students appropriately receive expert help on written assignments without overstepping the bounds of academic propriety*? Assistance comes from various sources: peers, parents, the university's writing center, a professional editor, another professor, even the professor in the course for which the assignment is being prepared.

It is impossible to devise hard-and-fast rules to cover this situation. The key to determining whether the assistance a student receives falls within the compass of ethical behavior is twofold:

1. Exactly how much help did the student receive? Is the substantive work still original?
2. Has the student openly disclosed the assistance to the professor in advance?

In most college classes, professors are likely to give the thumbs-up to students who ask whether they can get an editorial boost on their papers as long as such assistance is limited to nonsubstantive areas such as spelling, punctuation, grammar, and format. A notable exception are English classes where the mechanics of writing are an important intrinsic evaluation criterion, such as freshman composition.

Failure to disclose the outside assistance is not only unfair to the rest of the class from the standpoint of evaluation, it is also not in the best interest of the student receiving the help, because it can create an illusion of progress which melts in the crucible of a subsequent exam or graded essay that calls for the student to work independently. Similarly, the professor herself must often tread a thin line between providing *just enough* help, which teaches the student how to write better, and providing *too much* help, which breeds excessive reliance and, taken to an extreme, can even raise issues of equity in evaluation relative to other students who receive less assistance.

The issue extends to graduate as well as undergraduate levels. Ellen Fenwick, freelance technical writer and graduate student in the technical communication program at Boise State University under the tutelage of Mike Markel, sparked a firestorm of electronic discussion on TECHWR-L (see Chapter 12) when she raised the question of whether it was ethical for professional editors to assist graduate students on their theses or dissertations. Opinions varied widely, ranging from an "any help is cheating" position to an "anything short of rewriting the paper is OK" posture. Reports Fenwick:

> *The consensus seemed to be that it was ethical to assist, as long [as the] assistance was limited to normal editing functions (for example, spelling, grammar, sentence and paragraph structure, formatting). . . . Most concurred that the research, organization of ideas, and writing MUST remain the original work of the student.*[10]

Under this approach, editors felt fundamental problems on content or structure should not be fixed, but could suitably be pointed out for students to correct. A number of respondents felt learning to work with an editor was, in itself, a valuable learning experience for graduate students.

Fenwick points out that some companies offer ready-made research papers for a fee, even blatantly advertising their services—prompting states to enact legislation specifically prohibiting the practice on the grounds that it constitutes a breach of contract between the students and the university (namely, the students' tacit agreement to follow the rules, including the provision that work submitted for academic credit must be their own).[11]

Ironically, she also observes that the very same collaborative writing and "ghostwriting" that are commonplace in industry may run afoul of ethical issues in the specialized environment of the classroom. That's true. Unless the classroom assignment is itself collaborative, students would be well advised to check out *any* potential assistance with the professor *before* enlisting it. In fact, it would be only a slight exaggeration to say that *in cases of academic assistance, disclosure is nine-tenths of the law*. However, we would caution that even in the business world, there's a big difference between an editor or a colleague making an author's work better and the author cavalierly slapping a byline on a publication to which he has made no substantive contribution. Within the narrow confines of a contractual "ghostwriting" agreement, this practice is acceptable; otherwise it constitutes willful misrepresentation and plagiarism (see Chapter 5).

Fenwick concludes the TECHWR-L thread with useful ethical guidelines for a professional writer (or, for that matter, *anyone*) who is approached by a student for assistance on a writing project:

1. Be explicit with the student at the very beginning about what you ethically can and cannot do, and why.
2. If there is any doubt about the ethics of the work, have the student check with the professor. The student should be able to comfortably acknowledge the level of assistance you will be providing.
3. Do not contribute to or change intellectual content.
4. Do not do any original writing or any significant rewriting.
5. Where you see flaws or inadequacies in organization, logic, or expression, only point them out. Let the student decide on how to resolve them.

6. Explain all your editorial changes to the student, so that the student learns from them.[12]

Collaboration in the Workplace Yields Better Products

Collaborative writing is an integral part of the technical communicator's role in the workplace. Two situations most commonly call for us to serve that role:

- ❑ The job is too large for one person to handle alone (long reports, extensive manuals or documentation, major proposals)
- ❑ The communication requirement demands specialized expertise (the critical executive summary to a proposal, the "winner-take-all" marketing presentation that requires a narrated videotape)

Entire textbooks have been written on how to plan, write, and produce effective documentation and how to write winning proposals; that is not, of course, our purpose here. Our focus is the special professional and ethical requirements that accompany collaborative writing on such business documents.

Much of what we covered under collaborative writing in the classroom holds true for collaborative writing in the workplace as well. The difference is that *the stakes are higher in industry*. In the classroom project, the consequence of poor performance or late completion may be a letter grade for the participants; in business, the consequence may be a lost contract—and lost jobs. With higher stakes comes higher pressure. After all, if Susan's late inputs cause a college project to be late, Myron may wind up losing a few decimal points on his GPA. If Susan's late inputs cause a proposal deadline to be blown, Myron may wind up losing his job.

With higher pressure comes an ethical mandate for calm, professional comportment and clear, candid communication by all participants in the project. Early and clear definition of team and individual responsibilities, task orientation rather than ego involvement, diplomatic criticism and gracious receipt of same, active listening—all the benchmarks of successful classroom collaboration hold true in the business world as well. Let's take a look at the two most common collaborative writing projects in which technical communicators are apt to be involved: the software documentation manual (or online documentation) and the competitive proposal.

Early Involvement of Technical Communicators Means Better Documentation

The most important thing a technical communicator can do to improve the team effort on a software documentation project is to get onboard early. Documentation is too often an afterthought, tacked on just before marketing as programmers are working the final bugs out of the application. It falls to the manager of the technical communication department—in smaller companies, this may be one and the same as the technical writer who will be producing the documentation!—to convince management that documentation is an integral part of the product, one that must be planned from the beginning, developed as a cooperative venture among the programmers and the technical writer, subjected to thorough usability testing, and perfected before the software is released.

The key to carrying this message forward successfully is to convince management that *good documentation is good business*, because it reduces the number of service calls, increases user satisfaction, and breeds follow-on business. Numerous research studies have been conducted to quantify the business value of quality documentation; eight articles on the subject were compiled by guest editors Janice Redish and Judith Ramsey for the First Quarter 1995 issue of STC's *Technical Communication*.[13]

The presence of a technical writer on the software development team from project inception has major advantages:

- The technical writer's experience at document structure and planning ensures not only logical documentation but also a more orderly approach to document production.
- A close working relationship among programmers and writers allows programmers' technical knowledgeability to rub off on the writers, improving their understanding of how the product works. Better understanding makes for clearer writing.
- The writer can join forces with the usability specialist on the team—there had better *be* a usability specialist on the team, or the project is headed for problems too thorny for documentation to iron out!—and establish a "watchdog" presence during software development, looking out for the interests of the user.

While tongue-in-cheek, the cartoon does drive at the heart of the technical communicator's collaborative challenge: With apologies to normal, well-adjusted software programmers (*both* of them—the one in Idaho and the one in Sri Lanka*), if you think creative authors are temperamental, you probably have never met a brilliant software programmer! Mostly they dwell in cold, porcelain chambers deep within the inner recesses of computer laboratories, subsisting on Diet Coke and Twinkies, spinning their magic in FORTRAN and COBOL and other mysterious electronic tongues inaccessible to mere mortals. (*Just kidding!*)

Seriously, though, the editor-programmer relationship is very similar to the editor-author relationship discussed in Chapter 13. Here again, the key is

mutual respect. As technical communicators, we are well served by showing genuine interest in the expertise of our technical colleagues; not only is this essential to give us the knowledge needed to do our job properly, it builds a professional trust that can withstand the crucible of short deadlines, nasty bugs, and last-minute rework.

Competitive Proposals Impose the Acid Test of Teamwork

Proposal preparation center, editorial cubicle.

3:00 A.M.

Third day in a row.

Scott: *Got those galley proofs for the executive summary yet?*

Mary: Thought *I had them. Ross took them back for revision.*

Scott: *Again! Damn! Doesn't he realize that summary is four-color printing? We've only got 72 hours till* wheels-up*!* (Cool proposal talk for an airplane taking off to carry the proposal to its destination.)

Mary:	*Yeah, he realizes.*
Scott:	*So why did he take the proofs back for revision?*
Mary:	*He's the proposal manager, remember?*
Scott:	*Oh, yeah. . .*

Stories like this abound in the communication departments of major government contractors, where competitive proposals are the lifeblood of the business.

Yes, proposals are painful. But they're also important. (And, much as we hate to admit it, exciting!) Successful proposals win contracts that bring jobs and profit.

Given the critical importance of proposals to business success, it stands to reason that an intelligent company will allocate many of its top resources to the pursuit of new business through competitive procurement. That's where the collaborative writing comes in—industry can offer no better example of "writing by committee" than this feisty creature called the proposal. Indeed, that is the unique challenge of the species—how to breathe unity and coherence into a lengthy (often multiple-volume) document that has been written by a dozen or more individuals.

Two schools of thought exist on proposal management. There's the "old school," or "macho" approach, where the proposal manager becomes a cross between a drill sergeant and an air traffic controller. Under this approach, there is much fingerpointing, meetings often violate NBA decibel standards, and bodies pile up—but it *does* work.

Then there's the "new school" (see Self-Directed Work Teams, later in chapter), where the proposal manager becomes a cross between a basketball coach and a lead cheerleader. Under this approach, dress is casual, meetings are laid-back, and casualties are lower—and it *also* works, as long as there is sufficient structure.

That's the key—structure. There's no need for chest-thumping, head-butting, Type X management to produce a winning proposal. The "kinder, gentler" school of thought works better because it pivots on the same two basic elements we introduced earlier under classroom collaboration: respect and communication. In a high-stress environment that usually involves tight deadlines and heavy overtime, morale will be much higher under a genuine team approach that values the individual as much as it values the proposal win. Win probability will actually be higher, because *pride and cooperation produce a better product than do fear and intimidation.*

The trick is to retain the tried-and-proven controls of traditional proposal management within the new framework: structure without stricture, so to speak. Technical communicators have much to bring to this party; after all, project planning and communication are our forte! We can play a key role on a proposal team, not only through our traditional responsibility for producing the proposal itself—which obviously remains the critical function—but also by facilitating the internal communication that lubricates the gears of the proposal engine and cranks out the proposal with as little friction as possible.

Professional Association Initiatives Require "Writing by Committee"

Collaborative writing clearly has an integral role in the activities of professional associations, which are discussed in detail in Chapters 12 and 13. Most technical communication and scientific associations publish quarterly journals and monthly periodicals at a national level, as well as newsletters at the chapter level. Major conferences publish written *Proceedings*, including technical papers and workshop summaries. Coauthorship (discussed in the next section) is common on conference presentations and journal articles.

There's more. Committees, both local and national, work on specific projects and generate reports. Special Interest Groups (SIGs) and Professional Interest Committees (PICs) publish newsletters. More recently, World Wide Web sites and listings have been springing up at both local and national levels like mushrooms after a summer rain. All these communication projects demand cooperation and teamwork.

One would like to think that in all cooperative writing ventures, professional technical communicators would be paragons of selfless team-oriented communication, always putting task before personality, consensus before conflict, and team goals ahead of personal agendas.

Well, sor-r-r-ry! We've been there, done that, folks—have the tee-shirt! While it's true that technical communicators often possess exceptional talent for communication, they also run afoul of a basic axiom—it takes a fair amount of ego to be a good writer. The trick is learning to keep that ego in check when it comes to pursuing group goals.

Work on it. We have. We're not there yet, by the way—it's a career-long, a *life-long* process. But it pays off, both professionally and personally.

Writing for Publication Often Entails Coauthorship

This textbook would not be in your hands were it not for collaborative writing. Determined as both of us were to write a down-to-earth, practical guide to ethics and technical communication, the prospect was simply too intimidating for either of us to take on alone. How well we have done together, only you, the reader, can judge. But one thing is certain—working together, we *did* publish this book!

All the principles we have discussed in this chapter became very real for us as we worked the manuscript under an extremely tight publishing schedule. We'd like to tell you we pulled this off with no conflict or problems, but you'd know we were lying (and we've already come out against that in Chapter 4).

So, what words of wisdom do we have on coauthorship?

- ❑ Trust . . . always
- ❑ Candor . . . likewise
- ❑ Tact . . . well, usually!
- ❑ Strategic planning . . . up front
- ❑ Discipline . . . throughout
- ❑ Task orientation . . . and mothballing personal egos
- ❑ Communication . . . you'd sure better!
- ❑ A sense of humor . . . if you can't laugh, we guarantee you'll cry

Oh, one more thing:

- ❑ You have to be certifiably crazy!

Communication Specialists Can Be Community Leaders

Mastering the art of collaborative writing in the classroom and the workplace gives technical communicators the requisite skills to provide a unique leadership role in community organizations, where the ability to reach consensus and express that consensus in written or oral media is often key to accomplishing the organization's goals.

Examples are manifold, ranging from churchwork to Little League to jury duty. Jury duty? Sure. Mention that you are a professional communicator, and we can almost guarantee you will be elected foreman (apologies to our female readers . . . we must reject "foreperson" as a linguistic barbarism that not even our ardent egalitarianism and antisexism [see Chapter 10] can countenance). There may not be a written report involved in reaching a verdict, but the process is the ultimate exercise in reaching consensus! Little League? You bet. By-laws must be written and kept current. Schedules must be developed and published. Brochures must be designed and disseminated. Even parades must be planned and executed.

Church? Without mixing religion and technical communication, let us observe that whatever your spiritual beliefs and religious affiliation, the ability to listen attentively and communicate sensitively is bound to be an invaluable asset in your personal spiritual growth.

Having edited and produced proposals in the tens of thousands of pages, and also having struggled to compose three sentences of comfort in a sympathy card from a church congregation to a spouse grieving the loss of a partner of 50 years, we can tell you which one is harder. And also which one is more important.

Teaming in the Workplace is the Trend of the 1990s

In industry, make no doubt about it—*teaming is in*. But the notion that we are suddenly teaming just because we dutifully spout politically correct buzzwords like "integrated product-development teams" and "self-directed work teams," not to mention the implication that we were somehow *not* teaming before these concepts were invented, is fundamentally flawed.

Teamwork has always been an integral part of technical communication and it always will be. That is not to imply that integrated product-development teams and self-directed work teams are purely cosmetic, nor that they will be here today, gone tomorrow. Not so. Both concepts represent a substantive and fundamental change in the way we do business. Properly applied, both concepts—especially in combination—have the potential to permanently and profoundly change the complexion of the work force.

Accordingly, no discussion of teamwork in technical communication would be complete without a look at industry *Teams*, with a capital *T*.

We'll start by discussing how teamwork in technical communication applies within a traditional hierarchical management structure. Then we'll address teamwork in the context of multidisciplinary integrated product-development teams. Finally, we'll put on our "coaches' caps" and walk you down the primrose path of self-directed work teams. Be sure you stay on the marked trail, however—there are rogue elephants where we are going!

Teaming in a Traditional Vertical Hierarchy Can Bring Ethical Conflicts

Since feudal times, man (and back then it really *was* just man, not person!) has structured society and government in a vertical hierarchy: king, duke, earl, baron, knight, page, serf, horse, foxhound, falcon, woman (we're not sure of the exact order; if you care, take European History 101). The military works like that: general, colonel, major, captain, lieutenant, sergeant, corporal, grunt. Academe works like that: president, provost, assistant provost, dean, department chair, full professor, associate professor, assistant professor, adjunct instructor, graduate student, undergraduate student. Even the church works like that: pope, cardinal, archbishop, bishop, high priest, secondary priests, deacons, congregation.

In the business world, the traditional hierarchy translates roughly as follows: chairman of the board, chief executive officer (CEO), president, vice president, director, mega-manager, mini-manager, supervisor, task leader, salaried employees, hourly employees, services staff (e.g., cafeteria, janitorial). Technical communicators sometimes complain that they fall somewhere between hourly and services in the pecking order, although to some extent they have themselves to blame for that (refer to the discussion of "Doormat Syndrome" in Chapter 13).

Teamwork does have a place within a tightly structured vertical organization. Indeed, such an organization is more likely to wedge employees in ethical conflicts than an organization that is genuinely committed to *Teaming* with a capital *T* (see the later section on Self-Directed Work Teams). The challenge is for individual employees to hold to ethical values of fairness, honesty, and teamwork while ruthless colleagues are clawing their way up the corporate ladder (see Chapter 4).

How ruthless, really, is that ladder?

That brings us to the *Elephant Theory of Management Sensitivity*, which reduces the equation to two species: elephants and mice[14] (see Figure 8.2). Elephants are managers. They come in several different sizes. Mice are workers.

Figure 8.2 "The Facts of Business Life." Cartoon design by Steve Otte.

They come in basically one size. No offense, but in this formula most technical communicators are mice.

Technical communicators just entering the business world ought to be issued a laminated business card with Figure 8.2 on the front. On the back would be printed the Seven Rules of Survival for Mice in a World of Elephants, a fanciful and satirical look at techniques for handling management. Note that the tactics are presented in order of increasing desperation:

Rule 1. Hide. In the business world, this tactic is commonly referred to as "keeping a low profile." Thus, the neophyte technical communicator who persists in exiting the building via Executive Row every day to curry favor with the brass is inviting a 4:30 summons ("Oh, Janet, I'm glad I caught you. . ."), which has a way of leading to midnight oil back at the workstation. This is like the mice who boldly parade past the elephants. Such mice may be considered brave, or they may be considered stupid, but mostly they are considered dead (see arrow in figure).

Rule 2. Run away. Having detected the tell-tale trumpeting of rampaging rogue elephants, any intelligent mouse heads for the hills (or holes, as the case may be). Thus, the graphics specialist who overhears an executive in the hall boasting about the 75 color slides she is going to have to have by the next morning implements Rule 2 by rapidly developing a root canal and scheduling an emergency dentist's appointment. But what about the graphics specialists' colleagues who take the hit? Oh, well—life's a beast when you're a mouse.

Rule 3. Distract the elephant. This is risky, but it can be exercised as a last resort if the first two options have failed and the elephant's foot is on the rise. Thus, the technical editor who protects the print schedule by inserting a noticeable but easily correctable error in Chapter 1 to divert the proposal manager's attention from an unfixable error in Appendix J is applying Rule 3.

Rule 4. Offer the elephant a peanut. This can be very effective, *if* you have the right peanut. Candidate "peanuts" for technical communicators to keep in their survival kit include fancy color visual aids, parchment memo pads bearing the elephant's footprint, and framed lithographs that go nicely in large corner offices. The downside risk of this tactic is that elephants often get *addicted* to peanuts; thus, the mouse who depends on this technique for survival had better carry an ample supply.

Rule 5. Never try to shoot an elephant. The caliber of gun a mouse can carry cannot drop an elephant. It can only *wound* the elephant, which is not a good idea. The significance of this rule in the business world needs no further explanation.

Rule 6. Get the elephant to stomp on another mouse instead of you. Many quick-thinking mice owe their survival to rapid implementation of Rule 6 during elephant stampedes. However, technical communicators who apply this technique should remember the rule has a corollary:

Rule 6. Corollary: *Make sure the mouse you point to for surrogate stomping is a mouse you don't like—and definitely not a senior mouse, or rat, that you report to.* If a senior mouse is the only available alternative target, then be sure it is dead.

Rule 7. Get a bigger elephant to stomp on the elephant that is about to stomp on you. If all else fails, this tactic can be a life-saver. But those who exercise Rule 6 to save their mice hides would be wise to heed a most important caveat:

Rule 7. Corollary: *Never get between two elephants who are in the process of stomping each other.* Refer to Rules 1 and 2.

The above treatment is, of course, tongue-in-cheek (trunk?), but if you think about it, it is not without validity. A number of these self-preservation techniques conveyed metaphorically, if applied literally, would clearly constitute unethical behavior on the part of the practitioners.

Take the phony dentist appointment, for example; that's just like calling in sick to avoid overtime, which is dishonest and unfair to teammates who have to pick up the slack. And "peanuts" are a hop, step, and a jump away from bribery. The line between a simple favor and an unethical—in government circles, sometimes even illegal—gratuity can be rather fine. Five to ten dollars is the rule of thumb (see Chapter 9).

The red herring example on the proposal is tempting to justify on pragmatic grounds ("It was either that or risk blowing the submittal"), but that's situational ethics (see Chapter 2). After all, it is the proposal manager's job, not the technical editor's, to decide just how serious the error in Appendix J is and whether the proposal should be submitted with the error. The editor *owes* the proposal manager a clear explanation of the error, even if it reflects on the editor's performance. We specified that the error could not be corrected without blowing the print schedule; however, if alerted, the proposal manager might have been able to exercise another option such as pointing out the error in the contractual letter of transmittal that accompanies the proposal submittal, altering the company's commitment in summary fashion with backup data to follow, thereby avoiding significant financial exposure for the company.

And as for the diversionary mouse stompings, battle-scarred veterans of the corporate jungle will chuckle knowingly, but right behind the chuckle lurks a sigh. Those who have had their careers plateaued by the advance of unprincipled colleagues know that when you turn the elephants and mice back into people, it's not very funny at all.

Holding to your principles in such a world—now, that's teamwork!

Integrated Product-Development Teams Exercise Concurrent Engineering

Integrated product-development teams, or IPDTs, with their multidisciplinary make-up and their task-centered approach, have revolutionized the basic structure of many companies. IPDTs are built around the concept of *concurrent engineering*. Briefly stated, concurrent engineering calls for *all* aspects of product performance—technical performance, quality, producibility, reliability, maintainability, safety, life cycle cost—to be factored into its development right from the beginning, the initial design stage.

By comparison, before concurrent engineering, it was typical for engineers to rule the roost during the early stages of product development, with the other specialties being tacked on after the design was complete, almost as an afterthought. The result was often elaborate and costly designs that could deliver outstanding performance, but which could not be produced cost-effectively, or which were impossible to maintain due to inaccessibility of key parts. Back to the drawing board!

Concurrent engineering calls for *all disciplines* to be represented in the IPDT through *all phases of product development* ("cradle to tomb"), to ensure that characteristics such as safety and maintainability are "designed in" rather than added after the fact. Applied to quality, the concurrent engineering concept becomes Total Quality Management—*everyone* is responsible for product quality, from the first artist's concept of the product to the environmentally safe disposal of the last fielded product 30 years later (see Chapter 8).

Even within an IPDT structure, traditional functional affiliations (for instance, "Engineering," "Finance," "Manufacturing") are still likely to exist, which sometimes leads to bloody turf battles that can ensnare unwary team members—including technical communicators—in career-damaging political crossfires (like mice trapped between rampaging rogue elephants).

However, if senior management truly empowers the IPDTs, including the critical "power of the purse" (that is, responsibility for profit and loss), then team members know where their first loyalty is—to deliver a quality product to the customer and turn a profit. To do this, team members must set aside functional differences and power struggles, and cooperate.

When merit increases are tied to the team's performance, it's amazing how cooperative erstwhile competitors can become!

Self-Directed Work Teams Empower Employees—IF Management Is Enlightened

Self-directed work teams, or SDWTs, are similar to IPDTs (and, in fact, may well *be* IPDTs) in that they are focused on specific tasks. However, self-directed teams differ from traditional management approaches not only structurally, but also philosophically. The distinguishing characteristics of SDWTs are threefold:

1. Increased employee autonomy and "empowerment"
2. A facilitative rather than directive management style
3. A breakdown of the traditional vertical structure, reducing middle management

A company organized around self-directed work teams has a horizontal structure instead of the traditional hierarchy. Theoretically, at least (see *Dilbert*), less emphasis is placed on the "pecking order" and more on the product.

Even more important than the structural differences are the changes in style and philosophy. To be successful, a company-wide teaming initiative must have the firm commitment of management and—this is the hard part!—a genuine willingness on its part to turn a significant amount of decision-making authority over to the teams. Managers must be willing to change their styles from directive to facilitative, from supervising to "coaching." They must create a climate in which employees are not afraid to speak out, to assert themselves, and to take reasonable risks. If management is not willing to do this, if its commitment to teaming is only skin-deep, then it would be better not to implement the concept at all. A half-hearted, superficial sortie into teaming is worse than just staying in a traditional structure; if the initial pain of transition leads to a realization that "nothing has really changed at all" (that the elephants are still elephants and the mice are still mice, so to speak), then the outcome is not motivation and productivity, but disappointment and cynicism.

On the flip side, rank-and-file employees bear equal responsibility for the success of a teaming initiative. Workers must be willing to take on accountability for product success and to assume more leadership responsibility. If they aren't, management is more or less forced to revert to directive tactics to get the job done. In other words, it takes enlightened elephants and empowered mice.

A brilliant and delightful little book entitled *ZAPP! The Lightning of Empowerment: How to Improve Quality, Productivity, and Employee Satisfaction* by W. C. Byham and J. Cox[15] captures the spirit of self-directed work teams through inspired use of allegory. Byham and Cox create a fantasy factory where one department suddenly becomes possessed of a mystical "lightning" that sparks its employees to superior performance. The manager of another department that is beset with morale problems and poor performance is suddenly allowed to "see" the lightning that is energizing the other department while a dragon—symbolizing destructive old-school management techniques—is stomping around laying waste to workstations in the manager's own department.

When all is said and done, the moral of the story is for management not to "SAP" employees by failing to trust them, denying them the decision-making authority they need to do their jobs, refusing to listen to their ideas, and micromanaging them; but rather to "ZAPP" employees by trusting them, valuing their skills and opinions, genuinely *listening* to them, and—most importantly—giving them room to operate. That's the lightning that slays the dragon and turns out the widgets.

That's exactly what makes self-directed work teams successful—trust, respect, and open communication.

And that's where technical communicators can play a key role. We should, by our very training and expertise, be better prepared than most not only to survive but to *flourish* in a teaming environment—to lead and facilitate the process, to help people express their ideas, to encourage innovation, and to teach people to take risks when the situation demands it.

Like declaring independence against all odds.

No, teaming isn't that new, after all.

So let's hang together.

Summary

In technical communication, teamwork means the ability to work together effectively with clients, employers, and colleagues to develop and produce quality communication requirements that meet user requirements.

A common situation in technical communication that requires teamwork is the collaborative writing project: in the classroom and in the workplace.

Collaborative communication projects are included in most undergraduate technical communication programs. The keys to success are planning and cooperation: following a specific sequence of steps and applying the principles of group dynamics, especially respect and communication. Of particular importance is the ability to give and receive constructive criticism from a task orientation, rather than one of personal involvement.

Collaboration brings several advantages, including the potential to do more and better work by drawing from a broad base of skills and knowledge. However, it also entails risks, such as interpersonal conflict and inequitable division of work.

Collaborative writing is used in the workplace if the job is too big for one person or if it requires specialized expertise across a range of disciplines. Two lengthy projects that require team writing are software documentation and competitive proposals. In both cases, early involvement of technical communicators in document planning benefits quality. Proposals are often worked under unreasonable schedules; the key to success is to give the proposal team enough autonomy to encourage creativity and sustain morale, yet provide sufficient structure and clear direction to maintain quality and schedule.

Collaborative writing also comes into play in professional associations, in coauthorship, and in community organizations.

Technical communicators are called upon to exhibit teamwork in a wide range of business environments. The traditional vertical hierarchy tends to create conflicts between personal aspirations and commitment to team goals. Integrated product-development teams, or IPDTs, stress concurrent engineering, welding diverse disciplines to carry a product throughout its life cycle. Self-directed work teams, or SDWTs, call for increased employee autonomy (empowerment) and a facilitative rather than directive style of management.

Application Exercises

1. What is brainstorming? Where does it fit in the sequence of collaborative writing projects?
2. What are some strategies for dividing responsibility on a group project?
3. Explain the difference between task orientation and ego involvement (see also Chapter 13). Which works better on a work team? Why?
4. Why is it important to involve a technical communicator in the early stages of a software documentation project?
5. Why is it important to involve a technical communicator in the early stages of a competitive proposal?
6. List ways in which collaborative writing plays a role in professional associations.
7. List ways in which collaborative writing plays a role in community organizations and activities.
8. Define and exemplify concurrent engineering.
9. Explain how concurrent engineering relates to integrated product-development teams.
10. In what ways do self-directed work teams differ from a traditional business environment?

Exploration Exercises

1. Briefly describe a collaborative writing project in which you were involved. What went well? Why? What didn't go so well? Why? Compare your findings with the summary of advantages and disadvantages of collaborative writing. How might you improve the process if you undertook the same project again?

2. Give your idea of a good business meeting. Of a not-so-good business meeting. What key elements differentiate the two?
3. Are you a good listener? Why or why not? If not, what could you do to improve? If yes, what could you do to improve?
4. Have you ever coauthored a paper or report? If yes, how did it go? Why? If not, what would you look for in a partner? Why?
5 List your strengths as a team player. Then list areas in which you could improve as a team player. Be specific.
6. Apply the Elephant Theory of Management Sensitivity to an organization with which you are (or were) involved. Cite an example from your personal experience that shows one or more of the Seven Rules for Survival in operation.
7. Describe a vertically structured organization with which you are (or have been) associated. Suppose you were hired to convert it to a self-directed work team environment. How would you go about it? Outline and defend your plan of attack.
8. Read *ZAPP!* Write a short (2- to 3-page) reaction paper or make a brief (3- to 4-minute) oral presentation.

References

1. Fred D. White, *Communicating Technology: Dynamic Processes and Models for Writers*, (New York: Harper Collins, 1996), 42.
2. Paul V. Anderson, *Technical Writing: A Reader-Centered Approach* (Harcourt, Brace, Jovanovich, 1991), 677.
3. Lynn Quitman Troyka, *Handbook for Writers*, 3d ed. (Englewood Cliffs, NJ: Prentice-Hall,1993), 693.
4. Jules Harcourt, A.C. "Buddy" Krizan, and Patricia Merrier, *Business Communication*, 3d ed. (Cincinnati: Southwestern Educational Publishing, 1996), 77.
5. T.E. Pinelli, R.O. Barclay, M.L. Keene, J.M. Kennedy, and L.F. Hecht, "From Student to Entry-Level Professional: Examining the Role of Language and Written Communications in the Reacculturation of Aerospace Engineering Students," *Technical Communication* (Third Quarter 1995), 501.
6. L. Allen, "Introduction to Technical Writing," lecture notes, Metropolitan State College of Denver, Spring 1995.
7. Mike Markel, *Technical Communication: Situations and Strategies*, 4th ed. (New York: St. Martin's Press, 1996), 50.

8. Richard C. Weber, "The Group: A Cycle from Birth to Death," *NTL Reading Book for Human Relations Training*, NTL Institute, 1982.
9. White, *Communicating Technology,* 49.
10. Ellen Fenwick, Boise State University, "Ethics of Helping with Student Papers: Summary," wrapup of electronic discussion on TECHWR-L, World Wide Web, (2 January 1996) 8:36 A.M.
11. *Ibid.*
12. *Ibid.*
13. Janice Redish, Redish & Associates, Seattle, and Judith Ramey, University of Washington, guest editors of compilation of eight articles under the theme "Measuring the Value Added by Professional Technical Communicators," *Technical Communication* 42 no. 1 (First Quarter 1995).
14. Daniel W. Voss, "Editorial Training II," Martin Marietta Corporation, Orlando, FL, 1992 (now Lockheed Martin Corporation). Workshop at 38th Annual International Technical Communication Conference (STC), New York, NY, April 14-17, 1991. Paper by same title appears in *Proceedings,* pp. ET-154-157.
15. W. C. Byham and J. Cox, *ZAPP! The Lightning of Empowerment: How to Improve Quality, Productivity, and Employee Satisfaction* (New York: Fawcett, 1988).

Chapter 9 Overview

Conflict of Interest . . .

The Mission: **We must serve our clients and employers with integrity in a way that excludes considerations of personal advantage and promotes the interests of the company, as long as doing so does not violate the public good. We must conduct our professional duties in an equitable and even-handed manner.**

Chapter Roadmap

Definition and Focus. This section provides a philosophical framework for avoiding conflict of interest as a core value.

Beware Using Company Property. This section identifies the myriad of unethical behaviors that can result from lack of care in handling and using corporate property.

Kickbacks, Gifts, and Other Goodies. This section acquaints you with the ways that both the law and good ethics dictate that kickbacks, bribes, and similar means of influencing a professional's decisions and actions are unacceptable—yet sometimes gray.

Conflicts Arise from Divided Loyalties. This sections lets you distinguish between two types of divided loyalties: those of scope—an employer's interpretation of loyalty is much broader than that of the employee—and those of precedence—the company is only one of the numerous entities to which one owes allegiance.

Educators Face Divided Loyalties, Too. This section introduces some of the conflicts that arise in education, such as those related to grading, research, and selection of textbooks.

Conflicts Arise in Professional Associations. This section advises you of some of the conflicts faced in professional organizations, such as copyright release, reimbursement of expenses, and use of positions for personal gain.

Summary, Practical Exercises, and References

. . . Our Duty to Be Loyal and to Observe Fair Play

CHAPTER 9

Conflict of Interest: Divided Loyalties, Competing Priorities, and Conflicting Interpretations Make This the Grayest Area of All

Corporations have big meetings around dynamic and dangerous situations, but never even recognize small cancers inside the company.

—James Shannon,
former Executive Director,
General Mills Foundation[1]

Men who, in their business and social conduct, would not swerve from a line . . . of action, which was planned to represent that which was right and fair to all.

—Earl Sams,
writing about the qualifications of
the men first hired by J.C. Penney[2]

Definition and Focus

To research *Beyond The Bottom Line,* a study of the ways in which good ethics make good business, Tad Tuleja sent inquiries to scores of companies, ran requests for information in several prominent business publications, and interviewed executives from a dozen of the largest American corporations. The responses—"pounds of annual reports, codes of corporate behavior, news clippings, speeches, and correspondence" as well as hundreds of pages of notes—address all areas of business conduct.[3] It's interesting that having evaluated all of that data, the author comments on the "special, mounting concern" that industry is according this particular value:

> *When I first wrote to corporations asking them for information about their ethical standards, I made no specific mention of conflict of interest, yet virtually everyone who responded with an ethical code included something on this sensitive area . . . it is a dominant concern in today's ethically conscious boardrooms.*[4]

In explaining this tremendous amount of attention, Tuleja turns to the traditional definition of conflict of interest: "an engagement of an employee in activities that deflect his or her attention from the best interests of the firm."[5] Indeed, a vast majority of the corporate conduct statements we perused in our own research, as well as many of the other sources we consulted, use this same definition with only minor variations.

It occurred to us that this very definition may be at the heart of the furor surrounding this value area. Have you ever encountered a less specific definition? One more open to interpretation? One that has greater potential to be so all-inclusive?

"Activities" can cover an incredibly broad spectrum of human endeavors. "Best interests of the firm" can be equally broad, and frighteningly subjective, depending on who is determining exactly what constitutes those interests. Does "best interest" mean that employees are obligated only to avoid harming the company? Or are they obligated to promote the image and bottom line of the company? If the latter, need they do so only between 8 and 5, or must they do so 24 hours a day?

It is no wonder, then, that this value area is the one most likely to breed thorny ethical conflicts. It is the grayest of all.

We need to establish some solid ground, however, before we can venture into the gray—we need a definition. *Technical communicators are obliged to serve our clients and employers with integrity in a way that excludes considerations of personal advantage and promotes the interests of the company, as long as doing so does not violate the public good. We must conduct our professional duties in an equitable and even-handed manner.* It seems, then, that according to this definition, the public good is the top priority, with the company's well-being following a close second. But Chapter 11, which addresses social responsibility, examines what can happen when these two priorities collide. So we need to look further.

Well, it's easy to see that conflict of interest encompasses many other value areas we discuss throughout this book. Honesty is in the best interest of our employers and clients, as are legality, quality, teamwork, and professional growth. So what makes conflict of interest different? And what's left for this chapter?

What makes conflict of interest unique is that more than any of the nine other value areas, conflict of interest is the one with the greatest potential for head-on collisions between *personal* and *professional* points of reference. It is the value area that most frequently pits our *loyalties* one against another.

It is also the value area that opens us to the greatest threat, because the *perception* of conflict of interest is often as grievous as the reality—we have as much to lose if we are *perceived* to have violated a trust as we do by actually breaching one.

This sets a pretty serious tone for the rest of this chapter, doesn't it? Let's look at some of the ways in which conflict of interest can rear its ugly head. You'll find—as we did—that many of the subjects we explore and scenarios we present are among the most common ethical conflicts in the workplace. The behaviors in question are among the most widely practiced, yet all too rarely discussed, behaviors. And, chillingly, they are actions that people tend to indulge in day in and day out without even thinking about them.

In its most frequent guise, conflict of interest doesn't generally involve dumping millions of gallons of toxic chemicals that expose hundreds of people to life-threatening diseases, being flagrantly insensitive to the culture of an entire ethnic group, or flouting federal antitrust regulations (although, we readily agree these actions would, in fact, put at risk a company's best interests). Instead it is most often those small cancers to which James Shannon refers in the chapter-opening quotation that can lead to career-threatening misdeeds.

Beware Using Company Property

We have found that one advantage of living in a popular vacation destination is that numerous relatives and friends pay extended visits to explore Disney World or test their prowess on the ski slopes. During such visits, we expect our guests to use our bath towels, help themselves to the goodies in our refrigerators, place calls on our telephones, perhaps even drive our cars to see the sights. We consider these things the acceptable cost of hosting guests. However, we must confess that we'd be more than a little chagrined if, while we were at the office, our guests packed up the contents of our linen closets and kitchen cupboards into their suitcases and appropriated our cars to drive home, leaving behind hundreds of dollars of long-distance phone charges to boot! Call us ungracious if you want, but we would consider those expenses above and beyond any reasonable cost of hosting guests. Naturally, none of you would indulge in such impolite behavior any more than you would think of paying your personal bills from your host's checkbook!

Yet every day companies endure similarly inconsiderate behavior by employees who wouldn't dream of treating their hosts in such a manner, but who don't bat an eyelash at stealing from the company. Someone who would never make a 45-minute personal call to North Dakota from Mom's phone in Alabama may not hesitate to place that same call from the office. Someone who wouldn't dream of pilfering a $7.00 bath towel from her sister's linen closet might not think twice about taking home $7.00 worth of pens, pencils, paper clips, and computer paper from the office over the course of a year. Certainly, when all of the "extras" taken by all of the employees are added up, companies are justified in thinking the total extends above and beyond the reasonable cost of doing business. Employees who indulge in such behaviors might as well be paying their AT&T or Office Depot bills right out of their employer's checkbooks. (For another perspective on these issues, see Chapter 4).

The opportunities for unethical behavior in this regard are myriad. Smuggling supplies from work, placing long-distance personal calls from the office

"George, you're not ***stealing*** that ream of paper, are you?"

phone, using company letterhead for personal letters—and writing those letters on company time—are common occurrences. Now, we're not such ethical sticklers that we condemn these practices across the board, without exception. As we suggested in Chapter 2, who would want to work for a company that is willing to fire a valued employee for running four pages of personal papers through the copy machine? There will always be gray and there will always be room for judgment. But copying your personal tax return at the office and putting a company pencil in your briefcase are just the tip of the iceberg.

Expense vouchers can easily be manipulated, for instance. Most companies require receipts for big-ticket items like airfare and hotels, but not for the dozens of incidentals that crop up on a trip. How easy it is, then, to pad the expense report—just claim an extra buck for each tip, a few extra dollars for airport parking, and a couple of additional highway tolls. Or suppose the company specifies a maximum daily limit of $27.50 for meals, but breakfast is included in the cost of the conference you're attending, so you frugally spend only $15.00 each day for food (okay, so you like fast food!). Wouldn't that extra $12.50 per day come in handy? Piece of cake. Just claim the maximum meal allowance on the expense report, even though you didn't spend it. A frequent enough occurrence—and an unethical one.

"Two more company trips skimping on meals, and I'll have made enough on my expense reports to buy that blue silk coat."

And for those who aren't tempted by a mere $12.50 per day, there's the lure of easy money to be gained by manipulating or misusing billable hours. If you're charging $50.00 an hour to produce software documentation for a client, it sure could be tempting to include those coffee breaks or lunch hours on the final bill. And it sure would be unethical.

Of course, like other conflicts of interest, the misuse of company property is rarely as intentional or as black-and-white as these scenarios suggest. Mostly, it's gray. Those of us who wouldn't walk off with a company computer might, over the course of time, indulge in other "misdemeanors" that add up significantly. Those of us who would never bill a client for three hours we didn't work might, on a particularly chaotic day, be less than diligent about recording the amount of time we devoted to each project—and over the duration of an assignment the net effect may be the same.

Routine pressures and the fact that many such practices have become the norm make it difficult to distinguish the ethical implications of some conduct. Doesn't everybody do it? The lax culture that is either tolerated or even fostered by many organizations often leads to the perception that certain unethical behaviors are acceptable. Our responsibility, then, is not necessarily to establish stringent guidelines—four copies are okay, but five are not acceptable; two pens a year are fine, but take three and you're outta here—but to ensure that, first, we *do* think through the ethical implications of our actions, that we *do* apply value analysis. And second, to be sure that when we see shades of gray, we see *legitimate* rather than *chameleon* gray!

Kickbacks, Gifts, and Other Goodies

Perks can *make* a job—goodies like free coffee and popcorn, health club memberships, bonuses, and close-in, covered parking are strong motivators. In fact, many corporations now offer these extras routinely to entice and retain the best and brightest employees. Many other "perks," however, are not so above-board and, as such, can pose ethical dilemmas.

In general, both the law and good ethics dictate that kickbacks, bribes, and similar means of influencing a professional's decisions and actions are unacceptable—in many cases, even punishable if you're caught. For instance, Boeing and one of its sales executives are facing criminal investigations in Canada on suspicion of paying bribes totaling more than $1 million to sell planes to Bahamasair. The U.S. Justice Department may also investigate under the Foreign Corrupt Practices Act, which imposes stiff penalties for companies found guilty of making improper payments to win overseas deals.[6]

But here, too, there can be gray.

For instance, in many countries around the world government officials are paid minimal salaries in part because it is assumed that they will receive a considerable "boost" to their income through gifts from those to whom the officials can grant government favors. You want a permit to expand your Oldsmobile dealership and I want a car—in many places such reciprocity is not only legal and ethical, it is expected.

What if your company is doing business in one of these countries—is it then ethical (assuming it is legal) to do as the Romans do, so to speak? Tough question. Obviously a gray area with no easy answers.

Where does one draw the line on gifts? Let's say you send all of your printing to outside vendors. In the past year, you have spread the work around to three different companies, depending on the schedule and the specialty you required. So the holidays come around, and the guy from Joe's House of Ink brings you a bottle of white zinfandel. As the two of you chat, the rep from Blue Line Press stops by with a fifth of velvet-smooth, top-shelf, well-aged scotch. In the midst of introductions, the Marketing Manager from International Printers of Perfection enters bearing a bottle of 50-year-old cognac. If the value of the gift is the issue, you can probably accept the wine but not the cognac, and the scotch is iffy. If you deem all of the gifts acceptable, what happens in the coming year when International gets more business than Joe's? You may have perfectly legitimate justifications—like schedule, budget, and capability—for sending the work to one printer rather than another. Yet Joe might find it difficult to believe that the cognac didn't influence your decision. So should you accept no gifts at all?

Is a defense contractor "bribing" troops when it provides monogrammed baseball caps and coffee mugs to a graduating battalion that has just completed training in the use of its products? Lots of gray. No easy answers.

Many companies attempt to avoid these types of conflicts with guidelines; unfortunately, these can raise as many questions as they resolve. Avon says that "no employee shall accept . . . any gift, entertainment, or favor of a type that goes beyond common courtesies consistent with ethical business practices."[8] What, pray tell, is common courtesy? Are Avon's ethical business practices consistent with those of Exxon? Puget Sound's company policy says, "Acceptance of gifts, samples, entertainment, travel, or services by the employee . . . that create an obligation of favorable Company action is unacceptable."[9] What if I don't intend for my acceptance of dinner from a vendor at a convention to create an obligation, but my boss thinks it does? Other companies prescribe that employees may accept only gifts of nominal value. How do you set the limit, then? Is a $24.99 gift acceptable, but not a $25 one?

Again, we provide no hard-and-fast answers; we offer only value analysis and a set of core values on which to base the process. As we observed in Chapter 2, most of us have the capacity to distinguish right from wrong. We'd all be wise to minimize the influence of conflicting interests on our work and avoid compromising situations. Of course, if the dilemma is *really* gray and if the stakes are high enough, you could, as Professor Mary Guy suggests, ask yourself, "How would this look on the front page of the newspaper?"[10]

Conflicts Arise from Divided Loyalties

Most of us, companies and workers alike, agree that employees owe loyalty to their employers. Granted, in the current business environment, where each of us stands a fair chance of being "downsized" right out of a job at any time, that allegiance is not nearly as strong as it was in the days when one expected to retire from the first (and only) company he ever worked for after 40 years of service. Nonetheless, employment carries with it the expectation of a certain degree of loyalty.

Conflicts of interest arise in this area on two fronts. The first problem is one of scope—an employer's interpretation of loyalty is much broader than that of the employee. Does loyalty require dedication to any extent that management decrees?

The second is one of precedence—the company is only one of the numerous entities to which one owes allegiance. Can contending loyalties be prioritized? And if so, what is the order of supersession?

Often, divided loyalties hit hardest when they involve family and loved ones. For example, let's say a request for information that arrived on Monday must be in the print shop on Saturday. By the time you are ready to leave the office on Thursday, the document is still eight hours from being camera-ready. You are scheduled for vacation beginning on Friday. You plan to drive the 250 miles to attend your daughter's graduation and accompanying festivities on Saturday. The boss contends that no one else is available to pick up the pieces and meet the deadline. Of course, the circumstances need not be so dramatic. On more than one occasion you have demonstrated your commitment to working overtime during a crisis. You are equally committed to spending evenings and weekends with your two preschoolers. However, the new division manager puts in 11 hours a day, 6 days a week; she expects the same of the entire staff and routinely practices "management by walking around" between 6:00 and 7:00 every evening to see who still has nose to the grindstone. In both instances, the company's demands exceed what you believe falls within the legitimate realm of loyalty. In both instances, you must choose between being loyal to your family or to your employer. Prioritizing one's values and weighing the consequences to arrive at an ethical decision is stressful in such situations, to say the least.

Pursuit of one's professional development can also pose conflicts. For instance, suppose a recently hired software writer is strongly committed to the value of professional development. Unfortunately, the manager who writes his performance appraisal contends that hours spent attending conferences or meetings of professional organizations could be better spent cranking out more documentation—or perhaps helping the manager provide knee-jerk responses to the whims of the executives who write *her* appraisal.

The conflict arises when the writer has made a commitment to conduct an after-hours workshop for his professional association and the manager hits him with a time-consuming task passed down from her boss at 4:30. Real tough. And real world.

Another example—most corporations approve of (even encourage) attending graduate school when doing so enhances your contribution to the company. Often in such cases, supervisors are obligated to support the company policy by not making demands that interfere with class attendance. But what if your graduate program is geared not toward improving your performance on your current job but toward positioning you to change careers altogether? In this case, if the boss feels overtime is necessary, he may conclude that your insistence on attending class instead constitutes a conflict of interest.

Management is even less likely to support your personal pursuits at the expense of company demands, if those pursuits are as shiftless as craving leisure time to relax, unwind, and relieve stress or as enigmatic as writing the Great American Novel.

Speaking of writing, you run an even greater risk of encountering conflict of interest issues if the writing you do in your spare time is for personal profit. In the eyes of many employers, doing freelance work of any kind is considered moonlighting, which constitutes a potential conflict of interest. Most people who have a second job don't consider it a conflict of interest as long as they adequately perform the responsibilities of their primary position, they do nothing that harms the primary employer, and they do not work for a competitor of the primary employer. Policies about secondary employment vary from company to company. Many do, in fact, regard moonlighting of any kind as a potential conflict of interest, and so set stringent guidelines, require management approval of the outside position, or forbid it altogether. Adhering to the policy may not be sufficient to cover you, however, if your supervisor makes demands that hinder your ability to perform your secondary job as you should.

You can avoid conflicts—potential, actual, or perceived—by making your position on conflicting values and divided loyalties clear to your management right up front, and pursuing discussions that lead to a mutual understanding—*before* questions arise or problems occur.

Those who are self-employed face other types of potential conflicts of interest. The needs and demands of one client will often clash with those of another. Obviously, it is a blatant conflict to accept contracts from two clients to develop marketing materials for two competing products. That's a black-

and-white issue that is easy to avoid. But many gray issues lurk out there, and they're not as easy to solve. For instance, the schedules you've agreed upon with two different clients may suddenly change, making both manuals due on the same day. Meeting one revised deadline would be improbable; meeting both is nigh unto impossible. Both are long-time satisfied clients; neither is willing to settle for subcontracted work. Who gets priority? That's a delicate balancing act, indeed.

How does one walk *that* tightrope? The answers to some of these issues are fairly straightforward. Again we recommend that you minimize the influence of conflicting interests on your work. Just as it would be unwise for an attorney to represent both parties in a divorce, it simply doesn't make sense for us to place ourselves in compromising situations unnecessarily.

Also, be careful to completely disclose all personal interests that might unduly affect your performance or judgment. Fully inform clients of all conflicts of interest, including what your position will be if a conflict does arise, so everyone is clear about the consequences and alternatives. Finally, you can minimize conflicts if you make firm, contractual agreements with clients.

Other scenarios of ours are not so easily resolved. If you're lucky, some degree of compromise may be possible. Unfortunately, many divided loyalties result from situations for which no middle ground exists. In those instances, value analysis will help you determine the most acceptable course of action.

Educators Face Divided Loyalties, Too

The Dead Poets Society (Touchstone Films) portrayed a classic ethical dilemma in teaching—how does an institution uphold academic freedom when it conflicts with the desires of stakeholders? In the movie, John Keating's unorthodox classroom methods draw fire from the both the administration of the prep school where he teaches and the conservative parents of his students. Keating believes he has an obligation to teach critical thinking; his detractors believe he is encouraging reckless and self-indulgent behavior. In one scene, the dean admonishes Keating, saying, "the curriculum is set here, it's proven, it works. If you question it, what is to prevent them [the students] from doing the same?"

Keating replies, "I always thought the idea of education was to learn to think for yourself."

The dean's abrupt response and departing directive is, ". . . not on your life. Prepare them for college, and the rest will take care of itself."

The sad, and perhaps all too realistic, ending to the movie is that Keating refuses to comply and loses his job. Academic freedom is sacrificed to parental paranoia.

Academic freedom is but one of many ethical concerns in higher education. Observing fair play while maintaining standards in the classroom can pose several conflicts of interest for faculty. Suppose a capstone course is required for graduation and only one professor teaches it. He requires attendance because the course entails a significant amount of collaborative work among students. A single, working mother registers for the course in her final semester. Her job often demands significant overtime and her two children can be equally demanding of her time. She has a 4.0 grade point average and is a shoe-in for valedictorian. She has taken several other classes from this professor, none of them requiring attendance, and they have developed an excellent rapport. After the first class meeting, she approaches the professor requesting relief from the attendance requirement, citing the consistently high quality of her classwork and expressing her commitment to fulfill her obligations to her team members in producing the final project. Without the attendance requirement she likely would earn an A for the course; with it, she likely will not, ruining the perfect grade point average she has worked so hard (and sacrificed so much) to maintain. She feels she is being penalized because she is a nontraditional student. Certainly, the professor would be sympathetic. But is it a conflict of interest to suspend the attendance requirement for one student, even one who has proven herself with such stellar performance?

Yes, it certainly is. Standards are standards regardless of the circumstances. Any teacher who establishes absolute criteria for attendance, grades, or classroom behavior is ethically obligated to apply them fairly across the board, regardless of the situation. The professor who includes correct use of grammar as a measure of the grade on a paper for technical communication majors in Introduction to Technical Writing must do the same for the science and nursing majors in the class. It isn't fair to require that students in one discipline meet higher standards than those in another. The professor who does not require attendance of part-time, working students cannot require it of full-time, nonworking students. Clearly, the ethical obligation here is to establish standards that are appropriate to the course and fair to the broadest spectrum of students, then apply them equally.

These aren't the only conflicts of interest that arise in academe. Some faculty argue that the very system by which their performance is evaluated can create divided loyalties, particularly in the assessment area of professional development.

In academe, professional development has two significant components, research and publishing, both of which can be problematic for faculty. Research requires many hours spent in the library, laboratory, or field. Many of those hours are provided through release time (time for which faculty is paid but "released" from performing classroom duties), especially when the research is funded by grants. In other words, teachers are performing work that some contend enhances the stature of both the institution and the professor, but does not contribute directly to the students' education. To make matters worse, opponents of this practice argue that when faculty isn't in the classroom, someone else must be. That someone else is frequently a graduate assistant whose credentials can't be expected to measure up to those of the professor.

Now suppose a chemistry major from New Jersey and her parents have made substantial sacrifices to enable her to attend a university in California, a school they selected largely because two of the foremost chemists in the discipline are on the faculty there—what a learning opportunity! Unfortunately, both professors have 3/4 release time, funded by hefty grants from the National Science Foundation, so each teaches only one course a semester—a graduate course at that. And our chemistry major? In her first two years she is instructed only by graduate students and adjunct faculty.

Consumers expect faculty members to teach and be available to students. Colleges and universities capitalize on that expectation by shamelessly flaunting the reputations and accomplishments of their faculty in marketing materials. On the other hand, research is tremendously valuable—in fact, it is crucial—to the advancement of knowledge in all disciplines. Faculty members, then, are placed smack in the middle of a conflict of interest. Tenure, promotion, and raises depend in great measure on how much research they conduct—as does the continuing growth of knowledge and information in the field. But the demands of that research dictate that they spend less time in the classroom, away from the very students whose attendance makes the institution economically viable and their jobs secure.

Let's carry our example a step further. Suppose that as a result of their research, the California chemistry professors publish a ground-breaking textbook, for which a publishing house pays them an advance and royalties. Should the professors mandate the use of their book in the introductory chemistry classes offered on their campus each semester? Clearly the book is one of the leading publications available for classroom use, so it has much to offer students. But the authors receive royalties based on the number of copies sold. Is requiring their own text, then, a conflict of interest?

In this particular scenario, we think not. Teachers at all levels have an ethical responsibility to provide the best education for their students. That obligation entails using the best resources available, including textbooks. The ethical standard of measurement here, then, is the appropriateness of the text. If the professors' book is, in fact, as good as or better than any other they could select, it is not a conflict of interest for them to make it required reading. However, if an alternate book contains material that is more suitable for the course or content that is more complete, the teachers must forego the royalties (ouch!) in favor of the other text.

This is where the concept of up-front disclosure ties in. We addressed this concept in Chapter 8 when we discussed collaborative research and writing projects involving professors and students. The professors who select their own books as the required text of a course they teach would be well advised to avoid conflict-of-interest charges by making sure the book decision has the prior endorsement of the department chair. And if one of the authors also happens to *be* the department chair? That's what provosts are for. You get the point—putting the issue *on* the table avoids the risk of dealing with it *under* the table.

Conflicts Arise in Professional Associations

A brief word about conflict of interest as it pertains to professional associations is in order. At first glance, there would appear to be limited applicability. However, a number of potential areas for conflict of interest can arise. For this reason, some such organizations, such as STC, have developed specific policies addressing their concerns pertaining to members of the association in general and to senior leadership in particular.

Issues include copyright release on communication materials generated on a voluntary basis; reimbursement of expenses incurred on behalf of the organization; and use of conferences, meetings, or other association-sponsored events as a platform for personal gain.

Consider, for example, the professional leverage that is potentially available to an independent contractor who is elected to a national office within a major professional society. What, other than ethics, is to prevent her from parlaying a trip funded by the organization into a forum for personal business opportunities? And how about an author who uses a presentation at a national conference to hawk his published wares?

Like so many other scenarios involving conflict of interest, these examples pivot on two issues: intent and degree. It would be unreasonable, for example, to expect the author/conference presenter to make no reference to a book that addresses the same subject on which he is speaking. And it doesn't constitute venal capitulation to self-interest if the independent contractor/national officer happens to make a legitimate business contact while on society business. It's just like the difference between running those four sheets of paper through the office copier and running a ream and a half.

Gray, gray, gray!

As is the case in so many areas, it all boils down to personal integrity. And, yes, to a considerable degree, we must trust our elected officers in a professional organization, just as we must trust our colleagues, our professors, our supervisors, and our employers. As in most areas, the vast majority of individuals do merit our trust. Unfortunately, there are also exceptions. As Ronald Reagan was so fond of saying during détente:

Trust, but verify.

Summary

Variety is the touchstone of the conflicts we've examined throughout this chapter. Conflicts of interest can be the most difficult of all ethical dilemmas because they involve (1) differing interpretations of what is in the "best interest" of the company, (2) collisions between personal and professional points of reference, and (3) divided loyalties. This value area prescribes that we are obliged to serve our clients and employers with integrity in a way that excludes considerations of personal advantage and promotes the interests of the company, as long as doing so does not violate the public good. We must conduct our professional duties in an equitable and even-handed manner.

One conflict of interest is the use of company property. Many people who would not condone theft in their private lives do, in fact, steal from employers in a variety of ways, including use of company equipment and time for personal matters, padding expense accounts, and improperly recording time.

Although kickbacks and outright bribes are illegal, the legal practice of giving and receiving corporate gifts can open a can of worms, from an ethical standpoint. This can become a particularly thorny issue in the international marketplace, where some cultures define "bribery" in terms very different from Americans.

Opposing allegiances lead to conflicts of interest when the demands of the workplace contend with those of our families or with our personal needs.

The halls of academe also witness their share of conflicts of interest. Unfair application of standards, and the tension between the demand for faculty research and the obligations of the classroom are just two of the ethical issues that smolder on campus.

Professional associations are not exempt from the specter of conflict of interest. Concerns range from reimbursement of expenses to misuse of organizational events to serve self-interest.

Although conflicts of interest often mire those who are involved in a morass of ethical gray, a few fundamental guidelines will help ease the difficulties. When it is impossible to avoid such conflicts, you can minimize their impact by fully disclosing them—whether they are real, potential, or perceived—and make firm agreements on how you intend to deal with them, *before* they happen.

Application Exercises

1. Give three reasons why conflicts of interest are often more gray than other ethical dilemmas.
2. Why can a *perceived* conflict of interest cause as many difficulties as an *actual* conflict?
3. Provide an example of the inappropriate use of company property that is not discussed in the text.
4. What guidelines would you consider acceptable in terms of corporate gifts? Give two examples of gifts that meet your guidelines and two that don't. Compare your answers with those of your classmates.
5. In your experience, identify an instance in which your loyalties were divided. How did you resolve the issue?
6. Discuss the extent to which it is reasonable for employers to place demands on the number of hours employees work each week. How far do you believe an employer can reasonably "encroach" into an employee's personal time?
7. What is moonlighting? How does it differ from primary employment? What do you believe is an appropriate stand for a corporation to take on this issue?

8. Have you ever taken a class in which you felt the instructor did not apply the standards equally to all students? Describe the situation. Exclude identities. Should fairness be a measure that students are asked to assess in completing faculty evaluations? Why or why not?

Exploration Exercises

1. Visit the Human Resources or Ethics Department of a local company. Ask a company representative to provide specific examples of situations he would see as a conflict of interest.
2. Have you ever received a final grade that you believed was unfair? How did you handle the situation? How would you handle it today? With one of your classmates playing the role of the professor, enact a meeting in which you discuss the grade.
3. Working with several of your classmates, develop a list of conflicts of interest that have occurred on your campus recently. What was the central issue in each? How were they resolved? Are those resolutions satisfactory? Why or why not?
4. Meet with an administrator on campus. Ask her to explain the types of conflicts of interest that are common on campus and the process(es) by which different issues are resolved. Then write a code of conduct that addresses those issues.
5. Discuss with your classmates instances in which you have seen someone misuse or steal property on campus or on the job. Exclude identities. Discuss how widespread you think this practice is. Reach a consensus on the reasons why. Propose solutions.
6. Watch *Dead Poets Society*. Write a brief paper explaining how you believe the headmaster should have handled the conflicts in the school.

References

1. Kent Hodgson, *A Rock and a Hard Place: How to Make Ethical Decisions when the Choices are Tough* (New York: American Management Association, 1992), 46.
2. Charles Nelson and Robert Cavey, eds. *Ethics Leadership and the Bottom Line*, (Croton-on-Hudson, New York: North River Press, 1991), 313.
3. Tad Tuleja, *Beyond the Bottom Line: How Business Leaders Are Turning Principles into Profits*, (New York: Facts on File, 1985), xii.
4. *Ibid.*, p. 139.

5. *Ibid.*, p. 139.
6. *New York Times* via Internet, 6 February 1996.
7. Walter Manley, *Executive's Handbook of Model Business Conduct Codes*, (Englewood Cliffs, NJ: Prentice-Hall, 1991), 38.
8. *Ibid.*, p. 79.
9. *Ibid.*, p. 79.
10. Mary Guy, *Ethical Decision Making in Everyday Work Situations* (New York: Quorum Books, 1990), 80.

Chapter 10 Overview

Cultural Sensitivity . . .

The Mission: **In technical communication, the value of cultural sensitivity demands tolerance, understanding, and freedom from prejudice. It means embracing diversity, respecting rather than fearing differences, and reflecting that respect not only in our products but in our personal and professional behavior as well. To do so represents more than a professional value: It is a commitment to defend our birthright as human beings.**

Chapter Roadmap

Definition and Focus. This section provides a brief historical and social context to increase awareness of the need for cultural sensitivity as the world continues to shrink. The chapter translates this awareness into specific precepts to guide us in the culturally sensitive practice of technical communication.

Cultural Sensitivity—An Ethical Mandate. This section exemplifies the pitfalls that can undermine cultural sensitivity in our melting pot society and equips you with specific methods and guidelines for avoiding them.

- ❏ *We Must Avoid Stereotypes*
- ❏ *We Must Strive to Send a Multicultural Image*
- ❏ *We Must Not Stumble into Tokenism*
- ❏ *We Must Deal Gracefully with the Generic Pronoun and Avoid Sexist Usages*
- ❏ *We Must Recognize and Refrain from Sexual Harassment*

Cultural Sensitivity—A Business Imperative. This section explores the growing importance of intercultural awareness in the world marketplace. It explains how ethnocentrism is a barrier to intercultural sensitivity and presents practical examples of how to avoid problems and communicate effectively across cultures.

- ❑ *Overcoming Ethnocentrism Is Key to Intercultural Sensitivity*
- ❑ *Communicating Effectively across Cultures Demands Special Knowledge and Special Effort*

  Different Strokes for Different Folks

  A Pocket Guide to Survival in Intercultural Communication

Let's Practice What We Preach. This section may hit close to home. It challenges you to recognize and close any potential disconnects between your philosophical commitment to cultural sensitivity and your day-to-day practice of it, both inside and outside the workplace.

Summary, Practical Exercises, and References

. . . Our Duty to Respect the Growing Diversity of the Workplace and to Reflect it Both in our Products and our Actions

CHAPTER 10

Cultural Sensitivity: Not Just a Professional Obligation, But the Imprint of Our Humanity

And opening his mouth, Peter said, "I most certainly understand now that God is not one to show partiality, but in every nation the man who fears him and does what is right, is welcome to Him."

Bible, Acts 10: 34-5

I am an invisible man. . . . I am invisible, understand, simply because people refuse to see me. Like the bodiless heads you see sometimes in circus sideshows, it is as though I have been surrounded by mirrors of hard, distorting glass. . . . Nor is my invisibility exactly a matter of a biochemical accident to my epidermis. That invisibility to which I refer occurs because of a peculiar disposition of the eyes of those with whom I come in contact. A matter of the construction of their inner eyes, those eyes with which they look through their physical eyes upon reality.

—Ralph Ellison, *Invisible Man*

Definition and Focus

We've come a long way since 1947 when Ellison's *Invisible Man* was first published. But we're not there yet.

Our ability to get there will define our destiny as individuals, as a nation, and as a world community.

What was invisible in Ralph Ellison's childhood in the Old South became painfully visible in Selma, Alabama, in the 1950s, when African-Americans, galvanized by a strong but peace-loving leader, joined hands to throw off a centuries-old yoke of oppression. They began, innocently enough, by boycotting a bus system which operated under the racist law that compelled them to sit in the back of the bus.

And, yes, things *have* changed in the three decades since Martin Luther King, Jr., stood in front of 200,000 on the Mall in Washington, D.C., and delivered the unforgettable "I Have a Dream" speech that stirred the conscience of the nation.

Things have changed. But not enough.

Ironically, it is now prejudice itself that has become "invisible." Invisible, but still frighteningly operative. In an age of *political correctness*, vocal and violent racism has receded to the relatively small—but hardly to be ignored—lunatic fringe, those who wear their hate visibly, on their foreheads or their armbands. The more insidious enemy, the force that has the potential to rend the very fabric of our society, is the silent, faceless prejudice that lurks inside all of us, majority (whatever *that* is) and minority alike.

Abraham Lincoln warned us to look for "no trans-Atlantic giant" to step across the ocean and crush us. He told us that "all the armies of the world combined could not by force make a footprint on the Blue Ridge or take a drink from the Ohio." "No, if there is danger," he went on, "it is *from within*."

And that is precisely Ellison's point. The "inner eye" that rendered him invisible as a young African-American man growing up in a white society is the same inner eye that can, unless we are ever vigilant, blind us to the sensitivities of our fellow humans.

There is no more urgent imperative for the human race than to overcome the artificial barriers of race, culture, religion, gender, and sexual preference that have so long separated us—to shed the fears and prejudices that have set nation against nation, black against white, East against West, Arab against Jew, Serb against Croat. The list goes on.

Americans have a unique opportunity to break down the subtle and invisible barriers of prejudice, because we are a *melting pot* society. As technical communicators, we have a unique responsibility to contribute to that process—both within the nation and, increasingly, within the international marketplace.

In technical communication, the value of cultural sensitivity demands tolerance, understanding, and freedom from prejudice. It means embracing diversity, re-

specting rather than fearing differences, and reflecting that respect not only in our products but in our personal and professional behavior as well. To do so represents more than a professional value, it is a commitment to defend our birthright as human beings.

In this chapter, we will explore specific implications of cultural diversity for technical communication, both in the United States and in the world community. We will also consider how technical communicators can demonstrate cultural sensitivity through their actions as well as their communication products.

Cultural Sensitivity—An Ethical Mandate

Ours is, indeed, a melting pot society.

Ishmael Reed opens his essay "The Multinational Society" with an excerpt from the *New York Times*:

> *At the annual Lower East Side Jewish Festival yesterday, a Chinese woman ate a pizza slice in front of Ty Thuan Duc's Vietnamese grocery store. Beside her a Spanish-speaking family patronized a cart with two signs: "Italian Ices" and "Kosher by Rabbi Alper." And after the pastrami ran out, everybody ate knishes.*[1]

He closes it by suggesting that North America may well become an international cultural crossroads:

> *Or does North America deserve a more exciting destiny? To become a place where the cultures of the world crisscross. This is possible because the United States is unique in the world: The world is here.*[2]

It sure is. All you have to do is take a look around campus or your workplace, and you will have empirical verification of Reed's observation.

The numbers support the cosmopolitan trend. The classic Hudson Institute study, *Workforce 2000: Work and Workers for the 21st Century,* cited by Barchilon and Kelley (Arizona State), points out that nearly a third of those entering the workforce in the past decade and a half have been minorities:

> *Between the years 1985 and 2000, minorities will make up 29% of the new entrants to the labor force. This shift in minority representation in industry is one reason companies are encouraging diversity.*[3]

And Larry A. Samovar and Richard E. Porter, in their definitive text *Communication Between Cultures,* cite statistics that suggest that the United States' "melting pot" legacy is likely not only to continue but to accelerate:

Projections suggest that ethnic and racial minorities will make up about one-third of the population of the United States by the year 2000. [By 2050, the percentage will rise to 45 percent.][4]

Given such a trend, it clearly behooves technical communicators to take extreme care to be sensitive to cultural diversity in the products we generate and produce. Four ways in which we can do that are as follows:

- ❑ Avoid stereotyping
- ❑ Send a multicultural image, wherever possible to do so naturally
- ❑ Avoid tokenism
- ❑ Handle the generic pronoun with grace and avoid other sexist usages
- ❑ Recognize and refrain from sexual harassment

We Must Avoid Stereotypes

Humans don't handle ambiguity well. They have a tendency to want to shove unfamiliar things into little "pigeon holes" so they can get closure on them and be comfortable—never mind whether the item actually fits into the hole! That's how stereotypes come about. Samovar and Porter define a *stereotype* as "the perceptions or beliefs we hold about groups or individuals based on previously formed opinions and attitudes."[5]

Stereotypes are prejudices. They also illustrate the logical fallacy of false generalization (see Chapter 4); if we stereotype all members of a given ethnic group as possessing a particular characteristic, all it takes to disprove our preconceived notion is one exception.

Stereotypes are especially treacherous for technical communicators because they can creep into our writing and graphics without our conscious consent. As such, they are like little land mines of intolerance waiting to blow up and alienate our culturally diverse audience. The following paragraphs present four stereotypes to be on the lookout for.

1. **The glaring misconception that professional positions are occupied by men and clerical or paraprofessional positions by women.** A case in point is the medical text that repeatedly refers to the doctor as "he" and the nurse as "she." We all *know* there are female doctors and male nurses, but cultural expectations that date back to episodes of *Dr. Kildare*, *Ben Casey*, or *Marcus Welby, M.D.*, can subconsciously affect our pronoun selection without us even being aware of it—at least not in an initial draft. That's why we have to be on the lookout for such unintentional insults when we review our material.

   Another common example of gender stereotyping in business communication is the tendency to cast executives as males and secretaries as females. Note that this applies to *generic* references; it is clearly not stereotyping to refer to Alice as the secretary and Karl as the boss if Alice *is* the secretary and Karl *is* the boss! What if the company has all Karls for bosses and all Alices for secretaries? Well, that goes beyond the jurisdiction of the technical communicator to fix.

2. **The equally flawed notion that minorities occupy service or menial-labor positions while white Anglo-Saxon Protestants hold the management slots.** Consider the production brochure graphic that depicts the supervisor as a young white male in a business suit and the assembly line workers as a mix of minorities, females, and white males. Suppose the brochure uses a photograph instead of an illustration. Suppose, further, that there is only one floor supervisor and this person actually *is* a young, white male. What do you do then, falsify the photo? Surely not. Leave it out? And have a brochure spread consisting of all text? Use it as is, and risk implying a stereotype? That might be your only ethical option, but if you are forced to exercise it, you'd better look hard for a legitimate counterbalancing graphic that shows diversity to include elsewhere in the brochure. The tradeoffs involved for the editor in this type of situation can be thorny indeed. (See Sending a Multicultural Image and Avoiding Tokenism, later in the chapter.)
3. **The mistaken idea that senior citizens are infirm and mentally limited.** At first, you might be inclined to view this stereotype more as a deficiency in Western culture than a "land mine" for the technical communicator. But ask yourself this: When was the last time you saw a silvery-gray head in a piece of clip art depicting a management meeting? An ethnic and gender mix has become almost obligatory; age has received less attention.
4. **The misguided notion that persons with disabilities are fragile and require special treatment.** Special facilities to help overcome the disability, yes. Special treatment, no. A classic example of this stereotype in action is the natural tendency not only to talk *louder* to someone who is hard of hearing, but also to *oversimplify*—as though the hearing impairment implies an intellectual limitation as well. Another example is using the disability as a label rather than a condition (for example, use "person with a disability," which is factually descriptive, not "disabled person," which suggests the person somehow "broke down"; use "person with AIDS," not "AIDS victim"; use "child who is mentally retarded" not "mentally retarded child.") However, we should not let respect for those with disabilities lead us into timid euphemisms that are not only insulting, but obfuscate the message (such as referring to a child who is mentally retarded as "special").

   It is *not* stereotyping to make legitimate allowances for disabilities when we create communication products. Indeed, this is a communication responsibility. It falls in the same category as providing special facilities. Included are such important communication adjustments as the following:

Providing closed-caption titles for the hearing-impaired in videotapes that will reach a wide audience, especially if demographic audience analysis suggests a significant percentage of senior citizens (that's not stereotyping; that's medical reality).

Avoiding red-green contrast in online (and paper) documentation, to accommodate those who are colorblind. This is particularly important with safety warnings, which are often flagged with red. It's okay to use the red for the majority who are not colorblind, as long as you provide a redundant indicator such as a recurrent safety icon or, if you are using a multimedia platform, a warning beep.[6]

Printing a few church bulletins or theater programs in *oversized type* for older attendees whose eyesight is failing (this is respectful accommodation of a need, not stereotyping).

Including *Braille* location indicators underneath key directional signs in hallways, outside classrooms, etc.

When does a legitimate difference leave off and stereotyping begin? That can be tricky. Consider, for example, the studies that have pointed toward a male tendency to be task-oriented and competitive, and a female tendency to be person-oriented and cooperative, which is clearly relevant data for forming balanced collaborative work teams:

> *As the primary caregivers, women learn to value nurturing, connection, growth, and cooperation; as the primary breadwinners, men learn to value separateness, competition, debate, and even conflict.*[7]

Or how about the old saw that "boys are better at math skills; girls are better at verbal skills," a perception which is not, by the way, without some empirical basis. As long as we refer to such documented differences with scientific objectivity and sensitivity in our communications, there's nothing wrong with acknowledging them. Cultural sensitivity does not mean homogenizing legitimate gender differences into androgynous anonymity. At the risk of closing the section on stereotyping with a double stereotype:

As the French say, "Vive la difference!"

We Must Strive to Send a Multicultural Image

Whenever we have the opportunity as technical communicators to depict people in the workplace, we should be mindful of our responsibility to provide a genuine reflection of diversity in the workforce. One of the authors, who is in an interracial marriage and in the process has achieved diversification in the child workforce, vividly recalls a door-to-door Bible encyclopedia salesman [it was a sales*man*, by the way, not a salesperson!] who proudly displayed his wares on the coffee table, only to be shot down by a discerning six-year-old who asked rather pointedly, "Where are the little *black* children?" No sale.

Of all places not to have diversity!

Sending a genuine image of diversity in the workplace does not mean faking it or forcing it. The key is to be *natural* and *realistic*. Over the full run of a user's manual, for example, it would be appropriate for graphics depicting users to include females, minorities, and a person in a wheelchair. But it is overdetermined and patronizing to try to pack the entire spectrum of diversification into one picture! What you end up with will resemble Madison Avenue's first, clumsy attempts at diversity in advertising. Who can forget the politically correct fast-food restaurant commercials that featured six smiling, singing children: Caucasian, African-American, Mexican-American, Asian-American, Semitic, and Eskimo, who miraculously all arrived at the same restaurant at the same time.

We Must Not Stumble into Tokenism

Tokenism is a subtle form of prejudice. Practitioners may not even be aware they are engaging in it. Tokenism refers to self-conscious attempts to prove how unprejudiced one is that wind up sending exactly the opposite message.

The perfect example of tokenism was television's important but awkward shattering of the color line in programming in the 1960s. Can you remember the first sitcom with an African-American lead? [Careful, you're dating yourself now!] That's right: *Julia*, whose protagonist was an African-American nurse. If you remember the show, you will recall how soppily perfect Diahann Carroll's character was. Beyond that, every effort was made to purge her of any trace of genuine African-American culture. She was attired along conventional Middle American style lines, in a post-Barbie hairstyle. Julia was physically black but culturally white.

Comedienne Carol Burnett lampooned tokenism when her character self-righteously proclaimed her enlightened liberalism: "Some of my *best* friends are Negro."

Star Trek's scriptwriters made a frontal assault on tokenism in the episode "Plato's Stepchildren" when they called for a very real, nontoken, and passionate kiss between Captain Kirk and Lieutenant Uhura (for you non-Trekkies, Uhura was black). An unbelievable furor erupted at the studio. To make a long story short, the kiss scene had to be "watered down" before it could make the air; ratings-conscious executives feared the public outcry television's first interracial kiss would spark.[8]

Nichelle Nichols, the actress who played Uhura, remembers the producers lamely explaining to her the "political" difficulties filming the kiss might create. The producers even considered pairing Kirk off with Nurse Chapel and Uhura with Lieutenant Spock—evidently on the mind-boggling premise that it was okay for a black woman to kiss an alien but not a white man![9]

William Shatner was incensed. Yet for all his influence in Tinsel Town, Captain Kirk lost this one. He was ahead of his time.

When all was said and done, the scene was shot from behind in a manner that *implied* a kiss, but provided no direct camera angle and no evidence of actual contact. Thus, television's "first interracial kiss" turned into a token non-kiss—to be certain that all the bigots stayed tuned. Sad.

Another example of tokenism is needlessly calling attention to a person's color, supposedly in a positive context, just to show how "liberal" we are. For example, why is it necessary to mention in a sports story that the quarterback is black? Do we mention if the quarterback is white? Now if the quarterback is *female*, well, that's a legitimate news angle!

Another telltale sign of tokenism is the hypocritical tendency to fraternize publicly with minorities to maintain one's politically correct image while keeping the very same minorities at "arm's length" in one's private social circles.[10] In this scenario, the young African-American female engineer enjoys a circle of fawning managers at the office Christmas party, but the same young African-American female engineer turns up curiously absent from the invitation list for the New Year's Eve party at the local country club. Such *closet bigotry* is a perfect example of invisible prejudice in our society. It's hard to detect, harder to prove, and almost impossible to correct. We fear it is more widespread than people imagine.

How do we avoid tokenism in our technical communication? Study our motives. If we are genuinely committed to depicting diversification in the workplace, chances are our communication decisions will be legitimate. If we are more interested in parading forth our liberalism, we are likely to run afoul of tokenism—clumsily shoehorning a minority text reference or graphic in where it doesn't fit.

Does checking an annual report to make sure all minorities are represented somewhere constitute tokenism? No. Granted, the process may seem a bit artificial, but it may also be necessary to correct an inadvertent oversight that would send the wrong message. If we behaved this way socially (for example, conspicuously alternating minorities in the seating plan for a dinner party), we would be practicing tokenism. In editing a document, our role is different; nobody *sees* us counting up white and minority faces in the photos—they only see the finished document and the image it conveys *is* influenced by the representativeness of the photographs.

We Must Deal Gracefully with the Generic Pronoun and Avoid Sexist Usages

One of the most cantankerous characteristics of the English language is its lack of a gender-neutral pronoun with which to refer to the relative pronouns ("someone, somebody, everyone, everybody, anyone, anybody"). The problem, of course, is that these pronouns are always singular. It is grammatically incorrect to write: "Everyone has finished their report." This practice is becoming pervasive in spoken usage; in a couple of decades, it may even gain formal acceptance. It is *not* yet acceptable in written usage, however, which demands that professional communicators step up to the challenge of the generic pronoun.

For centuries, the solution was simple: Use the singular masculine pronoun for generic applications: "Everyone finished his report." Nothing could be simpler. Clarity, conciseness, and elegance were served.

Unfortunately, a little over 50 percent of the population was dis-served.

As a reformed practitioner and advocate of the generic masculine pronoun, one of the authors [care to guess which one?] finally ran face-to-face with how the other 50 percent of the population felt. The epiphany occurred during a graduate course in elementary education in which the textbook kept referring to the teacher as "she." Leave aside the fact that a significant majority of elementary school teachers are, indeed, female. "How do you know the teacher is a 'she'?" the erstwhile male chauvinist demanded. The author was converted.

Even though the abandonment of the generic masculine pronoun leaves some linguistic baggage to clean up, it truly *must* be abandoned. Why? What's the big deal?

The big deal is that using the masculine form for all generic pronoun references is a subtle form of sexism. It operates on the same principle as children's books in the 1950s that always depicted doctors as males and nurses as females. The young female child gets the subliminal message: "I can't be a doctor, because I'm a girl." The same message comes across with the generic masculine pronoun. The child gets the message: "If in doubt, it must be a boy, because boys are more important."

Sold. The generic masculine pronoun is history. Now what about the linguistic baggage?

By that we are referring to the admittedly awkward situations that can arise when you remove the generic masculine pronoun from the writer's linguistic palette:

> *Everyone of the students must complete his or her work before he or she takes his or her final exam if he or she is to pass the course.*

The sentence may be politically correct, but it's a safe bet most readers won't get more than halfway through it before throwing up their hands and moving on. There are three basic ways to deal with the situation:

1. Go ahead and use the *he/she, him/her,* and *his/her* pairs, but *only once* in a sentence!
2. Use the plural:

   All of the students must complete their work before they take their exam if they are to pass the course.
3. Rewrite the sentence:

   To pass the course, all students must complete their work before taking the exam.

The second option is the easiest and it is often the best. In the previous example, however, the third option is better, because conciseness and emphasis are served as well as cultural sensitivity.

Some well-intentioned but linguistically callow grammarians have attempted—futilely—to solve the generic pronoun problem by creating new, neutral forms such as *te, ter,* and *tem;* or *s/he*:

> *Everyone will please open* ter *book to page 507.*

All we can say in response to such barbarisms is to quote Hamlet's father describing his murder at the hands of his treacherous brother:

> *Oh, horrible! Oh, horrible! most horrible!* —*Hamlet,* Act I, Sc V

We believe that, in fact, the language *is* undeniably being murdered here and the perpetrators are language specialists, the very same who have a filial duty to protect it from such atrocities. Fortunately, we suspect that *te, ter, tem,* and *s/he* have about as much chance as catching on as the ill-destined two-dollar bill. You cannot legislate language use any more than you can legislate use of a particular currency. (Even less, in fact, in that the government could theoretically have removed all other forms of currency *except* the two-dollar bill, thereby forcing its use.)

Another way of handling pronoun references in a document is simply to alternate between masculine and feminine usage. At first glance, the *gender toggling* solution may seem objectionable on the grounds that it would be obtrusive, but in practice it can be quite natural. In witness whereof, we point out that this was our solution for this textbook. Reread a section at random, focusing on the use of *he* and *she*. See what we mean? You may not have even noticed; if you did, we doubt it was more than a passing observation.

By the way, we also made a conscious effort to balance the male-female scale when it came to doctors, lawyers, CEOs, and the like. The only place where we almost ran afoul of vestigial stereotyping was the lead to Chapter 12; in the first draft, the black-diamond skier was male. Since the wife of one of the authors is a better skier than he will ever be, the change was mandatory.

In fairness, we must note that the gender toggling approach is not universally accepted within the technical communication community. Some of our respected colleagues prefer a gender-neutral approach. So if you, like us, believe in gender toggling, be sure to check with your professor, employer, client, or publisher before converting your beliefs into oscillating pronouns!

Linguistic sexism is not limited to the generic pronoun. Noted grammarian Lynn Q. Troyka, author of *Handbook for Writers*, points out four other areas where sexism can creep into language:

1. Using "man" for situations where both men and women are intended:

   Not Man is destined for greatness.
   But Humans are destined for greatness.

2. Stereotyping jobs and roles by using *man* as a standard suffix:

   Not Chairman
   But Chairperson

3. Using expressions that exclude one sex:

   Not Mankind
   But Humanity

   Not Old wives' tale
   But Superstition

4. Using demeaning and patronizing labels:

   Not My girl will send you the fax.
   But My secretary will send you the fax.[11]

By and large, we endorse such non-sexist uses because of the cultural sensitivity they represent. However, we do feel that those who crusade against sexism in language sometimes go overboard, coining uses that are simply too awkward to be tolerated or taking umbrage at a use that is clearly poetic, not sexist:

- ❑ Do we really want our juries to be run by a "foreperson"?
- ❑ Are "man-eating" tigers really sexist?
- ❑ If we're running down the street, do we trip and fall in a "person-hole"?
- ❑ Shall all old Navy movies be edited to read "Person the decks!"?
- ❑ Must Neil Armstrong face EEO litigation for not saying "One small step for a human, one giant leap for humankind"?

We think not. Sometimes discretion is the better part of non-sexist usage.

"Oh, it's OK, Edith—the sign says it's a MAN-eating tiger!"

We Must Recognize and Refrain from Sexual Harassment

Before moving on to the international marketplace, let us consider an issue related to sexism that has commanded much attention in the workplace and on campus over the past few years—sexual harassment. Most companies and universities have developed strict policies on the subject. It used to be that telling an off-color joke was merely an exercise in bad judgment; now it can conceivably be a cause for disciplinary action, even termination of employment.

Like anything else, sexual harassment policies and the penalties that accompany them arose in response to a genuine need. Basic human respect, not to mention common sense, demands that we not use sex as a lever for advancement, as a means of humiliation, or as a topic for titillation in the mixed company we typically encounter in the workplace or the classroom. However, do we really need to live in fear that an innocent comment—even a genuine compliment—can spell a trip to the EEO woodshed and a major demerit on one's career ledger?

What constitutes legitimate sexual harassment? And what constitutes overreaction— abuse of an instrument that was created for a worthy purpose?

Here again, *shades of gray* cloud the lens. *One person's flirtation is another person's sexual harassment.* There are no absolutes in this area—short of outright physical advances or manifestly indecent behavior, everything is relative. It is fully possible, for example, for a small work team of both sexes to be comfortable with a steady diet of salty jokes, innuendo, and repartee, safely within the context of mutual human respect. But it is also fully possible for such a small work team to trample—either deliberately or inadvertently—the unstated sensibilities of a new member or even of employees within earshot who are *not* members of the immediate work group.

Discretion is the best advice we can offer. *If in doubt, err on the side of conservatism.* If you are debating whether to share an off-color joke or a racy comment, assume offense will be taken, even if it might not be. That's far better than assuming offense *won't* be taken when it actually is. If you're considering using words like "hunk" or "babe," reconsider.

Exercise discretion in your actions as well as your speech. Supervisors and professors: If you are meeting with a member of the opposite sex after hours, do so in a public place like the cafeteria or canteen. Don't leave yourself exposed to charges of sexual harassment; it takes only one incident, whether or not it has any basis in fact, to permanently damage your career. Students and employees, we give you the same advice. The chance of actually being harassed in a situation like that is fairly slim—but it is *not* zero. For your own protection, don't take the chance.

What does all this boil down to? It boils down to sensitivity and common sense. *Responsibility for avoiding sexual harassment is shared.* Unless the "harassing" behavior is totally out of line—such as physical advances or lewd and lascivious words or actions—the recipient owes the practitioner one warning: one clear, unequivocal statement that the behavior is not appreciated and will not be tolerated. This is true whether the behavior is telling dirty jokes, making excessive references to physical appearance, or making inappropriate physical contact (the lingering pat on the shoulder that stops just short of an unwelcome caress). The first time should bring a clear warning—an unequivocal sign that you take exception to the offender's words or behavior. That's communication.

If the behavior stops, that should be the end of it.

If it doesn't, that's sexual harassment. It is unethical and intolerable. And yes, it should be punished.

Cultural Sensitivity—A Business Imperative

Businesses that wish to survive into the next century are learning that to do so they must be competitive in the international marketplace, especially in rapidly developing Third World countries whose burgeoning economies are destined to provide a significant portion of the world's new business over the decades ahead.

Functioning successfully in the international arena requires two things:

1. Having a winning price (you can never get around this one!)
2. Exercising cultural sensitivity.

The second requirement is important not only for business success, but also because of the ramifications for international relations. The United States must undo the "Ugly American" image it has earned through the brash and insensitive actions of some U.S. citizens abroad—be they businesspersons, tourists, or military personnel—who have trampled on the cultural sensibilities of their hosts.

Sensitive, knowledgeable intercultural communication can go a long way towards improving our image abroad. Even more importantly, it can contribute to the pursuit of world peace. In his essay "Peace as an Ethic for Intercultural Communication," David W. Kale observes:

> *The guiding principle of any universal code of intercultural communication should be to protect the worth and dignity of the human spirit. . . it is unethical to communicate with people in a way that does violence to their concept of themselves or to the dignity and worth of their human spirit.*[12]

Applying this standard, Kale finds ethnic jokes, slurs, and racist slogans all to be unethical. Furthermore, he contends that the deliberate use of deceptive communication across cultures is unethical, as is the forcing of preferred political systems upon nations that don't want them.

Technical communicators can't do much about forcing political systems upon other nations, but we have much to contribute in the international arena when it comes to producing honest, culturally sensitive communication products. Before addressing specific guidelines for effective intercultural communication, let's take a look at the key factor that underlies the process: *ethnocentrism*.

Overcoming Ethnocentrism Is Key to Intercultural Sensitivity

Samovar and Porter define ethnocentrism as follows:

> ***Ethnocentrism:*** *the belief that one's own group or culture is superior to all other groups or cultures; the tendency to interpret other cultures according to the values of one's own.*[13]

The second part of the definition is as important as the first. The myth of cultural superiority is clearly arrogant; the distortion of another culture's values through the prism of one's own culture is much more subtle, but potentially just as damaging to intercultural relations. Indeed, Samovar and Porter state that ethnocentrism *"might well be the single major barrier to intercultural communication."*[14]

Let's return to *Star Trek* for a moment. In *Star Trek VI: The Undiscovered Country* (Paramount Pictures), one scene depicts peace negotiations with the long-hated Klingons, arch-enemies of the Federation. (Who can ever forget the Klingon cloaking device?) During dinner preceding the negotiations, Ensign Chekhov refers to the concept of "certain inalienable *human* rights" that could not be abrogated—whereupon the daughter of the Klingon chancellor replies tartly: "'Inalienable *human* rights'—the very *name* is racist!"

What a perfect example of ethnocentrism! Interesting, also, to note the philosophical progress from the days when TV's first interracial kiss had to be laundered to protect the ratings. By the way, Lieutenant Uhura was at the dinner table; certainly the irony was not lost on Nichelle Nichols.

Let's consider an example of ethnocentrism in the international business marketplace. United States businesspersons typically follow a straightforward, let's-not-mince-any-words, get-the-job-done approach. In many Eastern cultures, such an approach is considered rude. A more reticent, indirect style of communication is preferred. A U.S. businessperson blinded by ethnocentrism would be apt to view the indirect style as evasive rather than polite.

Young Yun Kim speaks specifically to the need for a bridge between East and West in her essay "Intercultural Personhood: An Integration of Eastern and Western Perspectives":

> *If we are to actively participate in this evolutionary process, the dualism inherent in our thinking process, which puts materialism against spiritualism, West against East, must be transcended.*[15]

In short, *we must overcome ethnocentrism.*

Where does ethnocentrism come from? We learn it. Albert and Triandis, in their essay "Intercultural Education for Multicultural Societies: Critical Issues," state:

> *Ethnocentrism tends to be characteristic of persons who have been exposed to little diverse information and to child-rearing practices which did not allow them to learn about value systems which differ from those of their parents.*[16]

Rodgers and Hammerstein captured that concept a bit more poetically in the first Broadway musical to take on a significant social message, the immortal *South Pacific*:

> *You have to be taught to hate and fear,*
> *You have to be taught from year to year,*
> *It has to be drummed in your dear little ear,*
> *You have to be carefully taught.*

Since ethnocentrism is a learned behavior, it can also be unlearned. For this reason, Albert and Triandis go on to maintain that "all children in a multicultural society should have the benefit of intercultural education."[17] They further contend that "intercultural education is necessary not only for students from diverse ethnic groups but also for their *teachers*."[18]

The key to overcoming ethnocentrism in our communication with other cultures, be they across the corridor at work or across an ocean, is to be honest with ourselves—to ferret out our hidden prejudices, drag them out into the light where they can be exposed, and thereby remove the cataract of ethnocentrism from the "inner eye" that Ellison referred to. Only then can we see other cultures, other peoples, with clarity and sensitivity—and act, speak, and write accordingly.

Before we can know others, we must know ourselves.

Assume, then, that we are genuinely committed to identify and eliminate all traces of ethnocentrism in ourselves. Assume, further, that we are reasonably successful in this venture. We have thus entered the realm of *cultural relativism*: That is, we accept other cultures on their own merits and refrain from imposing our own values and beliefs on others. Good. Does this then mean that we must accept every other culture purely on its terms, no matter what practices the culture includes?

Does intercultural tolerance include acceptance of the Middle Eastern punishment of encircling an adulteress within a stack of used tires, dousing the tires with kerosene, and setting them afire? Does it include laissez-faire endorsement of an African culture's practice of genital mutilation of infant females?

No.

Elvin Hatch addresses this issue in his essay "The Evaluation of Culture," as does Kale in the work cited earlier:

> *. . . there are some universal values on which we can build a universal code of ethics in intercultural communication.*[19]

Hatch asserts that *there are certain absolute standards against which all cultures may reasonably be judged*, without being guilty of ethnocentrism or intolerance. The core value underlying these standards is the *humanistic principle* which demands that *the well-being of people ought to be respected*:

- ❑ *We can judge that human sacrifice, torture, and political repression are wrong, whether they occur in our society or some other.*
- ❑ *Similarly, it is wrong for a person, whatever society he or she may belong to, to be indifferent toward the suffering of others.*
- ❑ *People ought to enjoy a reasonable level of material existence: We may judge that poverty, malnutrition, material discomfort, human suffering, and the like are bad.*[20]

Once we wander outside the parameters of the humanistic principle, however, cultural relativism kicks in. Thus, matters such as religion, diet, etiquette, sexual mores, and the like are areas where we should not judge nor meddle with the values and practices of other cultures.

Communicating Effectively across Cultures Demands Special Knowledge and Special Effort

How can we translate the theoretical concepts previously discussed into specific actions that will improve intercultural technical communication? Kale offers us a good jumping-off point with "four principles that should guide the actions of ethical communicators":

- *Address people of other cultures with the same respect that we would like to receive ourselves.*
- *Seek to describe the world as we perceive it as accurately as possible.*
- *Encourage people of other cultures to express themselves in their uniqueness.*
- *Strive for identification with people of other cultures.*[21]

Different Strokes for Different Folks

If we appreciate the uniqueness of diverse cultures we will be sensitive to *different styles of doing business.*[22]

In Britain, formality and conservatism characterize business relationships. Americans working in the United Kingdom on a joint venture international proposal soon discovered that the same direct, hard-driving business style that served them so well in the United States was not at all appreciated by their British hosts. And imagine the surprise of the American editors when they learned that the executive summary for a multibillion-dollar procurement would be conservatively worded, text-intensive, lean on graphics, formatted like a routine business report, and printed in black-and-white! Submitting a proposal of that magnitude in the United States without a slick four-color executive summary would be almost unheard of. But the marketing hard sell would have been the wrong choice for the audience—the British Ministry of Defense wanted "just the facts, Ma'am," with none of the marketing trimmings. The British editors prevailed; the executive summary was conservative—and the proposal was a winner!

In Germany, business and social relationships are sharply delineated. During business hours, it's strictly "Herr," "Frau," and "Fraulein"; after work at the beer hall, it becomes Fritz and Greta—but the two roles are not generally mixed. This, of course, represents a significant difference from U.S. business practice, where major contracts inked in the boardroom are often initiated on the golf course.

In Latin countries and in Mexico, a "laid-back," casual style is typical. Americans are relatively informal, but nonetheless hard-driving and schedule-oriented when it comes to business transactions. Punctuality, for example, is one Anglo-Saxon value that definitely made it across the Atlantic. American businesspersons operating in Spain or Mexico have to adjust to the concept that "siesta time" is very real; during the scorching heat of the midday, a business meeting is not likely to draw full attendance. Does this mean the Latins are less businesslike, less dedicated? No. Just different. The smart businessperson *adjusts*—how hard is it, after all, to schedule meetings early or late in the day?

In Japan, Korea, and other Eastern nations, courtesy and respect are valued more than aggressiveness and outspokenness. Direct eye contact, which is almost a necessity for credibility in the United States, tends to be offensive to an Eastern audience. That places a double communication challenge upon U.S. businesspersons visiting Eastern nations; not only must they be careful to avoid using strong eye contact themselves, but they must also overcome their own internal cultural tendency to interpret their hosts' aversion to direct eye contact as evasiveness or even dishonesty—which it isn't—rather than respect, which it is. Linguistic subtlety is another important aspect of East-West communication scenarios; namely, East is, and West isn't!

Suppose an American businesswoman is authorized by her company to commit up to $1 million—no more—of investment to a joint venture with a Japanese company. If asked by her hosts whether her company would be willing to put up $10 million, her answer, while polite, would definitely include the word "No." If the situation were reversed, the Japanese answer would likely be "well, that would be a challenge," "or that would require additional consideration." Savvy East-West communicators would recognize either of those answers as a "no"—about as firm as it gets. Without this knowledge, the American businesswoman could easily mishandle the situation, engendering ill will and endangering the business relationship.

A Pocket Guide to Survival in International Communication

The preceding paragraphs present good *general* guidelines for sensitive intercultural communication. But how about some *specifics*? Full treatment of this intriguing topic is outside the scope of this text—not to mention outside the expertise of its authors! We strongly recommend the works of two acknowledged experts in the field: William K. Horton, of Horton Enterprises in Denver, and the Terra Pacific Company of San Diego, specialists in translation and localization of documentation. Horton's works are a gold mine when it comes to intercultural aspects of graphics and iconography[23]; Terra Pacific provides a wealth of hands-on, real-world examples of the pitfalls one can encounter when "localizing" documentation by translating it into the language of a target region, usually in a different country.

"Henry, I don't think you asked him the way to the bus station!"

To whet your appetite for specifics, we offer a few "appetizers" culled from a Terra Pacific seminar[24] and a William K. Horton workshop.[25]

Avoid Using Idioms, Jargon, and Slang. Colloquial usages are almost impossible to translate. Picture the poor American tourists in a Middle Eastern nation poring over the pocket guide to Arabic trying to ask the way to the bus station and coming out with something which, loosely translated, means "go stick your head in camel dung." We lose more Mideastern tourists that way!

Seriously, however, if we describe a business venture to a non-American audience as a "piece of cake," we are likely to elicit not nods of approval but mystified stares (are they peddling a bakery, or what?) And imagine an attractive Eastern European visitor to the South Bronx being told she is "bad, man!" Then, when she takes offense, being offered the explanation, "No, man, you're cool. Like you're outrageous!" *If* she gets past "man" and "cool," she proceeds to look up "outrageous" in her pocket dictionary . . . well, so much for intercultural communication . . . You get the drift.

Forget Humor. The appreciative grin of a native reader in response to a clever turn of phrase may, when you place your work in the hands of a non-native reader, draw merely a puzzled stare. Don't use humor for cross-cultural communication. It simply won't translate.

Be Alert to Costly *Faux Pas*. The Chevy Nova never sold well in Central America. Puzzled marketeers finally found the answer in Spanish 101: Can you translate "no va"? That's right; it means "doesn't go." Oops.[26]

Watch Out for Different Meanings for the Same Phrase. Consider the example of American and British negotiators who agreed to "table" a thorny issue. The problem? Americans use "table" to mean "set aside"; the British use it to mean "put it on the table"—in other words, drive the discussion to closure. Hardly compatible meanings![27]

Be Extremely Careful with Body Language and Gestures. Nonverbal cues differ widely from culture to culture, and they can pack real social and emotional wallop. In Muslim countries, for example, the left hand is considered unclean. As a minimum, using it is impolite; in some circles, it is a major insult. Similarly, in the same cultures, showing the bottoms of one's shoes or feet is also considered insulting. As a consequence, graphics in documentation bound for Arabic nations must be checked to eliminate or deemphasize any depiction of the politically incorrect parts of the anatomy.[28]

The traditional American circular thumb-and-forefinger symbol for "OK" is a classic example of the danger of gestures in intercultural communication. In Japan, it's a symbol for money. In France, it means zero. And in Brazil, it's

an obscene gesture. And the raised hand, palm outstretched, that is so often used as an icon in manuals to indicate "Warning: Read this before continuing" sends an entirely different message in Greece: it means "go to hell."[29] While we will admit that even English-speaking readers of documentation sometimes feel as if they are being given this message (particularly when they feel as if it's written in Greek), we ought not be so explicit!

Be Mindful of Cultural Roles. A skilled intercultural communicator must be knowledgeable and sensitive to the social mores of the audience. A case in point is the sharp social differences in the role of women in Muslim nations, as compared to the West. The same linguistic and graphic conventions that would be considered obligatory in U.S. communications to avoid charges of sexism (such as referring to or depicting women as well as men in executive positions) would be extremely offensive to all-male Arab audiences. Also, it's important in depicting women in photographs or graphics to be sure they are modestly attired, in deference to Arabic customs governing female apparel. And any reference to or depiction of alcoholic beverages is strictly *verboten*.[30]

Exercise Caution in Translating Advertising. Two marvelous examples from Terra Pacific drive home the importance of being careful when using advertising slogans across the international marketplace:

> *The slogan "Come Alive with Pepsi" came over into Chinese as "Pepsi brings your ancestors back from the dead." Uh-huh, uh-huh!*
>
> *The Swedish vacuum company Electrolux once translated a perfectly good Swedish ad into English. The result? "Nothing sucks like an Electrolux."*[31]

Verify Graphic Conventions. A soap company advertisement for the Middle East featured the soap in the center, flanked by dirty clothes on the left and clean clothes on the right. The only problem was that people in the region read from right to left. As a result, the message they received was that the soap made clean clothes dirty![32]

Use Universal Icons instead of Text Whenever Possible . . . The familiar *No Smoking* icon is a perfect example of the efficacy of using widely recognized graphics in international communication. While "No Fumar" and "Defense de Fumer" are not all that difficult to translate, we are confident that a fair percentage of U.S. tourists abroad probably rely on the red bar and circle. Use of easily interpreted icons not only increases understandability, it reduces translation costs.

. . . But Choose Icons and Illustrations Carefully! Pictures are a particularly useful communication tool in areas of low literacy. But beware the possibility of misinterpretation: Witness the classic example of the baby food company that couldn't understand why it was experiencing abysmal sales in a developing African nation. It turns out people were interpreting the familiar picture of the smiling baby on the label to mean the jars contained ground-up babies![33] Jonathan Swift, be still!

Let's Practice What We Preach

> *"Lots of folks* talk the talk,*" runs the old evangelist's saying. "The question is, can you* walk the walk*?"*

In plain terms, can we as technical communicators demonstrate our commitment to cultural sensitivity not only in the products we create and produce, but also in our words and actions, inside and outside of the workplace?

Cultural sensitivity is not like a uniform you put on from 9 to 5 in the workplace. To be meaningful, it must reflect a deep and abiding personal commitment that is *never* relaxed, not in the workplace, not at home, not at Happy Hour, not even in the presence of others who do not share the same level of cultural sensitivity. Thus, the editor who carefully checks for ethnic balance in the photographs within the annual report, then turns around at Happy Hour and tells—or even laughs at—a disparaging ethnic joke is not only guilty of unethical behavior but also of hypocrisy of the worst sort.

Strong words. Yes. Important concept. Yes. In fact, we'd say a *vitally* important concept.

As communication specialists, our skills give us the ability to exert a major influence on human relationships. As a result, we have a unique and uncompromising responsibility to be leaders for intercultural harmony. We can be a voice for tolerance, we can be an active instrument of prejudice, or we can sit silently by—which supports prejudice by default.

There is only one morally acceptable alternative.

What, then, does it mean to be a *voice for tolerance*? How, specifically, do we *walk the walk*? Let's look at four situations where we have the opportunity to be leaders for human respect.

Refuse to Participate in Ethnic Jokes and Slurs

Suppose you are at Happy Hour with your boss and several coworkers. Your boss tells an ethnic joke whose punch line is a racial slur. What is the appropriate behavior? Assuming you do not share your boss's mirth at the debasement of fellow human beings, what should you say or do? Should you come right out and tell him, "I really don't appreciate that kind of joke. I find it offensive, and I wish you would not tell such jokes in my company"?

That would be ethically forthright and courageous, but realistically, how many of us are willing to lay our jobs on the line over an ethnic joke? Does anything less represent a moral cop-out? Isn't there anything between confrontation and endorsement? Yes, there is—what psychologists call *extinction*, or eliminating an undesirable behavior by consistently refusing to reinforce it. Stony silence can be a powerful antidote for racist behavior. If everyone greeted such behavior in that fashion, the practitioners would effectively be hung out to dry—right where they belong.

Resist Closet Homophobia; Do Not Discriminate or Disparage Based on Sexual Preference

Closet homophobia is an insidious form of invisible prejudice. For every violent homophobe who physically attacks a homosexual just on account of sexual preference, there are dozens of law-abiding citizens who would never dream of such a behavior who do not hesitate to join in a joke at the expense of gays. We would speculate that many sensitive individuals who are uncomfortable with ethnic or racial slurs fail to manifest the same level of sensitivity relative to those of differing sexual preferences.

Indeed, gays appear to some to be "fair game" for derision, since they supposedly are homosexuals *by choice* as opposed to being black by what Ellison refers to as an "accident to the epidermis." Leave aside the fact that recent medical research suggests that homosexuality may indeed have a biological basis, rooted in the hypothalamus. Compounding the situation are Biblical proscriptions against sodomy which, taken out of the historical context, are often used to lambast homosexuality as an *evil* behavior, even stretched to interpret AIDS as a just and fitting divine retribution for that behavior. Frightening, isn't it? (Zealous fundamentalists are quick to quote Leviticus 18:22, which refers to homosexuality as an "abomination." They rarely quote nearby Leviticus 11:12, which states that eating shellfish is an "abomination" [see Lying by Omission, Chapter 4].)

As a society, we have a long way to go before reaching cultural tolerance and sensitivity in the area of sexual preference. Indeed, we are fully aware that some of our readers may differ with us on this sensitive issue and we respect everyone's constitutional right to personal beliefs. But no one is entitled to use those beliefs as means of depriving others of their basic human rights. To do so is fundamentally no different than to immolate an adulteress in a ring of fire or to mutilate female babies.

We'll leave you with this thought: When we are genuinely culturally sensitive, we don't have to modify our behavior in the presence of minorities.

Let's say we are still growing and we *do* have to make a conscious effort to be sensitive. At least that's progress. Remember this: We can *see* the racial minorities in a room and we often *know* the religious preferences of those we work with day to day. The same is not true of gays. The person we hurt through a thoughtless comment may be a close coworker or even a trusted friend. Is that worth a cheap laugh? We think not.

Preserve the Line between Professional and Personal Relationships

Let's say you're a single instructor who has just taken a serious emotional hit—the sudden departure of your significant other. Suppose, further, that a very attractive member of the opposite sex (or, if you prefer, the same sex!) has shown an unusual and flattering interest in you, one that appears to extend past student respect for professor. Three students stay after class to ask questions. Do you find yourself deliberately answering the other two students' questions first to set up a one-on-one with your admirer? If so, be careful—that does not, in itself, constitute sexual harassment, but it *is* setting the stage for it. If you don't want to be struck by lightning, don't stand under a flagpole during a thunderstorm!

As an aside, does it then follow as the night the day that no professor can ever ethically enter a personal relationship with a student, that no supervisor can ever ethically enter a personal relationship with an employee? No. It may not be wise to court such a relationship, but there *is* such a thing as spontaneous combustion! We can all point to cases where precisely that has occurred—and where the situation has been handled with grace, equity, and decorum. We can also point to cases where the situation has *not* been handled well and has led to conflict of interest.

The key point is *not to mix* the relationships; once a personal relationship begins, there should no longer be direct reporting lines on the job or in the classroom (for example, a professor should not be grading his girlfriend; a supervisor should not be determining a merit increase or making a promotion decision for her boyfriend). Open disclosure must be made within the organization so that the necessary transfers can occur to prevent direct-line reporting between those involved in the personal relationship. Otherwise, both parties are leaving themselves open to conflict-of-interest charges, which can be very damaging, whether or not the charges have any basis in fact.

Respect the Line between Religious Beliefs and Professional Relationships

Devout practitioners of a given faith apply its principles in every walk of their lives, including their professional lives. No one can reasonably expect otherwise; how can we ask people to leave their spiritual convictions at the door when they step into the workplace or into the classroom? On the other hand, we *can* expect them not to confuse the church and the conference room, the pulpit and the lectern. The workplace and the campus are not suitable milieus for personal evangelical crusades, no matter how fervent our beliefs. The same cultural sensitivity that forbids discrimination or disparagement based on religion also demands we not seek to foist our own beliefs upon others, especially not if we attempt to do so with the leverage of authority.

What if *all* members of a small work group are members of the same religion? Is it not then appropriate to allow the line between professional and spiritual to be crossed? That's a tough one. Not without irony, we can find a close analogy in our earlier discussion of sexual harassment: Just as a salty joke among four consenting coworkers out of earshot of anyone else does not constitute sexual harassment, religious interpretation of a work-related development in the context of a community of like believers does not constitute unwanted evangelism. But the same caveat applies: The group of believers must exercise extreme caution not to impose its preferences upon a new member, a visitor, or others in the immediate work area. Thus, a quiet comment such as "we need to be good Christians and forgive this action" in a group of four Christian coworkers would not constitute cultural insensitivity; conducting a daily morning prayer session in an open work area would. If one Jewish person, Muslim, or atheist is offended, that is one too many.

It's about respect. It's about human dignity. It's about casting aside the silent veil of prejudice and replacing it with a smile of acceptance. It's about seizing the dream and moving it one step closer to reality.

> *Even though we face the difficulties of today or tomorrow, I still I have a dream. It is a dream deeply rooted in the American dream. I have a dream that one day this nation will rise up and live out the true meaning of its creed: We hold these truths to be self-evident, that all men are created equal.*[34]
>
> —*Dr. Martin Luther King, Jr.*

It's about millions of small, individual steps away from hatred and strife, toward global harmony and peace.

It's about vision. It's about courage. And it's about wisdom.

We close with words of wisdom from another great American leader, who only with the height of Anglo-Saxon arrogance can be termed a "minority." Indeed, not until the very word itself loses its meaning will we have stripped the invisible lens of prejudice from our inner eye. This great leader displayed clear vision as he called to a close one of the most gallant fights against injustice the world has ever seen, in order that no more blood be shed—on *either* side. To this leader, red and white had become *just human*. We can only hope that one day he will speak for all of us, from Idaho to Bosnia.

> *From where the sun now sets, I will fight no more forever.*[35]
>
> —*Chief Joseph*
>
> *Amen.*

Summary

In technical communication, the value of cultural sensitivity calls for us to respect and embrace diversity, both in our products and in our personal and professional behavior. To do so, we, like anyone else, must shed the invisible barriers of prejudice within ourselves that would set us apart from our fellow human beings.

To ensure cultural sensitivity in our communication products, we must avoid the pitfalls of stereotyping, tokenism, and linguistic sexism; and we must convey a multicultural image—*gracefully*, not clumsily, as if fulfilling a quota.

Linguistic sexism lurks in the use of the generic masculine pronoun ("everyone must finish *his* exam") and in other male-dominated usages ("business*man*"). The pronoun dilemma can be handled through judicious use of pairs (he/she), using the plural, recasting the sentence, or, in longer documents, alternate use of masculine and feminine pronouns (gender toggling).

Sexual harassment refers to the use of sex in the workplace or the classroom as a lever for advancement, as a means of humiliation, or in a manner that is offensive to at least one person present. It is a *relative* concept; one person's flirtation is another's sexual harassment. If in doubt, however, it is prudent to err on the conservative side; not only does this reduce the risk of inadvertently offending someone, it avoids the possibility of reputation-shattering allegations and possible disciplinary measures.

Exercising cultural sensitivity is an integral part of doing business successfully in the international marketplace. To achieve this, we must recognize and shed ethnocentrism: the belief that our own group or culture is superior to all others, along with the tendency to interpret other cultures through the distorting lens of one's own culture.

Cultural relativism refers to accepting other cultures on their own merits rather than trying to impose the values of one's own culture upon others. Cultural relativism covers anything that does not fall under an absolute humanistic principle which demands that the well-being of *all* people be respected. Thus, torture and other human rights violations cannot be accepted under the name of cultural tolerance, but a wide range of other customs, from religion to sexual mores, can—and should.

Different nationalities have different styles of doing business; success in international business relationships demands knowledge of and sensitivity to such differences.

Specifically, when localizing documentation by translating it into the language of a different country, it's important to apply the following guidelines:

- ❑ Avoid using idioms, jargons, and slang
- ❑ Do not use humor
- ❑ Be wary of costly linguistic mistakes
- ❑ Watch out for phrases that mean different things in different countries
- ❑ Exercise caution with body language and gestures
- ❑ Respect differing cultural roles, especially in male-female relations
- ❑ Be very careful in translating advertisements
- ❑ Verify graphic conventions
- ❑ Use universal icons instead of text
- ❑ Use graphics and icons carefully, watching for unintentional meanings

Cultural sensitivity is not a 9-to-5 phenomenon. Genuine commitment means 24 hours a day, 7 days a week, of tolerant human behavior, respecting the right of others to be different. Specifically, it means:

- ❑ Refusing to participate in ethnic jokes and slurs
- ❑ Resisting closet homophobia: not discriminating or denigrating based on sexual preference
- ❑ Observing the line between professional and personal relationships; avoiding conflict of interest
- ❑ Respecting the line between religious beliefs and professional relationships; not evangelizing in an inappropriate venue.

Application Exercises

1. Is it possible to be prejudiced and not even know it? Give a couple of examples to support your answer.
2. Define *stereotype*. List three stereotypes that are common in our society.
3. What is *tokenism*? Give two examples.
4. Revise these sentences to avoid linguistic sexism.
    a. A spokesman for the club announced plans for a holiday reunion.

b. Everyone in the class must submit his notebook for evaluation.

c. The doctor wanted to finish his rounds before the nurse went on her lunch break.

d. Man is the only species that murders his own kind.

e. "Please refrain from swearing when there are ladies present."

f. Tom Washington, the new black football coach at North Ridge High School, guided his team to the conference championship in his first year.

5. How would you define sexual harassment?

6. Where does this scenario move from an innocent exchange of office pleasantries into sexual harassment? Or does it? Explain.

   "Here's the fax you requested, Mandy. That's a really *nice outfit you're wearing. I really* like *what it does for you—or should I say what* you *do for it? (Time-out for full-body stare.) Hey, how about lunch today. . . "*

7. What is ethnocentrism? Where does it come from? Is it curable? How?

8. Compare and contrast the use of *absolute* versus *relative* standards to evaluate other cultures. Give an example of each.

9. Briefly describe the differing styles of doing business in the following countries:

   a. United States

   b. Britain

   c. Germany

   d. Mexico

   e. Japan

10. Why are body language and gestures dangerous to use in international communication?

11. List and describe three ways in which technical communicators can be a "voice for tolerance."

Exploration Exercises

1. Describe a personal prejudice that you have identified. Have you overcome it? If so, how? If not, do you plan to? How?

2. If you were (or have been) the victim of prejudice, would you rather face direct bigotry or hypocrisy and tokenism? Why?

3. Find and clip examples of linguistic sexism in newspaper and magazine articles and other publications.
4. Do you believe the recent furor over sexual harassment in the workplace is justified? Why, or why not?
5. Research the controversy leading to a Federal court order for Hooters restaurants to employ waiters as well as waitresses. What issues underlie the controversy? What is your personal opinion?
6. Describe a personal ethnocentric belief that you have identified. Now that you have identified it, what do you plan to do about it? Why? How?
7. Suppose your company has landed a major contract in a Third World nation whose repressive government still tortures political prisoners and commits other heinous human rights violations. You are assigned to a two-year tour of duty in that country—and along with the assignment, given a major promotion and a substantial salary increase. There is no danger to you personally; your company is in good standing with the government; indeed there are those who say some official palms were greased to land the contract. What would you do? Why?
8. Read William K. Horton's book on iconography, *The Book of Icons*, or attend a Terra Pacific workshop (or write for literature from them) on intercultural communication. Provide five specific examples of communication breakdown (linguistic, graphic, or social) that occurred as a result of lack of cultural sensitivity.
9. How much closer to Martin Luther King's "dream" are we today than we were when he presented his famous speech in 1963? Explain.
10. For years, Chief Joseph led the hopelessly outnumbered Nez Perce Indians in a valiant but doomed uprising against a treacherous government that had reneged upon a treaty by reclaiming much of the reservation it had granted the Nez Perce in Oregon and Idaho after gold and silver were discovered there. Rather than dig in for a last stand that would have claimed hundreds more lives on both sides and virtually wiped out his tribe, Chief Joseph delivered his immortal statement and laid down his arms. A century later, we ask ourselves, can we have peace without sacrificing justice? Can we have peace at all? How?

References

1. Ishmael Reed, "The Multinational Society," in *Bedford Reader*, 5th ed, X.J. Kennedy, Dorothy M. Kennedy, and Jane E. Aaron (Boston: Bedford Books of St. Martin's Press, 1994), 682.

2. *Ibid.*, 686.
3. Hudson Institute, *Workforce 2000: Work and Workers for the 21st Century*, Executive Summary for U.S. Department of Labor, Washington, D.C., 1987. Cited in "A Flexible Technical Communication Education Model for the Year 2000," Marian G. Barchilon and Donald G. Kelley, Arizona State University, *Technical Communication* (Fourth Quarter 1995): 593.
4. Larry A. Samovar and Richard E. Porter, *Communication Between Cultures* (Belmont, CA: Wadsworth Publishing, 1991), 304.
5. Samovar and Porter, *Communication*, 280.
6. William K. Horton, *The Book of Icons* (Salt Lake City: John Wiley and Sons, 1994).
7. Jeoffrey Chodorow, *Cases, Research, and Issues in Technical Communication* (New Jersey: Humanities Press International, Inc. 1990), 57.
8. William Shatner and Chris Kreski, *Star Trek Memories* (New York: Harper Collins, 1993), 282.
9. *Ibid.*, 283-285.
10. Samovar and Porter, *Communication*, 281-282.
11. Lynn Q. Troyka, *Handbook for Writers*, 3d ed. (Englewood Cliffs, NJ: Prentice-Hall, 1993), 421-3.
12. David W. Kale, "Peace as an Ethic for Intercultural Communication," *Intercultural Communication, A Reader*, 7th ed. Ed. Samovar and Porter (Belmont, CA: Wadsworth Publishing, 1994) 437-438.
13. Samovar and Porter, Communication, 315.
14. *Ibid.*, 285.
15. Young Yun Kim, "Intercultural Personhood: An Integration of Eastern and Western Perspectives," *Intercultural Communication, A Reader*, 7th ed. Samovar and Porter (Belmont, CA: Wadsworth Publishing, 1994), 424.
16. Rosita D. Albert and Harry Triandis, "Intercultural Education for Multicultural Societies: Critical Issues," *Intercultural Communication, A Reader*, 7th ed. Ed. Samovar and Porter (Belmont, CA: Wadsworth Publishing, 1994), 431.
17. Albert and Triandis, *Intercultural Education*, 425.
18. *Ibid.*, 430.
19. Kale, *Peace*, 437.
20. Elvin Hatch, "The Evaluation of Culture," *Intercultural Communication, A Reader*, 7th ed. Ed. Samovar and Porter (Belmont, CA: Wadsworth Publishing, 1994), 409.
21. Kale, *Peace*, 407.

22. Jules Harcourt, A.C. "Buddy" Krizan, and Patricia Merrier, *Business Communication*, 3d ed. (Cincinnati, OH: Southwestern Educational Publishing, 1996).

23. Horton, *The Book of Icons*.

24. Terra Pacific, Inc., San Diego and Los Angeles, CA, "Training Guide." Handouts from "Internationalization" seminar in Denver, CO, October 1994.

25. William K. Horton, handout from keynote address at STC's Rocky Mountain Chapter's Spring Conference in Denver, CO, 1993.

26. Terra Pacific.

27. *Ibid.*

28. Horton, handout.

29. Terra Pacific.

30. Horton, handout.

31. Terra Pacific.

32. *Ibid.*

33. *Ibid.*

34. Martin Luther King, Jr., "I Have a Dream," Lincoln Memorial, Washington, D.C., 28 August 1963.

35. Chief Joseph, upon surrender of Nez Perce Indians to federal forces after a valiant struggle, northern Idaho, 30 September, 1877.

Chapter 11 Overview

Social Responsibility . . .

The Mission: **In the context of technical communication, social responsibility is our duty to behave responsibly and ethically in using and disseminating information that affects the welfare of the public. It encompasses both serving the user—meeting his need for quality communication products—and protecting the user. It is also maintaining vigilance and advocating responsible behavior within our places of employment.**

Chapter Roadmap

Definition and Focus. This section provides the philosophical framework for social responsibility as a core value.

Our Duty to Serve and Protect. This section explains why our ethical obligation goes beyond merely *serving* those who use our communication products to also doing what we can to *protect* them.

We Must Serve the User. This section familiarizes you with ways to produce communication products that serve the readers' needs, including:

- ❑ *Think Critically.* This section suggests ways you can thoughtfully evaluate information for inclusion in your communication products.
- ❑ *Use Language Carefully.* This section introduces you to ways to use language in a socially responsible manner, such as using culturally sensitive language and avoiding doublespeak.

We Must Protect the User. This section familiarizes you with ways to produce communication products that protect the reader, including:

- *Ensure Accuracy and Completeness*. This section applies some of the AOC[7] principles of quality (see Chapter 7) as they relate to social responsibility, helping you determine when to compromise.
- *Analyze the Audience Thoroughly*. This section enables you to analyze the audience to determine how best to communicate in a socially responsible manner. It explains how levels of edit can be adjusted to meet requirements, it addresses the need to make information accessible to those with disabilities, and it explains why dangerous products demand meticulous documentation.
- *Accommodate Those with Disabilities*. This section discusses how and why our responsibilities go beyond the requirements of the law.
- *Document Dangerous Products*. This section examines our increased responsibility when users' lives and limbs are at stake.
- *Address Safety and Health Issues*. This section acquaints you with a variety of safety and health concerns to consider in preparing communications.
- *Evaluate Environmental Issues*. This sections familiarizes you with a range of environmental concerns to consider in preparing communications.

Summary, Practical Exercises, and References

. . . Our Duty to Preserve and Protect the Public Good

CHAPTER 11

Social Responsibility Means Serving the User Is Not Enough—We Must also Protect

In a free society . . . there is one and only one social responsibility of business—to use its resources and engage in activities designed to increase its profits so long as it stays within the rules of the game,which is to say, engages in open and free competition without deception or fraud.

—Milton Friedman[1]

Even the most private of private enterprises is an organ of society and serves a social function . . . it requires of the manager that he assume responsibility for the public good, that he subordinate his actions to an ethical standard of conduct.

—Peter Drucker[2]

Definition and Focus

Certainly, the opening quotations reflect contradictory opinions about corporate obligation to protect the public welfare. Perhaps part of the discrepancy stems from disagreement about the definition of social responsibility.

One school of thought contends that ethics and social responsibility are, in fact, two separate concepts with essentially different meanings. Robin and Reidenbach, for example, distinguish between them in this way:

> *Corporate social responsibility is the social contract between business and the society in which it operates. Business ethics, in contrast, require that the organization or individual behave in accordance with the carefully thought-out rules of moral philosophy.*[3]

Ethicist Walter Manley relegates social responsibility to second place behind ethics in evaluating a company's social performance. He defines it as "social obligations that transcend the economic functions of . . . profit."[4] Like Robin and Reidenbach, Manley associates social responsibility with specific issues, but links ethics with moral philosophy.

We agree that social responsibility encompasses duties that exceed profit—in particular, the duty to protect and preserve the public good. We argue, however, that social responsibility is not separate and apart from ethics. Instead, we believe with Peter Drucker that providing for the public welfare is virtuous and that we have an ethical obligation to do so.

In fact, a great many debates and case studies focus on the ethical dilemma faced by employees who are forced to choose between job security and taking some socially irresponsible action on behalf of an employer. Apparently, even Friedman would concur that individuals have an obligation to look out for the public good. In the essay previously cited, he comments that "There are no 'social' . . . responsibilities in any sense other than the shared values and responsibilities of individuals."[5]

So let's explore the nature of social responsibility as an *individually* held value. In particular, let's examine the ways in which technical communicators can exercise social responsibility. *In the context of technical communication, social responsibility is our duty to behave responsibly and ethically in using and disseminating information that affects the welfare of the public. It encompasses both serving the user—meeting his need for quality communication products—and protecting the user. It is also maintaining vigilance and advocating responsible behavior within our places of employment.* It is following the path of virtue when facing a choice between public well being and private interest.

Our Duty to Serve and Protect

Just as definitions of social responsibility differ in the workplace, so do company behaviors. For example, many companies pay commissions to their sales force. Others do not. Senior management at Digital Equipment Corporation, for one, believes that commissions can breed unethical behavior because they may prompt sales personnel to be more concerned with greedy self-interest than with client welfare. One manufacturer of over-the-counter medications,

Johnson & Johnson, withdrew its Tylenol products from store shelves immediately and voluntarily following the poisoning of the medication—a tragic, isolated incident for which the company was not to blame. Removal of the product cost the company millions, but the action has been frequently praised by press and public alike for the uncompromising ethical stance it represents. In sharp contrast, another company that produced asbestos suppressed evidence of the debilitating effects of the fibers for more than 30 years, allowing thousands of workers to suffer painful, frequently fatal illnesses that also drained families financially. Some firms fight charges of dumping toxic waste; others take voluntary action to clean up the environment. Clearly, the opinions of business leaders in this regard cover a broad spectrum.

As citizens, all of us are appalled when headlines decry the social irresponsibility of corporations. Just consider the Congressional hearings aimed at determining how much tobacco companies knew about—and hid—the addictive qualities of their products. So deeply felt is our disdain, so much a part of the American psyche, that Hollywood even immortalizes it in our entertainment. Movies like *Silkwood*, *Tucker*, and *Wall Street* dramatize the dangers—physical, financial, and otherwise—to which business subjects the public in the name of profit.

One of our favorite movies, the 1991 film *Class Action* (Twentieth Century Fox), vividly illustrates how corporations value profit over safety. A lawyer, played by Mary Elizabeth Mastrantonio, represents Argo, an automobile manufacturer, in a civil lawsuit. The nearly 150 plaintiffs are seeking damages for injuries suffered when their Argo-manufactured station wagons exploded into flames upon impact.

In preparing for the case, the lawyer uncovers references to a safety report, written before the car went into production, which identified the defect that caused the explosion. Argo executives buried the report and produced the flawed vehicles. When the lawyer asks her client why the company didn't simply recall the cars and retrofit the defective part, an inexpensive turn signal, an Argo executive explains the principle of actuarial analysis, or "bean counting" as he coldly calls it. The cost of retrofitting the cars was $50 million. Assuming an accident rate of *only* one explosion per 3,000 cars, the risk management expert predicted a total of *only* 150 deadly accidents at a cost of $30 million in damages, if everybody sued and won. "It's cheaper to deal with the lawsuits," the executive shrugs, than to "fix the cars." This harsh analysis points out all too painfully the point made by communication specialist Elvin Hatch that companies calculate the risk of producing dangerous products by assessing the cost of lawsuits relative to profit rather than by measuring the threat to human life.[6]

But wait, you say, isn't social welfare within the purview of legislators? Doesn't the law protect the public? Can't we fulfill our obligation to preserve the public good simply by following the law of the land? You may think so—you would certainly *hope* so; unfortunately, it just isn't always the case. There are, of course, hundreds of laws designed for occupational safety, environmental welfare, and consumer protection. Unfortunately, it is often possible to operate within the confines of the law without adequately protecting or providing for the public good.

Again, *Class Action* dramatizes some classic and, regrettably, all too common scenarios. The court orders the law firm representing Argo to turn over the potentially damaging safety report to the plaintiffs' attorneys in accordance with the rules of disclosure. The defendant's lawyers follow the "letter of the law" by handing over the report, but they also mislabel it and bury it in a mountain of paperwork so it cannot possibly be uncovered before the trial begins.

When a young, conscientious attorney questions the tactic, a partner replies, "You're just defending your client."

"But what about the victims?" the young lawyer queries.

"They should have gotten better lawyers," is the casual, off-handed, and icy response.

And indeed, the codes of ethics of the bar associations require that lawyers provide their clients with the best possible defense. How, then, can society balance the right of individuals to the best legal defense with the right of the public to be safe from dangerous products?

Unfortunately, we need not rely on Hollywood to provide examples. The real-life case of the Ford Pinto illustrates the economics of profit versus liability.

In the 1970s, Ford designed and produced the Pinto in an effort to quickly enter the burgeoning and profitable small car market. The company's well-documented, internal tests proved that the Pinto's gas tank could rupture in collisions at speeds as low as 20 miles per hour, resulting in potentially fatal fires. A safer gas-tank design did, in fact, exist. However, Ford officials neither retrofit the cars with the safer tanks nor warned consumers about the danger. Why?

During the court proceedings of the numerous lawsuits filed against Ford as a result of Pinto accidents, company officials explained the reasons for their decisions. They believed that all of their "socially responsible" options—warning customers, retrofitting the gas tanks, or delaying production—would have had a "negative impact on share price."[7] In other words, Ford's decision-makers believed they were acting in the best fiscal interest of the company and its shareholders by selling defective vehicles.

By the way, Ford did compensate many plaintiffs, but not *all* of them. Moreover, the company was cleared of all *criminal* charges in relation to the Pinto cases; indeed, during the trial on charges of criminal homicide, the court did not admit the evidence documenting that Ford officials were aware of the dangerous flaw.[8]

So we can't count on companies, laws, or courts to protect us as fully as we would like in every instance. That's why we must rely on one another as individuals. That's why our ethical obligation goes beyond merely serving those who use our communication products. We must also do what we can to protect them. So let's examine in detail how we do both—protect and serve.

We Must Serve the User

The value of good, clear technical communication finally made it to the silver screen! A set of written instructions figures prominently in one dramatic scene of *Apollo 13* (Universal City Studios) in which NASA engineers instruct the astronauts how to assemble a CO_2 filter from a collection of mismatched materials aboard the spacecraft. The instructions serve the crew well, enabling them to construct the life-saving device, just seconds before the CO_2 in the lunar module reaches lethal levels.

We rarely associate our daily documentation chores with such high drama, but perhaps we ought to. If the bacon grease on your stove catches fire, you certainly want the fire extinguisher that you bought but never looked at to have concise instructions that you can quickly and easily grasp. If you're replacing the brakes on your car, you want well-written instructions

"What do these instructions mean?"

that enable you to complete the task properly so your vehicle does, in fact, stop at the next lowered railroad-crossing gate you encounter. If you're igniting the pilot light on your gas furnace, you want explicit instructions that enable you to follow the procedure without sparking an explosion.

Just as we expect to purchase quality products that pose no threat to our safety, we also expect to be provided with instructions for their safe operation. In fact, liability laws dictate that manufacturers provide documentation that is adequate to ensure the safe operation of their products. Obviously, not everything we document is as weighty as the examples we've cited. But as environmental communication specialist Laurel Grove points out, "communicators are uniquely positioned to do good in the cause of environmental ethics."[9] We would add that our unique position is not limited to environmentally related issues but extends to all areas of social responsibility.

Of course, the best means of serving our users is to provide quality communication products. In Chapter 7 we characterized quality as *defining and performing to standards that best serve the user's unique communication requirements*. In short, we must provide the best possible quality within the constraints of cost and schedule. Doing so may, indeed, require difficult tradeoffs among cost, schedule, and quality. However difficult, though, we cannot sacrifice the tenets of quality on the altars of cost and schedule if doing so in any way threatens the safety, health, or well being of our users. Two of the AOC^7 principles of quality in technical communication (see Chapter 7) are particularly crucial to protecting the public good—accuracy and completeness. We'll look at these more fully later in this chapter. A third tenet, objectivity or fairness, also merits special mention in a discussion of social responsibility. If we are to be socially responsible, we must present facts impartially, untinged with any biases—our own, those of our colleagues, or those of our employers.

The range of circumstances here is vast and varied. For instance, how many times when editing a technical document have you queried a subject matter expert about a complex technical term or obscure acronym and acquiesced when he glibly replied, "Don't sweat it. Everyone will know what that means"? You know this guy, he's played straight with you before, so you figure you can accept his judgment. If the document happens to be a report for an engineering journal, leaving the terminology as is may be appropriate. But what if the document is a set of instructions for repairing brakes on a school bus—and the mechanic who will use it has an eighth grade education? Are you then justified in accepting without further question the expert's bias toward the technologically elite? If dozens of school children are on that bus when it must make an emergency stop at a busy intersection, wouldn't you be glad you persisted in changing or defining the questionable terminology to ensure the audience, the verbally limited bus mechanic, understands it? As Grove says, we "speak the language of technology and that of the lay person," so we "can translate the concerns of both sides" to ensure that our communications reflect no biases that impede understanding.[10]

Let's look at another example. You are the editor on the technical volume of a competitive proposal for construction of a low-cost apartment development in a former Soviet-block country. As you read through the building specifications, it occurs to you that all of the paint used in the apartments will be lead-based, lowering construction costs and increasing profit. A chat with someone in the legal department reassures you that the use of lead-based paint is legal in the target country; your employer is obeying the letter of the law. But you're bothered by the knowledge that lead-based paints are potentially dangerous, particularly when used in an apartment complex that is likely to be home to children. What should you do? Reread Chapter 2!

The point is that the use of lead-based paint is outlawed in the United States for the safety of the consumer. Is it then ethical for U.S. companies to use the same dangerous product in other countries? Doesn't doing so reflect an ethnocentric American bias against citizens of less technologically advanced countries—as if somehow their health and safety are less important than our own? (Consider the cultural issues we raised in Chapter 10.) Shouldn't protection of public health extend equally to all ethnic and demographic areas? Are we not obligated to raise such issues, to ensure that they do not remain hidden in a mountain of specifications, like the safety report in *Class Action*?

Think Critically

Another way in which we can serve the user is by thoughtfully evaluating and critically analyzing information in our companies. As Grove explains, "Our advantage is that we are accustomed to sorting through issues, separating rhetorical wheat from hyperbolic chaff, and making information clear."[11] Because technical communicators often work in advanced, state-of-the-art industries, that hyperbolic chaff is sometimes expressed as a dependence on technology rather than people. Paul Dombroski of Ohio University cites the *Challenger* explosion as a disastrous failure of the belief that technology can substitute for personal responsibility:

> *. . . we cannot legitimately expect that increasing amounts of objective technical information will always resolve ethical questions. The disaster could not have been prevented by collecting more technical information, because technical information does not determine fully either its own meaning or its own ethicality.*[12]

In other words, no amount of additional information could have saved the Shuttle astronauts unless someone accepted the ethical responsibility to set aside cost and political issues and scrub the launch. Unfortunately, no one did.

As the communications link between experts and lay audiences, between scientists or engineers and users, we must ensure that reliance on technology doesn't create a black hole in which the needs of our users get lost. Grove explains that technologists "shield themselves by ignoring the human, social dimension and take refuge in facts and procedures."[13] For example, in dealing with the spread of the virulent hunta virus in the southwest United States, consideration of the Native American culture was a significant determinant in the success or failure of any medical procedures designed to eradicate the disease. Many of the behaviors that played a role in spreading the virus were linked to religious rituals and beliefs. Simply telling people not to indulge in those behaviors would have been ludicrous, since these actions were part of daily religious practices that the Native Americans feel they could not in good conscience abandon for any reason. Similarly, the use of condoms to slow the spread of AIDS cannot be effective in a culture in which religious beliefs prohibit the use of any method of birth control. It is up to us to critically analyze all of our communications to ensure that *all* issues, both technical and human, are addressed.

As an employee of a company or a client, we are expected to present the best possible image for the employer who is signing our paychecks. To do so without question, however, may be to shirk our social responsibility. When our employers give us information to disseminate, we must assess more than its communicative impact. We must assess its social impact, as well.

Is the product we're documenting in any way unsafe or unhealthy? Do the raw materials used to manufacture it unnecessarily deplete natural resources? Is the process used to manufacture the product in any way wasteful or harmful? Does that process result in any injurious byproducts? Can the product be disposed of safely when its useful life is completed? Can it be recycled? Is it biodegradable?

If we see the potential for danger to the public welfare, we must take the initiative to question and act. We rarely have the authority to change company policy in these areas; however, we *do* have the obligation to raise the issue for open corporate discussion.

Use Language Carefully

Having evaluated and analyzed the information we are to disseminate, we can further exercise our ethical imperative by using socially responsible language. Doing so takes many forms.

We can, for instance, take care to be culturally sensitive in our choice of words. Using nonsexist words, sending a multicultural image, and dealing empathetically with intercultural communications are all steps in the right ethical direction. (For a thorough discussion of our duty to use language that expresses respect for increasingly diverse audiences, refer to Chapter 10.)

Also, we can ensure that our communications present our messages clearly. Where others camouflage meaning, we should elucidate it. Recall Groves' admonition to separate the linguistic wheat and chaff when the next document you're assigned to edit is rife with *doublespeak*. Are shoppers well served if we allow a food distributor to market "distressed produce" instead of "bruised fruit"? (Yes, we've really seen that one—and they get even *more* absurd!) Are your readers well served, for instance, if you allow a health care pamphlet to advocate "nutritional avoidance therapy" instead of a "low-calorie diet"? Are employees well served if we print a policy manual warning against "inventory shrinkage" rather than "employee theft"?

In his definitive work, *Doublespeak*, William Lutz, English Department chair at Rutgers University, brings George Orwell's term up-to-date:

> *Doublespeak is language that pretends to communicate but really doesn't. It . . . makes the bad seem good, the negative appear positive, the unpleasant appear attractive or at least tolerable. Doublespeak . . . avoids or shifts responsibility . . . conceals or prevents thought; rather than extending thought, doublespeak limits it.*[14]

Lutz identifies at least four different forms of doublespeak:

1. Euphemism
2. Jargon ("specialized language" used within a trade or profession)
3. Gobbledygook or bureaucratese ("piling on words")
4. Inflated language ("used to make the ordinary seem extraordinary")[15]

What prompts such unclear use of language? Many things. An attempt at political correctness may cause a bureaucrat to call discrimination "disparate negative import." Consideration of a sensitive ego may prompt a woman to send her friend to a store that sells clothes for "full-figured" rather than overweight women. Shrewd marketing strategy may lead a high-end construction company to build "customized culinary centers" rather than common kitchens. An effort to improve public opinion—or delusions of grandeur—may entice a car dealer to advertise "preowned vehicles" or "experienced automobiles" rather than the much-maligned "used cars." And we're all familiar with situations in which simple lack of education, plain unawareness, or blatant arrogance are at the root of such flawed usage.

Silly as these examples are, they are still fairly innocuous. Yet allowing them to appear in print or on the CRT will not serve the readers' best interests. Use of convoluted language, for example, can be costly. A 1994 *Denver Post* article carried a report about the savings realized when a federal agency rewrote the rules for use of citizen's band radios into plain English.[16] The number of questions from the public dropped so dramatically that five full-time employees who answered such calls could be reassigned to more productive tasks. Clearly, the rewrite saved users of CB radios as much confusion, frustration, and time as it saved taxpayers money.

Unfortunately, there are more dire motives for using doublespeak, ones that have more serious consequences.

We Must Protect the User

The era of the Vietnam "conflict" introduced baby boomers to the ugly face of doublespeak. Facing rising opposition to the war effort, the Department of Defense, and even the Oval Office, sought to assuage public opinion via the deliberate misuse of language. Although they couldn't prevent the media from reporting news of the war, officials *could* attempt to manipulate the complexion of those reports by using words that deliberately obscured the truth. Whereas Americans might protest the loss of pilots in crashes, we were less likely to object to "prematurely terminated flights." After all, we could hardly protest what we didn't understand. When David Brinkley read Pentagon briefings that U.S. troops were "servicing targets" using "coercive diplomacy" we weren't as horrified as we would have been had we been told our pilots were "killing Vietnamese villagers" by using "napalm bombs." The *New York Times* lashed out at this kind of governmental doublespeak in a 1973 editorial, saying, "The Pentagon made semantic history when it coined the abominable phrase 'protective reaction strike.'"[17]

You get the point. And if you're familiar with George Orwell's *1984*, you're more than a little dismayed.

Language is, indeed, thought. Manipulate the language and you can manipulate the thought.

The *New York Times* agreed. The 1973 editorial went on to say, "deliberate misuse of the English language to disguise rather than communicate facts" had so alarmed the nation's English teachers that the National Council of Teachers of English (NCTE) had established a Committee on Public Doublespeak.[18] The *Times* reported the "enthusiasts of this timely [linguistic] reform movement even [suggested] an annual Orwellian award to the worst example of Doublespeak."[19]

In fact, the NCTE passed two pointed resolutions at its 61st annual meeting in 1971:

> ***On Dishonest and Inhumane Uses of Language***
>
> *Resolved, That the National Council of Teachers of English find means to study dishonest and inhumane uses of language and literature by advertisers, to bring offenses to public attention, and to propose classroom techniques for preparing children to cope with commercial propaganda.*
>
> ***On the Relation of Language and Public Policy***
>
> *Resolved, That the National Council of Teachers of English find means to study the relation of language to public policy, to keep track of, publicize, and combat semantic distortion by public officials, candidates for office, political commentators, and all those who transmit through the mass media.*[20]

Sad to say that despite the best efforts of NCTE and others, more than 20 years later little has changed. In Grenada, for example, soldiers participated in "predawn insertions" rather than invasions. And Desert Storm brought a whole new cavalcade of linguistic atrocities. The *Quarterly Review of Doublespeak*, edited by Lutz, in its call for nominations for the 1991 Doublespeak Awards, lashed out at the verbal offenses that proliferated during the Persian Gulf War:

> *During this "armed situation," massive bombing attacks became "efforts." Thousands of war planes didn't drop tons of bombs. . . [instead] "force packages" "visited a site." These "weapon systems" didn't drop their bombs on buildings and human beings, they . . . "degraded," "neutralized," "attired," "suppressed," "eliminated," "cleansed," "sanitized," "impacted," "decapitated," or "took out" targets; they didn't blow up planes,*

> *tanks, trucks, airfields, and the soldiers who were in them, nor did they blow up bridges, roads, factories and other buildings, and the people who happened to be there.*[21]

Nor is the deception confined to the military. Legislators "enhance revenue" rather than "raise taxes." Department of Energy (DOE) officials talk about "energetic disassembly" instead of a "nuclear meltdown." In fact, instead of telling the public that some rather serious accidents happen routinely at nuclear power plants (a frightening thought in light of Chernobyl!), DOE casually invented the mind-boggling term "normally occurring abnormal occurrences" (not nearly so frightening!).[22]

Manipulate the language and you manipulate the response to it. Manipulate the thought, and you manipulate the subsequent behavior.

So why all the fuss? Clearly, tracking, publicizing, and combating doublespeak can be a full-time task today. We agree with Dr. Lutz, though, that communicators have an ethical obligation to resist—in fact, doing so is an effort to protect the public welfare.

But we hear you protesting that you're not likely to write speeches for politicians or press releases for the Pentagon or DOE. So how does this apply to you? There are several ways.

First, rhetorical subterfuge is common outside the beltway, too. Carefully evaluate your company's and your clients' communications, and ruthlessly eliminate doublespeak. Social responsibility compels us to protect our users from linguistic dishonesty. In other words, if the factory in which you are employed is about to close its doors, laying off thousands of workers and decimating the city's economy, don't write a press release saying the company plans a "volume-related production schedule reduction" (as did General Motors).

But don't stop there. On and off the job, our unique expertise gives us both the opportunity and the responsibility to be a voice for reason and critical thought. Thus, when we see through the deception of local politicians referring to "enhanced vehicular highway revenue," we owe it to our mystified neighbors to dash off a letter to the editor of the local newspaper querying the city fathers as to whether they are contemplating raising road tolls or instituting speed traps. (Stylistic aside: We contemplated using "city mothers" here in the interest of avoiding sexism, but we found ourselves perched on an even more precarious linguistic precipice. And "city parents" . . . well, you see our dilemma. See Chapter 10.)

Ensure Accuracy and Completeness

The perils of poor communication are well documented, but who is responsible if a message is misunderstood? Clearly the burdens of accuracy and completeness fall in large measure on the communicator.

Suppose you receive camera-ready copy of an operations manual for a "quick" final review. The engineer leading the project hands it over with abject apologies for not involving you sooner, explaining that a ridiculously short schedule and limited budget forced his team to produce the document on-the-fly. Now there really isn't enough time to make significant changes, but his supervisor suggested you be allowed to "give it a look, just to be sure we haven't done anything humiliating, like a typo on the cover." As you skim the pages, you're bothered by a nagging suspicion that some safety procedures may have been downplayed or overlooked. You decide to do some checking. A well-informed colleague confirms what the lead engineer told you. This is a new project for a very demanding client with a very large checkbook. Any delays could result in loss of the contract and would certainly jeopardize future work with this client. And yes, it does seem that this procedure should mention the safety concerns, but the situations involved are not life-threatening. What now? Adding the safety information would cost additional time and money, perhaps jeopardizing the project altogether. Do you risk the success of the contract?

There are certainly circumstances in which we can afford to trade off content for cost and schedule. Sacrificing a full-color cover for a less expensive

black-and-white version is an acceptable tradeoff. Sacrificing information that can prevent injury to meet a schedule is not—not if we are to accept our social responsibility.

Recall our advice in Chapter 7 to "get questions answered." If answers aren't forthcoming, leave out any questionable material as long as the omission does not flaw the document. If the omission *does* flaw the document, however, as in this case it does, you are ethically obligated to protect the user by delaying the schedule until the flawed, incomplete documentation is corrected.

Clearly, part of the solution to this dilemma is to change the way the company does business. Communicators can help prevent this kind of scenario and prepare more socially responsible products if we are involved in the early stages of planning and developing projects. Advocating such a process serves our users' interests as much as our own.

Analyze the Audience Thoroughly

Much of what we advocate here depends on audience analysis—the foundation of our communication products. How often have you heard that audience analysis is necessary to determine what information to include and how best to present that information? Axiomatic as this premise is, it bears repeating here in the context of social responsibility. Communicators cannot possibly preserve and protect the public good—that is, the good of the audience—if we do not understand who that public, or audience, is.

The extent to which we conduct our audience analysis will depend on the availability of resources. If time, money, and staffing are no object, we may be able to spend several months on-site in Paris observing first-hand how our new clients, a French telecommunications firm, will use our products and documentation. Unfortunately, such unlimited resources are rare—in the authors' combined 40 years of experience, neither of us has yet managed to land an assignment analyzing a Parisian audience, although our documentation has indeed been used in France! How, then, do we decide when enough is enough? How do we know that we've analyzed the audience sufficiently to protect the public from misunderstanding our message?

Again, we turn to our definition of quality communication as serving the users' unique needs. Consider editing, for example. Historically, editing meant catching every error and improving every sentence. Then some of us got smart and began advocating levels of edit—adapting the thoroughness of the edit to suit the resources available, the purpose of the document, and the needs of the users. Adjustable levels of edit are now the norm. Today, some insightful communicators are advocating a similar process for audience analysis—levels of analysis defined by the amount of information and degree of detail you gather.

Accommodate Those with Disabilities

It is imperative to keep in mind that our audience increasingly contains those whose abilities to use our communication products are in some way impaired. As the workforce grows older, for example, many users of our business-oriented communications will be less mobile, will have less acute sight and hearing, and will suffer other impediments of aging. Also, people with impaired vision, hearing, motor skills, and mental capabilities are increasingly moving into the professional mainstream. In response to these developments, the Federal Americans with Disabilities Act, which began to take effect in 1992, mandates that employers make reasonable accommodations to eliminate discrimination in hiring, training, and promoting employees with physical or mental disabilities.

Obviously, technical communicators play a significant role in upholding this law. After all, we document the software and hardware, write the policies and procedures, and develop the training that are at the very heart of the mandated accommodations. We believe our obligation is more far-reaching. We cannot be satisfied with merely complying with the law. We need to identify all ways in which we can make our communication products as accessible as possible to as many people as possible.

Nor can we limit our efforts to business-related products. Fortunately, those with special needs are also gaining greater access to the social as well as professional mainstream. As a result, communicators are obligated to assist in providing support in social as well as professional endeavors.

What do these obligations entail? First, as part of our audience analysis, we must learn how impairments of all kinds affect our users. Then, we must design communications that suit those users. We must be sufficiently familiar with a variety of media to make use of those that enable users to control how they receive information. We must be sufficiently knowledgeable to adapt page and screen design. We must understand special learning needs. In other words, we have to learn enough to ensure that the products we develop enhance rather than detract from special users' abilities to avail themselves of information.

So, let's assume you're documenting a product as innocuous as a vacuum sweeper. The information you collect about the users is likely to be limited to their ability to understand communication. If your schedule is short, your budget is minimal, or your resources are otherwise limited, you could justifiably spend less time and money analyzing the audience and allocate more resources elsewhere.

Dangerous Products Demand Meticulous Documentation

But what if you're documenting the use of a saber saw or a handgun cleaning kit? A procedure for disposing of toxic substances or turning off a dangerous piece of equipment in an emergency? What if you're communicating the risk of children swimming in a lake polluted with high levels of mercury? Or the precautions necessary to follow after a tanker truck spills volatile fuel in the streets of an ethnic neighborhood where English is not the first language? What if your documentation is likely to be used by a blind child or wheelchair-bound paraplegic? If you minimize audience analysis in these circumstances to save time or money, you're shirking your social responsibility. In these circumstances, you need to analyze the audience in greater detail—demographics, cultural data, perhaps even psychological factors.

We cannot ensure that our messages will *never* be misunderstood. We do have the social responsibility, however, to take every precaution and seek every measure to analyze the audience as fully as necessary to minimize the likelihood of misunderstanding. And the more serious the consequences of misunderstanding are likely to be, the harder we must strive to prevent it.

Many of the scenarios concerning social responsibility arise in relation to health, safety, and environmental communications, so let's look at these in greater depth.

Address Safety and Health Issues

Earlier in this chapter we discussed a scenario involving the inclusion of (or lack of) safety procedures in a manual. That is one of the obvious concerns of socially responsible technical communicators. All of our communications should incorporate the information necessary to ensure safe operation of the products we document. Are there less obvious safety concerns?

What about policies and procedures within our own companies? Do they safeguard employees? Often they are written by subject matter experts who concentrate on the best interest of the company. Security experts, for example, develop procedures and policies to protect company property. If the procedure for accessing the building after hours puts employees in harm's way by making them stand in a darkened, recessed entry while punching in a lengthy code, a socially responsible editor should identify this as a concern during the review of the written procedure. Our editorial obligations are not limited to identifying proper comma placement.

Do the training procedures followed in your company include all information necessary to safeguard employees? Do they cover all potentially hazardous situations? Do they properly balance the needs of the employees with those of the corporation? Do they accommodate employees who have special needs? As advocates for the readers, even when those readers are our colleagues, we are responsible for evaluating the accuracy and completeness of the material.

And what about health-related concerns? The dire consequences of distributing inaccurate or incomplete material are obvious if you develop communication products for a health care provider or pharmaceutical company. If you don't work for such a firm, however, are you exempt from ethical considerations involving social responsibility and public health? No such luck!

If you write operations manuals for an automobile manufacturer, you must be concerned about vehicle pollution (as you certainly *must* be as one of many who breathe the emissions). If you illustrate packaging for food products, you must be concerned about the contents of those packages. (Just how many "rat hairs" *are* acceptable in a hot dog?) If you edit instructions for a manufacturer of agricultural products, you must be concerned about the chemicals that go into those products. (Just what *is* in that ground water that filters into your well?)

Most companies have inspection procedures to ensure the healthiness of their products and quality management personnel whose job it is to certify the procedures are followed. Unfortunately, such procedures are only as good as the people who conduct them and people miss things. So we must take the initiative to point out problems when we see them. "That's not my job," is not an ethical excuse for failing to do so.

Even more important, many such inspection procedures, as is the case with those who test the contents of the hot dog you ate for lunch, inspect only for exceptions to government-mandated quality standards. "Now wait," you protest, "you surely don't expect me to single-handedly change government-mandated standards?" Well, not exactly. (Of course, we *do* expect you to be politically active—but that is another book!) Standards, even laws, *can* be changed (although not single-handedly) through persistent lobbying and lawsuits—both routes that often require a great deal of funding, always consume an extensive amount of time, and invariably entail articulate, forceful communication, which just happens to be our forte. We're not advocating that you spend your days at the statehouse or in court. But as a company employee, and therefore a company stakeholder, who is on-site every day, you are in an ideal position to lobby for socially responsible behavior within your corporation on as timely a basis as possible.

Evaluate Environmental Issues

Environmental concerns encompass many of the same issues we've already explored, but many more are specific to the discipline. For instance, governmental requirements to inform the public about certain environmental activities and to involve the public in decision-making impose added requirements on communicators who desire to be socially responsible. Decisions must be made about how much information to release and when to release it. Often companies tend to disseminate as little environmental data as possible and to hang on to it as long as possible before sharing it with the public. As a technical communicator responsible for presenting the information, you may not be in a position to make or influence these decisions; however, you *are* likely to be in a position to ensure that the concerns are raised in appropriate quarters.

Since the 1960s when Rachel Carson published *Silent Spring,* public awareness of environmental issues has mushroomed. The public expects businesses to be responsive to environmental concerns and demands laws to regulate companies that are not adequately responsive. The government has enacted laws to protect the air, water, land, ozone, and other natural resources.

The public clamors for more; industry lobbies for less. Environmentalists assert that the laws aren't stringent enough; opponents protest that they go too far at the expense of the economy. To be sure, there *are* difficult tradeoffs to be made. Any of us who flicked on a light switch to read this book forfeits the right to blindly attack electric power plants that burn fossil fuels. And if our jobs depended on a paper mill that was 95 percent pollution-free rather than 100 percent pollution free, our environmental purism might be tempered with economic pragmatism. But as communicators trained and experienced in gathering and interpreting information, we must at the very least research, evaluate, and express the positions of both factions fairly and accurately.

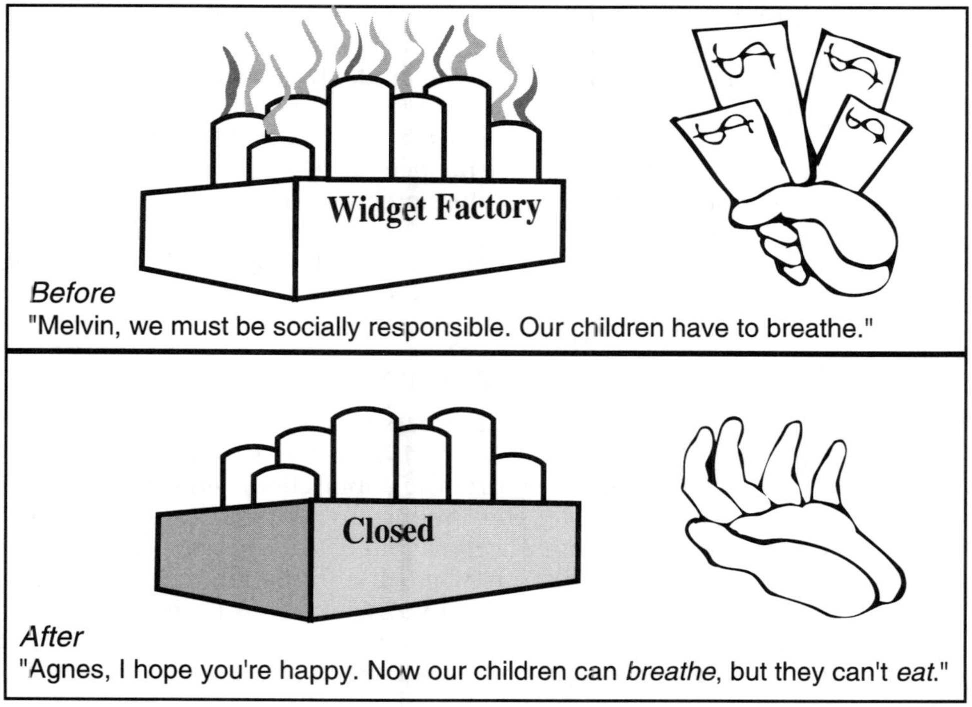

Other ethical issues surface when the public is involved in environmental risk assessment and evaluations. Many state and federal regulations require that both government agencies and private organizations solicit public input and keep the public informed about potential or existing hazards or impacts on the environment. Whether the concern is the safe disposal of nuclear waste at a former bomb factory in Colorado or the impact of building a golf course near the Florida Everglades, the public has a legal right to know and speak its peace.

Whether your employer embraces this obligation or merely abides by the minimal requirements of the law, dealing with the public in areas of environmental risk or impact poses ethical concerns for the communicator. What are your options?

You could simply pass along to the public whatever information you are given by your superiors. Or you could evaluate that information and modify it to best suit the needs of your audience. By now you know what we recommend. Communicators have the social responsibility to facilitate the timely communication of information and to tailor messages to audience needs. In fact, we would suggest you go one step further and function as a resource by which the public can communicate needs, opinions, and concerns back to your employer. As advocates for the audience, we have an obligation to ensure that public involvement is a vital part of our company's communication plan.

Summary

There are two schools of thought concerning social responsibility. One contends that it is not an ethical concern because it is outside the scope of morality. The other holds that securing and safeguarding public welfare is indeed virtuous and that we have an ethical obligation to do so. This chapter is predicated upon the authors' belief in the latter.

We cannot always depend on the law, the courts, or corporate America to fully protect us. We must rely on each other as individuals. As technical communicators, we are obligated to behave responsibly and ethically in using and disseminating information that affects the public well being.

To do so, we must first and foremost serve the user. We must ensure that our communication products are accurate, complete, and objective. To do so, we must conduct adequate audience analysis, evaluate all of the data we are given for publication, and use language that is both clear and socially responsible.

We are also obligated to protect our users. Doing so entails ruthlessly eliminating any attempts to manipulate their thoughts through the misuse of language (for example, doublespeak). It also involves zealously guarding the quality of communications related to health and safety—even at the expense of budget and schedule.

But we must not stop there. We must look beyond our own communication deliverables to the products our employers produce and to the materials and processes used in their production. Although our positions are related to writing rather than manufacturing, editing rather than designing, or illustrating rather than purchasing, it is unacceptable to overlook a potential danger to the public just because identifying it does not fall within our job descriptions. We must advocate social responsibility within our companies as well as within our publications departments.

Application Exercises

1. Explain the difference between the two prevalent definitions of social responsibility.
2. Explain the role of accuracy in serving and protecting the user.
3. Explain the role of completeness in serving and protecting the user.
4. Explain the role of objectivity in serving and protecting the user.
5. Define the term "socially responsible language." List five examples.
6. What is doublespeak? Give two examples. Why should we be wary of it?
7. Explain the importance of audience analysis in the context of social responsibility.

Exploration Questions

1. If you are employed, interview someone in your company's legal department. Identify any health, safety, or environmental regulations that govern the production of your company's products.
2. If you are not employed, visit a local company and proceed as directed in #1.
3. Watch *Class Action*, *Silkwood*, *Tucker*, or *Wall Street*. Write a brief paper (2 to 3 pages) describing the movie's stance on social responsibility.
4. Clip an article (or a few articles) from a newspaper, magazine, newsletter, or other printed material. Write a brief paper (2 to 3 pages) in which you cite instances of doublespeak and recommend more socially responsible language.

References

1. Milton Friedman, "The Social Responsibility of Business to Increase Its Profits," *Ethics, Leadership, and the Bottom Line*, eds. Charles Nelson and Robert Cavey (Croton-on-Hudson, NY: North River Press, 1991), 245.
2. Peter Drucker, "The Responsibilities of Management," *Ethics, Leadership, and the Bottom Line*, eds. Charles Nelson and Robert Cavey (Croton-on-Hudson, NY: North River Press, 1991), 241.
3. Donald P. Robin and R. Eric Reidenbach. *Business Ethics: Where Profits Meet Value Systems* (Englewood Cliffs, NJ: Prentice-Hall, 1989).
4. Walter Manley, *Critical Issues in Business Conduct: Legal, Ethical, and Social Challenges for the 1990s* (New York: Quorum Books, 1990), 16.
5. Friedman, "The Social Responsibility."
6. Elvin Hatch, "The Evaluation of Culture," *International Communication: A Reader*, 7th ed., eds. Larry A. Samovar and Richard E. Porter (Belmont, CA: Wadsworth Publishing, 1994), p. 411.
7. Dale Cloninger, "Managerial Goals and Ethical Behavior," *Financial Practice and Education* 5 no. 1 (Spring/Summer 1995): 57.
8. *Ibid.*
9. Laurel Grove, "Speaking for the Environment," *IEEE Transactions on Professional Communication* 38 no. 3. (September 1995), 166.
10. *Ibid.*
11. *Ibid.*
12. Paul Dombroski, "Can Ethics be Technologized? Lessons from Challenger, Philosophy, and Rhetoric," *IEEE Transactions on Professional Communication* 38 no. 3 (September 1995): 146.
13. Grove, "Speaking," 166.
14. William Lutz, *Doublespeak* (New York: Harper & Row, 1989.), 1-6.
15. *Ibid.*
16. "Pentagon stands tall in doublespeak," *The Denver Post*, 5 October 1994.
17. *New York Times*, 12 December 1973.
18. *Ibid.*
19. *Ibid.*
20. National Council of Teachers of English, 61st Annual Meeting, 1971.
21. William Lutz, "The First Casualty," *Quarterly Review of Doublespeak* XVII: 4 (July 1991): 1.
22. *Doublespeak*. Films for the Humanities. Princeton, NJ, 1993.

Chapter 12 Overview

Professional Growth . . .

The Mission: **As technical communicators, we aggressively pursue our own professional growth through self-development activities that keep pace with the latest advancements in our profession. Such activities include formal education, membership in professional associations, in-house training, on-the-job training, and professional networking. We seek and accept candid evaluations of our professional performance and use such feedback to define activities that will promote our professional development.**

Chapter Roadmap

Definition and Focus. This section addresses the critical importance of professional development as a means of keeping pace with the information explosion. The chapter explores several avenues through which we can pursue such development.

Formal Education Programs Provide Depth and Breadth of Curriculum. This section reviews several alternatives you can pursue to further your education, including on-campus degree programs, distance learning, and skills-oriented courses to meet specific job requirements.

Professional Associations Enhance Career Development. This section describes numerous professional growth opportunities you can gain by joining a professional association.

- *Local Chapter Meetings Explore Diverse Topics*
- *Seminars and Conferences Are a Valuable Forum for Growth*
- *Professional Literature and World Wide Web Sites Offer a Gold Mine of Information*
- *Specialized Subgroups Serve Specific Needs and Interests*

In-House Training Brings Education to the Workplace. This section presents formal educational opportunities that are often available right in the workplace, including in-service classes, seminars and workshops, videotapes, and computer-based training.

On-the-Job Training: Cornerstone of Professional Development. This section looks at various strategies to maximize cross-training and professional growth opportunities within the performance of job responsibilities. It provides specific suggestions on how to ensure that training gets priority, even in the face of conflicting job deadlines; it explains mentoring; and it suggests ways in which electronic media can be used to promote professional development.

Networking Harnesses Expertise to Solve Problems. This section illustrates the near-limitless growth opportunities available by exchanging information with professional colleagues, especially through electronic media such as e-mail, electronic bulletin boards on local area networks, and the Internet.

Candid Feedback on Performance Spurs Growth. This section discusses how you can seek and use performance feedback to accelerate your professional development. It offers some advice on how to benefit from annual performance appraisals—even when you don't agree with everything your supervisor has to say! It also addresses feedback from clients, professional competition, and peer reviews.

Summary, Practical Exercises, and References

. . . Our Duty to Provide Products and Services That Will Best Serve Our Users

CHAPTER 12

Professional Growth: Staying Ahead of the Information Avalanche

A black-diamond skier navigating ungroomed territory near the topmost ridge of a mountain range dislodges small chunks of ice. The skier whooshes on down the slope, oblivious to the miniscule impact of her deed upon the environment. The chunks of ice slither and tumble down the slope, picking up fresh wet snow, gathering momentum. Moments later, the skier hears a rumble, growing to a deafening roar. She glances back over her shoulder. Avalanche! The very word chills an alpinist's heart. Now it's no slalom—it's a race between life and death.

By one estimate, the total knowledge base of the human race doubled from 1 A.D. to 1800. That base doubled again in just three one-hundredths of the time, from 1800 to 1860. And the trend has continued to accelerate, with the period of time needed to double the sum total of human knowledge ever diminishing. By 1991, it was estimated at just 23 months! It has been predicted that by the year 2000, human knowledge will double every 35 weeks. By 2010, the period will be down to 2 weeks. By 2018, the world will double its knowledge base every 8 hours! It's mind-boggling. Just like an information avalanche, the knowledge explosion continues, spurred by technology.

—Motivational speaker Dennis Mannering[1]

Definition and Focus

The exponential growth of human knowledge, the technology explosion, the global electronic proliferation of information—all have far-reaching implications for technical communication as a profession and for technical communicators as professionals.

Sometimes it seems as if a new version of our software applications hits the market every other day. Just when we finally get comfortable with our tools, they change—they get better. And if we want to stay gainfully employed, we'd better get better right along with them!

We wind up feeling like the black-diamond skier trying to outrace the avalanche. The skier, at least, has years of training and consummate skill on her side as she flees from danger. In the face of the information avalanche, what do we, as technical communicators, have going for us as we flee from corporate downsizing? We have professional development.

In today's environment, professional development is not just a value or an aspiration, it's a necessity. Those who value and practice professional development in technical communication will have a distinct advantage over those who don't. Twenty or thirty years ago, it was sufficient for a technical writer to possess a sound basic education in science or engineering along with a command of the language. That, plus the knowledge gained on the job, could generally carry writers through their careers. Not any more. The difference is not in the *core skills*—our aptitude for technical disciplines, our ability in written and oral communication—but in the *tool skills*—our proficiency with the instruments that enable us to translate our core skills into a finished product, whether that be paper, diskette, a World Wide Web site, an interactive multimedia presentation, or CD-ROM. Technical editors or illustrators who failed to trade in their blue pencils or ink pens for a cursor five years ago are already at a severe professional disadvantage today. Tomorrow, they will be dinosaurs.

In most technical communication positions today, it's no longer sufficient to be just a technical writer or editor or just a technical illustrator. *Electronic publishing* has blurred the edges of our traditional areas of specialization. For example, both editors and technical illustrators are generally expected to possess at least basic competency with electronic page composition in addition to their specialty skills. Increasingly, those who create and storyboard *interactive multimedia products* are also expected to share in the hands-on production of those products, which requires proficiency in the software applications that translate creative concepts into electronic reality. And design of *World Wide Web sites* has, in less than a year, migrated from the arcane domain of a small cadre of "techies" into the basic tool set of many technical communicators. It's often not enough to develop the content of the Web site; the job often demands at least basic familiarity with the HTML (Hypertext Markup Language) programming that creates such sites, as well as a working knowledge of the rapidly proliferating software tools that speed the rather tedious HTML coding process.

As technical communicators, we aggressively pursue our own professional growth through self-development activities that keep pace with the latest advancements in our profession. Such activities include formal education, membership in professional associations, in-house training, on-the-job training, and professional networking. We seek and accept candid evaluations of our professional performance and use such feedback to define activities that will promote our professional development.

Before proceeding into the several arenas of professional development, let us pause briefly to consider this question: Is a technical communicator who steadfastly refuses to participate in any form of professional development behaving unethically? Not in the same sense as we would describe a technical communicator who steadfastly refuses to be honest. Honesty is a higher value than professional development, a societal core value. Dishonesty hurts others; lack of professional development hurts primarily the practitioner.

But we believe that professional growth nonetheless deserves a place in the hierarchy of values for technical communicators; without it we would soon find ourselves lacking in the basic tool skills that enable us to serve our employers, our users, society, ourselves.

In other words, if we are *not* committed to the value of professional development, we may soon find that we have seriously diminished our ability to exercise other critical values of our profession, such as our commitment to provide quality products and to demonstrate social responsibility. We may *want* to translate these values into action, but we may lack the skills to do so. And what good are ethical motives without the means of putting them into action?

Formal Education Programs Provide Depth and Breadth of Curriculum

The term "formal education" tends to conjure up images of a young couple strolling along a sun-drenched pathway next to an ivy-draped brick building with a stack of textbooks under their arms—perhaps with a pipe-puffing professor seated on a nearby bench pondering Hegel's Dialectic or Einstein's Theory of Relativity. You know, the type of picture you find on the cover of a college catalogue.

But for every idyllic case like that, there are a dozen scenarios of working parents rushing around with a fast-food burger in one hand and a textbook in the other, trying to wedge in dinner with the kids between work and an evening class at the local community college.

Fortunately, a diversified array of educational avenues has evolved to keep pace with our carousel world. So, if you are not in a position to dedicate yourself to four years or more of concentrated education in a campus setting, despair not! You can still benefit from the depth and breadth of curriculum that are built into a formal degree program.

Evening and weekend classes are available at community colleges and many four-year institutions. Increasingly, graduate programs are being tailored to the schedules of working people. In addition, distance learning programs are multiplying at a prodigious rate, reflecting the growing demand for a "flex-time" approach to formal education. And if your development need is geared toward a specialty skill, you can acquire or refine that skill without enrolling in a formal degree program. Numerous options exist, as discussed in the following sections.

Whatever course of formal education you choose, it's important to note that many employers will reimburse you for the cost of your tuition and books. Not only will you acquire skills that improve your value to your employer, but also the very fact that you are showing the initiative to develop yourself professionally sends a message that often opens doors of opportunity and accelerates career development.

Degree programs in technical communication are sprouting on college and university campuses like mushrooms after rain. The Society for Technical

Communication now lists more than 200 undergraduate degree or certificate programs in technical communication—an enormous increase over the seven programs in existence in 1970. Many universities now offer graduate programs as well. Moreover, the exponential growth trend is expected to continue. Information is at the heart of the world's technology explosion; technical communication is the art and science of improving information. It is a rapidly growing career with both breadth and depth of opportunity. That's the reason for the growing impetus to provide formal preparation at undergraduate and graduate levels.

Distance learning programs bring formal education opportunities to those whose work or family responsibilities will not permit enrollment within traditional programs. They use several different media:

- Television
- Videotapes
- Multimedia, off a server, on diskette, or on CD-ROM
- Correspondence courses
- World Wide Web sites

We know a student who also works full-time who makes Wednesday his class day. He starts with a three-hour, one-class-per-week community college course right after work. He then goes home and takes a second class on television from 9 to 10 P.M. Sure, Wednesday's a long day. But this way he can work enough overtime to stay in his boss's good graces and still get to play softball. Homework? That's what weekends are for.

World Wide Web sites offer an exciting new avenue for education. In just the few weeks that this chapter was taking shape, we found posted to an electronic technical communication list (see TECHWR-L, later in this chapter) three different references to Internet-based classes pertaining to technical communication: at Dartmouth, Columbia, and Dakota State University. Almost certainly, this just scratches the surface of what is *presently* available—let alone what will be available by the time you're reading this. For example, Netscape extends the Global Academic Network, a consortium of schools offering dozens of electronic courses under the auspices of MIT.

Professional growth through education does not necessarily imply enrollment in a formal degree program. *Skills-oriented courses* taken "a la carte" can help *meet specific job requirements*. If your professional development need is geared to a specific skill, such as a particular computer application that has become critical to your job performance and in which you lack the requisite proficiency, consider a number of educational alternatives within the community: *community college courses, evening courses at high schools and vocational educational centers*, and, last but not least, *the library*. As a taxpayer, you paid for these centers of learning—why not use them?

The cost of community college courses is usually a couple of hundred dollars; evening classes at high schools and voc-ed centers are usually just a fraction of that. Moreover, your employer may be willing to pick up the tab for skills courses even if they are not part of a formal degree program, both because they are directly related to your productivity on the job and, frankly, because the price tag is much lower—meaning the short-term return on investment is high.

Once the sole province of hard-back books, libraries—be they public or at the local high school, community college, or state college—have become multimedia learning centers complete with videos, CD-ROMs, and live Internet hookups. So even if you don't have your own computer workstation with which to access online educational opportunities, help may be no further than a mile or two away. The cost? It's practically free.

The message is simple: If you wish to pursue your education, whether it be through a formal degree program or through selected skills development courses, don't let scheduling difficulties stop you. Be creative. Shop around. Where there's a will, there's a way.

Professional Associations Enhance Career Development

Another valuable means of promoting your professional development is through membership in professional associations relevant to your field. Numerous professional associations exist in areas directly or indirectly related to technical communication. A representative list includes the following:

- ❑ The Society for Technical Communication
- ❑ International Association of Business Communicators
- ❑ Association of Teachers of Technical Writing
- ❑ Public Relations Society of America
- ❑ International Society of Technical Illustrators
- ❑ International Interactive Communication Society
- ❑ Professional Photographers Association
- ❑ Society of Broadcast Engineers
- ❑ Business Marketing Association
- ❑ International Television Association
- ❑ American Medical Writers Association
- ❑ Association of Professional Writing Consultants
- ❑ International Print Management Association
- ❑ Society of Motion Picture and Television Engineers
- ❑ IEEE Professional Communication Special Interest Group

Contacts for these and other organizations are readily available at college or community libraries—or, increasingly, through World Wide Web sites, which can be located using a Web search engine (see Electronic Networking, later in the chapter). Some of the associations (STC, for example) also have student chapters centered at community colleges, colleges, and universities.

Membership fees for professionals typically run from under a hundred to a few hundred dollars a year; student rates are generally much lower. True, the cost is more than pocket change, but over the long run the benefits to one's career are incalculable. Employers are often willing to cover part or all of the cost of memberships in a professional society if you can provide good rationale for the return on investment in terms of your versatility and productivity in the workplace.

Professional associations offer numerous opportunities for growth:

- ❑ Local chapter meetings
- ❑ Conferences and seminars
- ❑ Professional literature and World Wide Web sites
- ❑ Specialized subgroups
- ❑ Networking (discussion later in chapter)

Local Chapter Meetings Explore Diverse Topics

Local chapter meetings explore a variety of topics relevant to technical communication. STC chapters of the Society for Technical Communication, for instance, meet monthly on subjects ranging from usability testing of software documentation to interactive multimedia to professional ethics. It's generally not necessary to belong to an organization to attend a chapter meeting; usually all it takes is a small non-member's fee. So if you see a meeting notice on a subject of interest to you professionally, check it out. Rather than join several organizations, it may make sense—not to mention be more cost-effective—to pick one that comes closest to covering the full range of your professional development needs, then supplement that affiliation with a "smorgasbord" approach that gives you the best of several other worlds.

Seminars and Conferences Are a Valuable Forum for Growth

Seminars and conferences sponsored by professional organizations offer another valuable forum for professional development.

Seminars focus on specific subjects and skills. They usually range from one to three days in length and provide intense, in-depth training. They can be rather pricy—tuition ranges from under a hundred to over a thousand dollars—but they often yield the greatest tangible gains in the shortest space of time. Once again, employers will often pay for such seminars, particularly if the content relates to an immediate and specific business requirement.

Conferences, which vary from local to international in scope, are more broad-band in coverage, drawing experts to conduct technical sessions covering a wide range of subjects in a variety of formats (lectures and presentations, hands-on workshops, demonstrations, panel discussions, informal rotating discussions called progressions). Some conferences have a thematic focus; others cluster activity into several major areas of interest such as writing and editing, visual communication, online documentation, research and development, and management issues. Most conferences are scheduled to facilitate *streaming* by participants; that is, selecting the majority of sessions from one area of interest, with a smattering of sessions from other areas. That way, it's possible to gain both general exposure to the field and specific training and development in an area of personal need or interest.

A valuable byproduct of conferences is the published *Proceedings*, a compilation of the papers presented at the conference. Not only is the *Proceedings* a useful reference that enables attendees to take home more than just notes from sessions of interest, but it also allows them to cover two or more bases in the inevitable event of scheduling conflicts between sessions of interest. (Choices, choices!)

Professional technical communicators should consider the possibility of not just *attending* conferences but *presenting* at them. Before you retreat into humility ("Oh, what could *I* have to contribute; I'm no expert!"), just remember the tongue-in-cheek definition of a consultant in industry: "Anyone who is knowledgeable on the subject and comes from far enough away that we have to send planefare to get him here."

If the prospect of preparing a presentation for a national conference is daunting, you can always begin by presenting first at local chapter meetings and small in-service training sessions at work. Then, as you become more comfortable with the process, you can gradually expand your reach to regional conferences, larger in-service training sessions at work, and so on. Next thing you know, you'll be presenting to a roomful of colleagues at a national conference without batting an eyelash!

Nor are conference presenters limited to seasoned professionals. Typically, sessions at conferences are intended for a range of audiences from novice to expert, so all levels of expertise are represented and welcome in presenters as well. For example, at the 42nd Annual Conference of the Society for Technical Communication in Washington, D.C, a group of students from Metropolitan State College at Denver presented a progression on how to conduct a successful student publications competition. And at the 11th Annual Florida Technical Communication Conference in Orlando, a panel of students from Mercer University delivered an informative and insightful session on the value of networking in the job placement arena.

Not only do presenters share their knowledge with attendees at the sessions they conduct, they also bring home a wealth of information to their colleagues by attending other sessions themselves. The authors have presented at numerous conferences for the Society for Technical Communication and we can vouch for how professionally rewarding these experiences have been—not only from what we have taught others, but from what we have learned ourselves. It's a two-way street.

Professional Literature and World Wide Web Sites Offer a Gold Mine of Information

Professional associations also publish *literature* containing a wealth of useful material for self-development. Typically, such literature includes a periodic chapter newsletter addressing local issues and promoting chapter activities, a monthly national magazine with a range of features and articles, and perhaps a quarterly technical journal devoted to scholarly research (such as the IEEE's venerable *Transactions*). Such material is generally archived at university libraries.

Just as we exhorted you to consider the possibility not just to attend conferences, but to present at them, we urge you not just to *read* professional journals, but to consider *publishing* in them. Not only will you be sharing your professional expertise with others, you will find no better way to accelerate your own development—not to mention your reputation. There's nothing unethical about receiving credit, remuneration, or advancement for one's legitimate professional contributions.

And while we're at it, let us talk about writing a *book*; we can modestly lay claim to some experience in the writing and publishing arena! Seriously, the authors do feel we have, in our own way, made a contribution to the growth of the profession through this book. So, if you have the urge to publish, we urge you to *go for it*. Tape a little calendar to your computer and discipline yourself to write just one page per day. In a year, you'll have a book. Success boils down to one word: *start*.

Everyone's jumping on the juggernaut of the Internet and professional societies are no exception. Many now have sites and listings on the *World Wide Web*, making it possible to exchange information in real time electronically. (For a detailed discussion on the use of the World Wide Web for professional development, see Electronic Networking later in the chapter.)

Specialized Subgroups Serve Specific Needs and Interests

Within the membership of many professional associations are subgroups that focus on specific areas within the profession. Known variously as Special Interest Groups (SIGs), Professional Interest Committees (PICs), or other titles, such organizations within organizations are particularly valuable resources for professional development because, by the very nature of their focus, they often yield immediate gains in job performance. The Society for Technical Communication, for example, currently includes more than a dozen professional interest committees at the international level covering subjects ranging from online documentation to multimedia to marketing, supported by feeder special interest groups working the same subject areas at the chapter level.

In-House Training Brings Education to the Workplace

Most companies have an in-service training program of some kind. Large corporations often have a separate department dedicated to personnel training and development. Consider the criticality of training to manufacturing, for example. Let's say a company wins a large production contract from the government to supply a more energy-efficient ventilation system for government warehouses. To win the bid, the company had to shave its profit margin to the bone. To deliver the product on time and still turn a profit, the company will have to do everything possible to accelerate the learning curve of employees on the production line; minutes, even seconds, saved in repetitive assembly operations add up to spell the difference between profit and loss.

When it comes to technical communication, the relationship of training to cost-effectiveness is not as straight-line as it is in manufacturing. But here also, indeed in every facet of a company's operation, in-service training of employees can be critical to maintaining and improving productivity.

Several mechanisms exist for implementing formal in-house training:

- ❑ In-service classes taught by experts within the company
- ❑ Seminars and workshops conducted either by in-house experts or by external consultants
- ❑ Videotapes
- ❑ Computer-based training

One suggestion. If you take company courses, keep records. Certificates of completion are generally given out at the end of courses. There is generally no transcript, per se, as you have with college courses, but it certainly wouldn't be hard to compile one if you keep records on the courses, including date of completion. A healthy list of self-development courses, especially in tool skills such as computer applications, looks impressive on a resume—not only for the specific skills it reflects, but, even more importantly, for what it says about your initiative in going out and acquiring those skills.

On-the-Job Training: Cornerstone of Professional Development

As valuable as formal in-service training programs are, probably the most significant form of professional development an employee gets is what he or she learns right on the job—on-the-job training, or OJT for short. Who among us cannot cite a supervisor or an expert colleague who has helped us master the basics of a new job or an unfamiliar software application? The key elements of effective OJT are:

1. Clarity
2. Simplicity
3. Structure
4. Respect

Achieving *clarity* requires the expert to look at the situation through the learner's eyes—the hallmark of gifted teachers. This is analogous to the technical communicator's mandate always to consider the communication process with an eye to the audience.

Simplicity tells the expert what *not* to teach. For example, if the learner only needs to understand the basic word processing element within a software package to get the job done, it is counterproductive to burden her with irrelevant specialty functions like text-to-table conversions, Greek equations, and the like.

Structure refers to a systematic, "building blocks" approach—one that does not suddenly skip a half dozen steps and call upon the learner to run before he can walk.

Respect is the most important of all; it means *genuinely sharing* expertise in the spirit of mutual professional growth rather than parading it as a form of self-aggrandizement and intimidation. One of the marks of a true professional is the ability to provide expert guidance in a supportive, nonthreatening fashion that motivates rather than demoralizes the recipient.

Yet as important as OJT is, it is sometimes derailed by business imperatives:

> *Training and developing technical writers and illustrators is an integral part of building and maintaining an effective technical communications organization. However, this important initiative often collides with pressing business communication requirements such as proposal deadlines. Training can always wait; the deadline never can.*[2]

Training Demands Procedures and Priority

Making sure that training receives due attention in the contending priorities of the workplace is an important management responsibility. In developing new employees, it's good to have a well-defined procedure spelling out the steps in the training process, including a timetable for completion. If closely monitored by management, the procedure becomes a watchdog defending the important—but easy to postpone—training process from the ravages of pressing deadlines. Feedback to the employee on job performance should come early and often, beginning with informal sessions and working into formal product reviews—and the feedback should be honest (see the discussion of personnel appraisals in Chapter 4). Accountability for a new-hire's development should be shared equally between the employee and the supervisor.

Cross-Training Keeps Organizations Viable

On-the-job training certainly isn't limited to new employees. A robust program of personnel development including aggressive *cross-training* is critical to the health and long-term viability of any organization. A technical communications group is no exception. Yet, when confronted with a critical job requirement, management generally keeps turning to its experts to get the job done well and on time.

> *There is nothing inherently wrong with this; specialization is not, in itself, detrimental. However, we must distinguish between specialization which is professionally rewarding to the employee (as well as productive to the organization) and "treadmilling" an individual in excessive repetition of a particular assignment. Not only does such treadmilling frustrate the specialist, it also denies other editors the opportunity to develop in that area. Cross-training has value in maintaining morale as well as giving management maximum flexibility in staffing—an option which can be critical in meeting pressing requirements such as proposal deadlines.*[3]

Strategic Staffing Decisions Share the Pain and Boost the Gain

So how do we avoid "burning out" our experts without turning our novices loose on the job? And if we do the latter, how do we maintain quality? The answer is to make strategic staffing decisions that balance the fulfillment of job requirements with personnel development. One effective strategy is apprenticeship—assign someone who is just learning a particular skill along with an expert to work a job together. That way, the job gets done, the workload is shared, and the novice moves down the road to expertise.

In large organizations, this strategy can be facilitated by the use of a database program such as Microsoft Excel with skills sort capability. "By exercising basic search and extract functions, management can quickly identify . . . experts [to teach]. . . and inexperienced personnel who require training, to guide personnel development and cross-training initiatives.[4]

Self-Directed Work Teams Close Vertical Doors but Can Open Horizontal Ones . . . If We Grow with the Flow

The growing movement in the workplace away from traditional supervisor-employee relationships into self-directed work teams (see Chapter 8) has given added importance to cross-training for career development. The breakdown of the traditional hierarchical structures within corporations has chopped the traditional career ladder down into more of a step stool. There are fewer opportunities for vertical ascent into management; thus, career-conscious employees now must "grow horizontally" by broadening their range of skills. To maintain morale, which is critical to success in the team approach, it is incumbent upon management to provide more than lip service to cross-training—it must provide genuine opportunities. Strategic staffing decisions, coupled with individual initiative, are key to success in this vital area.

Cross-training implies management involvement, because it pivots on staffing decisions that may not be the individual employees' prerogative to make. There are several other methods of informal on-the-job training over which the individual professional has more control. All these methods depend on free flow of information concerning the traditional skills and ever-expanding technological tools of our craft.

Typical vehicles to facilitate informal on-the-job training include *internal user groups, informal classes, working lunches,* and *mentoring.* The first three all represent the imposition of structure, albeit informal, upon the ongoing process of information exchange that occurs naturally in the workplace. Put professionals together over coffee and doughnuts, over lunch, in an after-work seminar, and good things happen. Dedicate such sessions to a specific topic and even better things happen. The best approach is to provide just enough structure to give focus, while allowing enough latitude to encourage innovative thinking and open sharing of ideas.

Mentoring pairs experts with novices to promote growth. Linda Kaplan of Sun Microsystems, Inc., in Mountain View, California, eloquently traces the etymology of the terms "mentor" and "protege":

> *The word* mentor *has its origins thousands of years ago in Greek mythology, in the tale of Odysseus. When Odysseus was away from home for many years, he entrusted his son, Telemachus, to his friend and advisor, Mentor. While Odysseus was gone, Mentor served as a guardian, teacher, and father figure to his young protege. . . . In modern-day terms, mentors are influential people who significantly help you reach your major life goals. . . . The word "protege" is taken from the French verb,* proteger, *to protect, and is used here to denote the people who are the recipients and beneficiaries of the mentors' wisdom.*[5]

Currently enjoying a surge of interest in the corporate world, mentoring is an extremely valuable method for promoting professional growth. It applies to technical communication in three basic scenarios:

1. Within Industry: Pairing senior personnel with new hires or junior personnel
2. Within Academe: Pairing graduate students or seniors with underclass students
3. Across Industry and Academe: Pairing practicing professionals with students

The two key requirements for establishing a mentor-protege relationship are (1) to pair a skilled (senior) professional with a novice and (2) to make sure that the relationship is outside formal channels of supervision and evaluation. Cross-training among peers is not mentoring per se; mentoring implies a senior-junior relationship.

In industry, it's very important that the protege not report to the mentor within the organization; a supervisor who is responsible for an employee's personnel appraisals and career advancement cannot also fulfill the role of confidante and advisor. A mentor-protege relationship requires a special trust and candid communication; indeed, one of the subjects the protege is likely to want to discuss is the relationship with the supervisor!

In academe, the mentor-protege relationship needs to be outside the traditional professor-to-student framework. A professor certainly can be an effective mentor, but not if the student is—or will be—in his class during the period of mentorship. Issuing grades, just like giving personnel appraisals, is incompatible with a mentor-protege relationship. That is not to say a supervisor or a professor cannot also provide guidance and counseling; indeed, they can and should. However, a mentor-protege relationship is a formal alliance that encompasses years and, as such, should be independent of the "chain of command," be it corporate or academic.

Professional associations can combine the merits of mentoring and networking by publishing a *skills matrix* along with the local chapter directory to encourage exchange of ideas and give novices a list of experts whom they can consult for help in specific areas. In memberships that combine student and professional populations, ice-breaking strategies can be used at meetings to *pair professionals with students*. Something as basic as planned seating can encourage networking and spark informal mentoring that might otherwise be prevented by the natural human tendency to gravitate toward one's peer group.

An online skills database fosters electronic networking. Large organizations can facilitate this process by publishing an online skills matrix so employees will readily know where to turn for expert advice in a given area. One useful mechanism for such a matrix is an online information database such as a HyperCard stack that leads users to in-house resources and expertise in critical skills areas.

Combined with electronic networking, an interactive database of this nature is invaluable in promoting professional development. It is particularly useful in organizing informal training activities such as after-work forums, seminars, working lunches, and so on. Electronic bulletin boards on local or wide area networks (LANs and WANs) are another excellent means of exchanging professional information

Having employees train each other is a win-win situation; not only does such on-the-job training build an organization's skills base, but it also contributes to teamwork and to the morale of its employees.

Networking Harnesses Expertise to Solve Problems

Networking is an important—indeed indispensable—means of professional development. Exchange of information among professionals can occur through physical meetings such as conferences, seminars, and casual hallway exchanges; via letter or telephone; or, increasingly, via electronic communication.

E-Mail Is an Efficient Way to Resolve Technical Questions

Not even a decade on the scene, *e-mail* has become a staple of business communication, because it frees the communicators from endless bouts of "phone tag" and infuriating Voice Mail "trees" that lead the caller through a labyrinth of blind alleys and eventually back out the same way she came in.

Suppose you are working a job and you lack a specific piece of information, such as the meaning of an obscure acronym. If the job is due in print in two hours, you pretty much have to use the telephone or invest some shoe leather. But if the editing cycle extends through the rest of the week, you can simply fire off an e-mail inquiry to one or two colleagues who may have the information you need, receipt the message as a reminder that you need the information, and go back to your editing. A few hours later, tomorrow perhaps, the answer appears on your screen. You learn something, the job moves along, and not a minute has been wasted. Next time, on a different project, you can return the favor.

Internet Puts a World of Resources at Your Fingertips

The meteoric ascent of the Internet as a viable business communication tool has given rise to an already vast and rapidly expanding network of professional user groups and listings. These global electronic forums are an outstanding resource, both for professional development and for resolving specific problems on the job.

An excellent example of such a forum is an Internet listing for technical communicators called TECHWR-L. Over a thousand technical communicators, popularly known as *Techwhirlers*, subscribe and post to this rapidly growing international list, which is the brainchild of Eric J. Ray at Oklahoma State University. The beauty of this type of tool is that you can literally bring a world of resources to bear on a specific problem, in near real-time.

Suppose it's 4:30 P.M. and you run afoul of an irksome software bug. You can work into the evening vainly wrestling with the bug, you can make a bunch of phone calls but never track down an answer, or you can simply post an inquiry to the list and go home. The next morning, when you open your e-mail and check the list, you are likely to receive a half dozen responses to your inquiry. Not all will be helpful, of course, but there's a good chance at least one of them will give you exactly what you need to solve your problem. And it's all free. However, if you benefit from such a service we feel you incur an ethical obligation to help out others when their questions fall within your areas of expertise.

A particularly valuable tool on the Information Highway is the powerful *search engine* such as HotBot, Alta Vista, Savvy Search, Web Crawler, Excite, Lycos, and others which can, with staggering rapidity, sift through globally disseminated terabytes of data and zero in on the most arcane topic imaginable. We once searched for "oolite," a geological formation resembling a pearl, and within 15 seconds, got a *hit* deep within the posted proceedings of an obscure regional geology conference. Amazing! Try it.

Candid Feedback on Performance Spurs Growth

Professionals who are committed to their own growth and development routinely seek honest feedback on their performance.

We can hear the cynics, jaded by years of political skirmishes in the hallowed hallways of corporate America, rush to object: "What you mean *seek*, Kemo Sabe? It will come to you." There's truth in that objection; sometimes it does seem as if we get all the constructive—or not-so-constructive!—criticism we can handle unsolicited, without looking for more! A case in point is the annual employee review, a nearly universal business practice whose popularity often runs somewhere between that of a migraine headache and a root canal without anesthetic.

Performance Appraisals . . . To Grin and Bear

The dreaded annual performance appraisal, to which raises and promotions (and layoffs and demotions) are tied, has been with us for a long time and will be with us for a long time to come. And we would be naive if we adopted a "Pollyanna" attitude toward the process. We are keenly aware of the political realities of the workplace; we carry the scars to prove it.

But political realities notwithstanding, it is a serious mistake to retreat into a totally negative mind set regarding feedback on one's professional performance. After all, without candid feedback, how are we to know where we can improve? How are we to fix what we don't know is broken? There's nothing more beneficial to career development than an honest, constructive, specific appraisal of one's performance.

If you are fortunate enough to be in a work environment where you receive that kind of appraisal, count your blessings. But even if your appraisal is a mix of legitimate criticism and some political garbage that raises your ire, it's wise to try to separate the wheat from the chaff. That requires mental toughness, but it can be done. Dodge the political bullets, paying lip service where you have to, but don't let your irritation at the groundless or marginal criticism blind you to the valid observations.

A good approach is to list all the legitimate growth areas that have been identified and draw up a specific plan of action to develop your ability in those areas. Consider all the alternatives we have explored in this chapter, be it an evening class down at the vocational education center, a series of instructional videotapes from your company education office or the public library, or a distance learning course over the World Wide Web.

Client Feedback . . . If You Care

Feedback on one's professional performance is not limited to the annual appraisal, of course. Client feedback on our communication products is an essential part of our business; without it, we cannot be assured of providing the customer satisfaction that keeps the paychecks coming.

User satisfaction is the key to success in business. It is, therefore, a wise practice to develop a simple but inclusive client feedback form and encourage those who use our communication products to let us know where we have served their needs effectively and where we could improve our performance.

If we do not know, we cannot grow.

Not all client feedback must be captured in written form. Even a casual hallway chat can yield valuable insight into what made a project tick or what caused it to go awry. A quick e-mail exchange can serve the same purpose.

Professional Competitions . . . Let You Share

Another form of professional critique available to technical communicators is the professional competition, such as the annual events sponsored by STC and IABC. Experienced judges review and evaluate entries spanning a wide range of communication product categories (such as promotional brochures, newsletters, software manuals, and technical illustrations, to name just a few), providing valuable feedback to those who enter and significant national recognition to those who earn awards.

Following the judging for such competitions, copies of the publications submitted for evaluation can often be made available to technical communication faculty as an invaluable aid to classroom instruction. Moreover, winning entries are generally displayed at regional and national conferences. Instructors could easily contact the editor of an outstanding corporate newsletter, for example, and request a subscription for classroom use. It's hard to imagine such a request being turned down.

Peer Reviews . . . If You Dare!

Product evaluations are not limited to client satisfaction forms or professional competitions. Rare is the communication product that cannot be improved by a thorough critique by a fellow professional. The time for such a critique is *before* the product is finalized—whether that means going to press, adding the final narration to a videotape, or pressing a CD-ROM.

It's never easy to invite criticism of one's work. The authors are all too familiar with that fact, having just finished exposing our "great American novel of ethics" to the slings and arrows of our professional colleagues before the final editing cycle and committing it to press. Editors are the most dangerous people in the entire world to give your work to; they are, by their very

nature, critical. Fresh copy is to editors as fresh meat is to piranha. Editors view a professional critique—and rightfully so, by the way—as a mandate to find the blemishes, not wax eloquent over the beauty spots. We've worn that hat too, when asked to review others' work, and we've pulled no punches!

But when the pain subsides, the product improves. There's no question about it—the constructive criticism we received from peer reviews of this book has helped us improve the product for you, our users.

So set aside your ego and open yourselves up to the world of professional expertise available to you. Your reward will be not only a better product, but also another step forward on the never-ending path of professional development.

Which comes in handy the next time you hear the rumble of an avalanche.

Summary

Faced with an information explosion and ever-accelerating technology, today's technical communicators must depend upon ongoing professional growth and development to maintain the basic skills they need to do their jobs. Several different avenues exist to support this important process.

Formal education includes undergraduate and graduate degree programs in colleges and universities; distance learning via television, videotapes, correspondence, or Internet; and skills-oriented courses at community colleges, high schools, vocational education centers, and libraries.

Professional associations offer a variety of alternatives for professional development, including local chapter meetings, conferences and seminars, professional literature, World Wide Web sites, and networking.

Many employers, particularly large companies, offer formal in-house training in the form of company-sponsored courses of study, seminars and workshops, videotapes, and computer-based training programs.

On-the-job training is an important part of professional development. Orientation and training of new-hires works best if disciplined by a detailed procedure that defends training initiatives against the pressure of job deadlines.

Systematic cross-training of all employees, not just new-hires, is critical to the long-term survival of an organization, especially in today's environment of self-directed work teams. Electronic databases and skills sorts can be used to support strategic staffing decisions that balance personnel development with the fulfillment of job requirements.

Other forms of on-the-job training include internal user groups, informal training sessions such as working lunches and after-work seminars, and mentoring. A useful tool for facilitating such professional exchanges is an interactive online guide to in-house resources and expertise in specific skills areas.

Networking is a particularly valuable method for professional development. It can occur through physical meetings, by letter or telephone, or through electronic forums such as e-mail, internal company bulletin boards, and World Wide Web users' groups and listings. The latter have significant potential because they literally bring a world of expertise to one's fingertips. Search engines are particularly useful.

Candid feedback on job performance and communication products is an essential part of professional development, because it identifies specific areas where growth is needed. Mechanisms to provide such feedback include annual performance reviews and product evaluations such as client satisfaction forms, professional competitions, and peer reviews.

Where the Buck Stops . . .

While there are many incentives for professional development and diverse paths by which to achieve it, *ultimate responsibility for growth resides with each individual professional. You make your own future.*

Application Exercises

1. List and briefly describe four ways in which a technical communicator can pursue his or her professional development through formal education.
2. List and briefly describe three types of formal in-house training programs commonly found in large companies.
3. List and briefly describe three ideas by which informal on-the-job training can be encouraged.
4. Cross-training has significant benefits to an organization. Likewise, failure to cross-train carries significant penalties. Summarize each.
5. Why shouldn't a supervisor serve as a mentor to an employee or a professor serve as a mentor to a student?
6. Suppose a colleague comes to you having just received considerable criticism in a performance appraisal. Some of the criticism appears to reflect a personality conflict between your colleague and your mutual supervisor. However, other areas of criticism are directed at your colleague's lack of proficiency in a software application that is critical to career growth in your department. Your colleague has just come out of the boss's office and is incensed, knocking papers off the desk. You quickly take your colleague aside for a coffee break, out of earshot of the rest of the group (and the boss!).
   a. What is your ethical obligation to your colleague?
   b. Describe how you would approach the conversation over coffee.

Exploration Exercises

1. Identify a specific skill that is important to your professional development as a technical communicator and find two or three available resources at work, in academe, or in the community that could help you develop that skill.
2. Repeat #1 for two other skills.
3. Actually implement one activity that you discovered in #1 or #2.
4. Browse sites on the World Wide Web relevant to the professional development of technical communicators. List and briefly critique three of them. Be sure to include the uniform resource locators (URLs); they're those tags that usually begin with "http//:"!
5. Visit your campus or community library. Identify and briefly describe three resources there that could promote professional development in technical communication.

6. Obtain a course catalogue from a local university, community college, or community adult education program. Highlight courses relevant to professional development in technical communication.
7. If you're employed full time, interview whoever is responsible for your company's in-house training program. List and briefly describe three courses or other activities it sponsors that could promote your professional development.
8. If you're a full-time student, visit a local employer and proceed as directed in #6.
9. Interview the director of the in-service education program for your county's public school system. Briefly describe three activities the school system sponsors to promote the professional development of its teachers, and in each case suggest how such an approach might be applied to professional development within technical communication.
10. Design the main menu of a HyperCard™ stack summarizing the critical skills areas in your job or within your degree program.
11. Without getting shot or run off the property, obtain copies of the form used for employee appraisals at three different companies. For each form, list two or three things you like about it and two or three things you don't like about it, with a brief explanation. Be prepared to share your findings with the class or work group.

References

1. Dennis Mannering, closing keynote address at 40th Annual Conference of Society for Technical Communication, Dallas, TX, June 1993.
2. Daniel W. Voss, "Editorial Training," Martin Marietta Corporation, Orlando, FL, 1988 (now Lockheed Martin Corporation). *Proceedings* from 37th Annual International Technical Communication Conference (STC), Santa Clara, CA, May 1990, ET-58-61.
3. *Ibid.*
4. Daniel W. Voss, "Three Whiz-Bang Online Ways to Promote Professional Development in a Whiz-Bang Online World," Martin Marietta Corporation, Orlando, FL, 1994 (now Lockheed Martin Corporation). *Proceedings* from 42nd Annual Conference of the Society for Technical Communication, Washington, D.C., April 1995, 164-167.
5. Linda Kaplan, "Corporate Mentoring," *Proceedings* from 38th International Conference on Technical Communication (STC), New York, 1991, ET-185.

Chapter 13 Overview

Advancing the Profession . . .

The Mission: As technical communicators, we have an obligation not only to pursue our own professional growth, but also to assist our colleagues in their professional development and to work together to enhance the reputation of our profession.

Chapter Roadmap

Definition and Focus. This section identifies our commitment to advance the profession. To do so, we first must define the terms *profession* and *professionalism*. Then we must translate these definitions into our day-to-day performance as technical communicators, both collectively and individually.

What Is a Profession? This section presents six criteria for a profession as defined by the Society for Technical Communication. It can increase your appreciation of technical communication as a profession. It will also give you insight into the hotly debated issue of credentials and certification.

What Is Professionalism? This section discusses six forms of commitment that characterize professionalism. It can help you define and refine your own professional image.

The Image of the Profession Depends on the Performance of Its Practitioners. This section explains how, when it comes to professional image, a profession is only as good as its weakest link.

We Must Uphold Professional Standards. This section can broaden your perspective on how to deal with difficult issues on when to hold the line on quality and ethics.

Professional Relationships Demand Mutual Respect. This section provides tips on how to avoid "ego involvement" and how to deal with the thorny issue of incompetence in a colleague.

Avoiding the "Doormat" Syndrome: How to Earn Professional Respect. This section challenges you to seek out and earn professional respect—and offers specific suggestions on how to accomplish that.

Summary, Practical Exercises, and References

. . . Our Duty to Respect and Assist Our Colleagues and Enhance the Reputation of Our Profession

CHAPTER 13

Advancing the Profession and Enhancing Professionalism: A Charter We All Share

Definition and Focus

The opening cartoon depicts all-too-familiar words for technical communicators the world over. Even the popular satirical business cartoon strip *Dilbert* picked up on our oppression with the creation of a classically diffident and downtrodden technical communicator.

One is reminded of comedian Rodney Dangerfield's theme: "I can't get no respect."

Just what do technical communicators have to do to command a higher measure of professional respect? Act like a profession.

As technical communicators, we have an obligation not only to pursue our own professional growth, but also to assist our colleagues in their professional development and to work together to enhance the reputation of our profession.

Is advancing the profession truly an ethical concern?

That's a legitimate question. Is the technical writer who fails to share a new technique with her coworker behaving as unethically as the employee with his fingers in the company till?

Put in those terms, well, no. But as we pointed out in the introduction to Chapter 12, professional growth gives us the skills we need to *act* upon our ethical values. Similarly, our commitment to advancing the profession is inseparable from our commitment to quality and teamwork—both key ethical values of the profession.

If we don't work together to enhance our skills, we will not flourish as a profession. And if we are *not* willing to meet this responsibility, we are *not* exercising teamwork, nor are we serving the goal of providing the best possible quality to our users, which is our fundamental charter. So it does, indeed, behoove us to include service to our profession and to our colleagues in our ethical values—and to put that value into practice.

To do that, we have to understand what we mean by a "profession" and by "professionalism." What makes a profession? What do we mean when we say, "That's unprofessional," or "She's a real pro"? And just how do we as technical communicators, "act like a profession"? Let's take a look.

What Is a Profession?

The Society for Technical Communication conducted an extensive study to define what criteria govern whether a given field of endeavor—in this case, technical communication—qualifies as a profession and also to get a handle on what we mean when we use the term "professionalism."

The results were published in that organization's quarterly journal in 1993, authored by Terry Skelton, manager of the STC Professionalism Committee, and Shirley Andersen-Hancock, assistant to the STC president for professional development.[1] We'll summarize the findings and see how technical communication stacks up against other professions such as medicine and law.

The STC Professionalism Committee came up with six criteria for a profession:

1. Specialized education and preparation
2. A body of knowledge acquired through research
3. A unique and indispensable public service
4. Autonomy in work practices
5. Ethical practices according to enforceable codes
6. Commitment to public service and social responsibility

Let's consider each briefly in turn.

It Demands Specialized Educational Preparation

Doctors go to medical school. Lawyers go to law school. And technical communicators—well, our situation has not historically been the same. But our evolving profession *is* rapidly gaining momentum in the area of formal educational preparation; both undergraduate and graduate degree programs in technical communication are growing at a prodigious clip (for a detailed discussion, see Chapter 12). Just as doctors and lawyers follow a prescribed

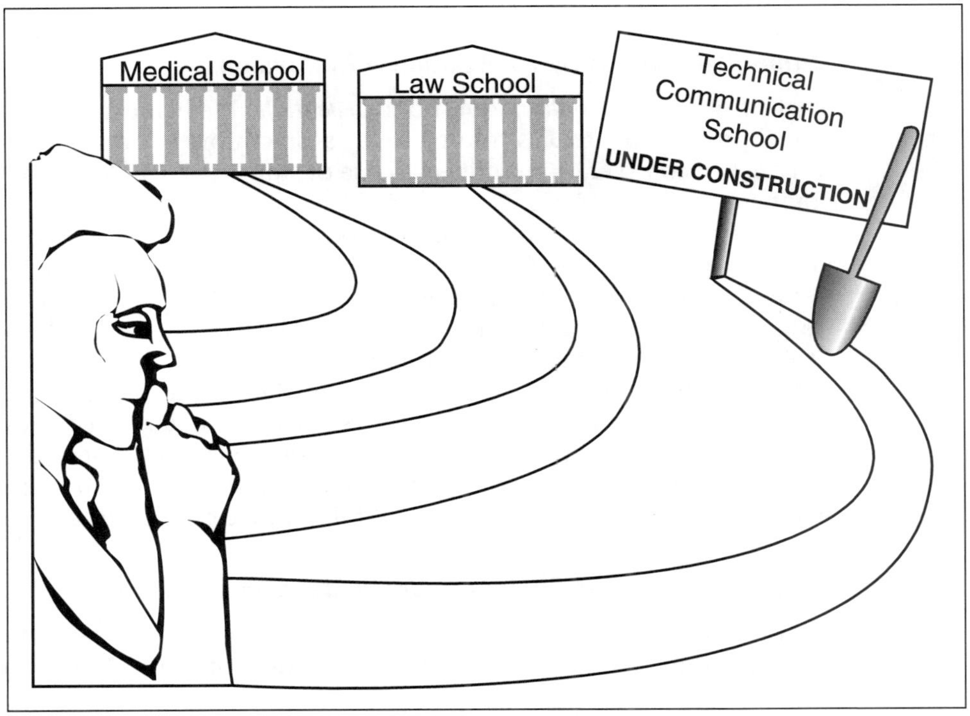

course of study, so, too, do technical communication students pursue a disciplined regimen. Instead of "Basic Human Anatomy" or "Torts and Liabilities," we enroll in "Writing and Editing" or "Research Methods"; the point is, a core curriculum exists.

The difference, of course, is that there are no licensed practitioners of law or medicine who did not attend an accredited medical school or law school. In technical communication, on the other hand, degree programs have come fairly recently upon the scene. Before 1980, there were very few. Thus, the veteran practitioners of our profession are unlikely to have a degree in technical communication; they are more likely to have pursued their educational training in engineering, English, journalism, or the like. Some don't even have a degree; they have translated core communication skills into professional proficiency through direct, on-the-job experience.

Does that mean technical communicators are less "professional" than doctors or lawyers? No. It's important to distinguish between the terms "professional" and "profession." Professionalism grows from individual performance and personal ethics; it can be applied to any field of endeavor. A profession is defined by specified criteria. Using the criteria set forth by STC, a major professional association within the field of technical communication, one would certainly have to acknowledge that both medicine and law, as well as a number of other professions, have more rigorous academic standards governing entrance into the profession.

Credentials: Gateway to Practice a Profession

That brings us to a related question: *credentials*. Medicine and law are both *licensed* professions; that is, to qualify as a practitioner, a doctor or lawyer not only has to complete a prescribed course of study, but also must pass a rigorous competency exam (the dreaded medical boards and the bar exam). *Licensing* is the strongest form of credentialing; a notch below in rigor we find *accreditation* (a term usually reserved for institutions) and *certification* (for individuals).

Licensing is a matter of law; an unlicensed pharmacist who dispenses drugs can—and probably will—go to jail. Licensing is generally reserved for professions whose practice is intimately connected to the public good, where personal safety and well-being could directly be affected by incompetency. Certification is commonly used in fields where a broad base of knowledge and competency is required for successful performance, but where there is a less direct link to the immediate public welfare. Thus, a financial planning counselor or a computer professional would likely be certified rather than licensed.

Certification: A Hot Subject in Technical Communication

Right now, technical communication has no formal procedure for certification, let alone licensing. The diversity of our profession, the absence of one single association such as the American Medical Association or the American Bar Association, and mixed and sometimes heated feelings about the subject have combined to keep certification from being universally accepted and implemented. STC conducted an in-depth survey of its membership on the subject in 1987 and came up with the conclusion that there was insufficient interest within the membership to reach the break-even point on the cost of developing a valid test instrument (a costly, specialized process that requires expert assistance).[2]

Further, there was evidence of some resistance, particularly from senior practitioners who had already earned their stripes in the school of hard knocks. Now, STC is revisiting the certification issue, aware that the lack of a formal credentialing process could, compared to tightly regimented fields such as medicine and law, lessen the image of our profession. To achieve certification, however, technical communication will have to vault a number of significant hurdles:

- ❑ Reaching consensus among several professional associations on what constitutes the unique knowledge base and distinguishing competencies of our profession
- ❑ Accommodating both the wide range of specialties within the profession and the broad spectrum of disciplines to which it applies
- ❑ Developing and endorsing one standardized testing instrument
- ❑ Gaining recognition and acceptance of certification from employers and clients
- ❑ Avoiding legal pitfalls such as antitrust or discrimination suits by candidates for certification who do not pass (or do not take) the test and third-party tort liability suits by clients aggrieved by poor or unethical performance by a certified practitioner
- ❑ Solving the "grandfather" issue of senior practitioners who might have difficulty with a standardized test, particularly if there's a high focus on computer tool skills—which there obviously should be
- ❑ Keeping the test instrument current and valid in an area of rapidly accelerating technological growth
- ❑ Covering the cost of developing, administering, and maintaining the test[3]

You can readily see why the issue is so controversial!

It Rests upon a Body of Knowledge Acquired through Research

The body of knowledge underpinning technical communication crosses multiple disciplines:

- ❑ Composition, reading, and rhetorical theory
- ❑ Principles and conventions of graphic and instructional design
- ❑ Management theory
- ❑ Systems theory
- ❑ Cognitive psychology[4]

That's naming just a few. While technical communication does require *depth* in certain key areas, such as writing, editing, and rhetorical analysis, it is unique in its demand for *breadth* because it is applied to so many different disciplines. In this respect, technical communication is very different from medicine or law, where depth is critical. Not only is the coursework leading up to a basic M.D. or law degree focused within the discipline, but in both professions, there is also a growing trend toward specialization, requiring even greater depth within a certain area.

How about research? As technical communicators in industry are well aware, the company that does not allocate a portion of its resources to stay on the cusp of emerging technologies rapidly loses its competitive edge and ultimately drives itself out of business.

The same holds true for technical communication as a profession. Research means new ideas, new methods; these, in turn, mean better performance, more value added. In today's fiercely competitive marketplace, those who stand still are left behind. If we cannot offer our clients more than they can purchase for themselves in a sophisticated software package, why should they seek—and pay for—our services?

In medicine, research is the foundation upon which so many hopes are pinned—who among us isn't hoping for a dramatic breakthrough that will tame forever the scourge of cancer, drive AIDS to its knees, or lift a palsied child to her feet? The body of knowledge from medical research is vast and growing. Importantly, technical communication is very much a part of that process. Electronic archiving and real-time sharing of medical data offer great hope for the future, and organizations such as the American Medical Writers Association are in the thick of the fray.

In law, the very briefs and case histories that govern the outcome of trials become the collected body of research for the profession, providing the vital foundation of precedence that is the cornerstone of American jurisprudence.

Are we vigorously pursuing "pure research" directly within the field of technical communication (as opposed to our frequent role in shaping the content of research within *other* disciplines)? The answer is a resounding "Yes." One need only peruse the IEEE's *Transactions*, STC's *Technical Communication*, or ATTW's *Technical Communication Quarterly*, among many others, to see that our profession is definitely building an ever-wider base of theory and research on subjects ranging from iconography to usability.

A number of professional organizations encourage such research through grants and stipends. Skelton and Andersen cite nine specific instances of such research, much of it in this case sponsored by STC grants, ranging from a classification scheme for levels of edit to a model for developing and implementing graphic user interfaces.[5]

On the second criterion, then, technical communication stands on firm ground.

It Provides a Unique and Indispensable Public Service

That could be trickier. No reasonable person would accord a technical communicator the same degree of immediate indispensability as a paramedic applying a tourniquet at an accident scene.

But that doesn't mean our profession does not have a significant impact upon—and therefore a commensurate ethical obligation to serve—the public good (see Chapter 11). Occasionally, proficiency in technical communication can extend to a life-and-death situation. The vital importance of clear, unambiguous instructions for proper handling of radioactive waste would be a good example. And while most comma splices don't lead to meltdown, there are other ways in which clarity can relate to personal safety.

As an example, a technical communicator at a recent conference told of an instance in a non-English-speaking nation where poorly translated instructions for self-injection with insulin led directly to a patient death. One cannot help but wonder whether the relatively small market and the relatively high cost for quality translations were factors in the inadequate documentation. And, if that indeed were the case, wouldn't the person who commissioned the inferior translation be just as morally responsible for the death of that patient as a surgeon who came into the O.R. under the influence of alcohol and lost a patient?

But even in the vast majority of cases where technical communication is not directly linked to a life-and-death situation, it is unquestionably linked to the public good. Consider poorly documented software, for example. That might not *kill* anyone—though we have caught wind of a few litigations involving justifiable homicide on this issue—but it certainly leads to user distress, diminished productivity, and possibly even interruption of important public support functions.

The STC study discerned three indicators of professional status:

1. Public recognition of work
2. Elevation of status
3. Remuneration[6]

While technical communicators have not generally attained the status of doctors and lawyers in terms of public recognition, we are definitely coming up the curve in this arena. Much of our scientific writing and editing is ghost-writing for authors who get the byline, but at least those whose names appear on the articles definitely know, generally appreciate, and often depend upon our expertise. And our researchers are beginning to amass a formidable body of accredited original research in their own right.

As to remuneration, studies by various professional societies have indicated that while technical communicators are not standing toe to toe with physicians and attorneys in this regard, we're generally holding our own within the professional business and technical workforce.[7]

Perhaps most importantly, any parent who has ever struggled on Christmas Eve with a heap of bicycle parts and an instruction manual that bears the dreaded caveat "Some Assembly Required" will voice a silent hymn of praise to the proficient, albeit invisible, technical writer!

It Demonstrates Autonomy in Work Practices

"Autonomy in work practices suggests that professionals in a field have gained such respect as to be allowed by society to influence the technical content of their work as well as the scope and conditions of the service they perform."[8] It implies a degree of technical specialization beyond the reach of experienced laity, as well as the existence of specific criteria governing admission to and advancement within the profession.

When it comes to autonomy, technical communication clearly lags other professions, such as medicine and law. The specialization is there, but the entrance criteria are wobbly at best (see previous discussion of certification). A significant subset of our profession, independent contractors, do have more day-to-day autonomy than technical communicators employed by corporations, but in both cases our actions are largely client-driven. Historically, our voice within corporations has been rather meager; however, as Skelton and Andersen observe in the STC Professionalism Committee's report, "the advent of interdisciplinary work teams and the 'total quality' ethic in business suggest greater opportunity for participative decision-making for all professionals, including technical communicators. So we should experience increasing autonomy."[9]

It Follows Ethical Professional Practices according to Enforceable Codes

Technical communication has no one code of ethics, because there is no one body governing its practitioners. Exploration Exercise #3 in Chapter 3 tells you where to find the ethical standards of four associations related to our profession: STC, the American Medical Writers' Association (AMWA), the International Association of Business Communicators (IABC), and the Association of Professional Writing Consultants (APWC).

All of these sets of standards are cogent expressions of the values espoused by their respective organizations, and they provide useful guidelines that members can apply in attempting to resolve ethical conflicts within the workplace. In no case, however, can the standards really be called a "code," certainly not an "enforceable code."

Enforcement implies some form of sanction for a breach of ethics. A doctor who commits an ethical breach can be suspended from practicing medicine; a lawyer can be disbarred. That's because the American Medical Association and the American Bar Association wield enormous professional clout with their memberships—and, significantly, they are the *only* recognized organizations for their profession. This is not so with technical communication. A technical communicator who commits an ethical breach faces no meaningful sanction unless his offense happens to break the law. Sure, he may be censured by STC or IABC or AMWA—*if* he even belongs to one of those organizations—but he is still free to continue practicing the profession as if the breach had never happened.

Clearly, then, as a profession we are three steps removed from having a strong, uniform, enforceable code. First, we would have to establish certification for the profession, with agreement among all cognizant professional associations as to what constitutes that certification. Second, we would have to generate a uniform code of ethics that became part of the certification process. Third, we would have to enforce the code by suspending or revoking the certification of any practitioner who commits a breach of ethics. That would require both time and cooperation.

However, the absence of one uniform, enforceable code of ethics certainly does *not* relieve technical communicators of our obligation to be personally committed to ethical professional practices in discharging our responsibilities to our employers, our clients, and the public. In fact, the more aggressively our organizations and our practitioners pursue this imperative, the greater professional esteem we will enjoy.

It Reflects Commitment to Values of Public Service and Social Responsibility

Technical communicators definitely do have a commitment to the public good and a charter to demonstrate social responsibility. The STC Professionalism Committee report sees at least three specific avenues through which our public service and social responsibility are served:

1. Using socially responsible language
2. Facilitating the timely communication of information
3. Putting the public good above special interest (that is, serving as public advocates)[10]

We are committed to clarity, accuracy, and usability in the communication products we produce. When we meet that commitment, we improve communication, boost productivity, and serve both the economy and the user. The public at large is fed up with government and corporate *bureaucratese* and linguistic *gobbledygook*—witness the drive for plain English laws for both government and commercial transactions. Where others obfuscate, we should enlighten. Where others confuse, we should clarify. Where others serve a vested interest, we should serve the public good. The first two are fairly easy to serve; the latter can be more difficult in that we often serve both the corporate bottom line and the needs of the user. When those two objectives dovetail, our course is clear. When they collide, we may find ourselves in a position of ethical conflict.

The question of how our commitment to the public good relates to our obligations to our employers and clients is explored in Chapter 11. Ultimately, we should serve the user before both our employers and ourselves:

> *Regardless of the special interests of companies and agencies that employ technical communicators for corporate gain, or technical communicators' own motivations, our work ultimately embodies the social responsibility of helping science and technology meet the needs of individual users and communities of users.*[11]

What Is Professionalism?

Professionalism is an individual thing. The same STC study that identified the six criteria governing a profession came up with six additional criteria that individual practitioners can apply to assess their own level of professionalism. It calls for commitment to:

1. The profession
2. A professional calling
3. Organizing the profession
4. Education
5. The service ethic
6. Achieving professional autonomy[12]

It Is Commitment to the Profession

Professionals work for more than a paycheck. Commitment to a profession means a lifetime of preparation, practice, and continual professional development (see Chapter 12). A professional is constantly seeking to advance the state of the art, learn new techniques, find new solutions. A professional goes the extra yard—not for the promotion, not for the money (although desiring these things is not in itself unprofessional), but primarily because the client needs it or simply because the job calls for it.

In medicine, commitment to the profession means the doctor who takes the extra few minutes with patients while making her rounds on Christmas, while her family waits at home to open their presents.

In teaching, commitment to the profession means the professor who takes the extra few minutes with the struggling student after class, cutting into time needed to conduct research for his book. It also means the adjunct evening instructor whose car sits alone in the parking lot at 11:00 o'clock at night, for the same reason.

In technical communication, commitment to the profession means the editor who doesn't just hack away at the work but takes the extra time to show interest in an author's philosophy and aspirations, developing a relationship of trust to cushion the impact of the rewrite the editor knows the work needs. Commitment to the profession means the software documentation specialist who does the extra rewrite that gives the last full measure of clarity to a complicated procedure. Commitment to the profession means the technical illustrator who goes back and makes all the line weights consistent, even though she knows the client will never notice the difference.

Commitment to the profession means doing the job right because it's there to do, not because somebody is making you do it.

Commitment to the profession is pride.

It Is Commitment to a Professional Calling

Skelton and Andersen cite several other authors and derive a consensus: Our commitment to advance the calling of technical communication is measured in our efforts to systematically advance the field and to standardize the practice of our profession to serve the public good. This goes beyond the commitment to the individual job and the individual client previously discussed.

In law, commitment to the professional calling is measured in the efforts of attorneys to exploit the new electronic media to improve archiving and increase the accessibility of the vast amounts of past legal documentation that underpins current cases. It goes beyond the current trial.

In dentistry, commitment to the professional calling means giving priority to a never-ending imperative to exploit new technology to improve treatment and lessen patient discomfort. It goes beyond the current root canal.

In technical communication, commitment to the professional calling means setting up an electronic bulletin board for sharing state-of-the-art techniques with ever-improving hardware and software to ensure that not only we ourselves but also our fellow communicators are armed with the latest tools to serve their users. It goes beyond the current document.

Commitment to a professional calling is vision.

It Is Commitment to Organizing the Profession

Obviously, a key part of organizing the profession is active membership in professional organizations. As technical communicators, we can advance our collective interests by joining publicly recognized associations such as those listed in Chapter 12.

Through these organizations, we can work with colleges and universities to define minimum entry-level standards for technical communicators. We can work with employers on how to continually improve the performance standards of practicing professionals. We can encourage research and development in our field. We can sponsor and support training initiatives.

As a profession, we also must confront the issue of credentialing previously discussed. We do not currently certify nor regulate our practitioners in any manner. And while certification carries with it a host of issues previously described, the lack of it reflects something less than solidarity in the organization of our profession.

Commitment to organizing the profession is reaching a standard of professionalism.

It Is Commitment to Education

Our professionalism as technical communicators carries with it both the realization that our education is a never-ending process and the commitment to act on that realization. Chapter 12 describes the many avenues open to us for ongoing education, both in formal degree programs and through external and in-house training.

A significant number of technical communicators are employed as full-time faculty in the growing number of undergraduate and graduate degree programs in technical communication. These individuals will play a vital role in the growth and advancement of technical communication as a profession in the years ahead. Their ability to mold graduates who are well-versed in the basics of writing, editing, and rhetoric and proficient in the increasingly critical computer tool skills will have an enormous impact on the status of our profession. If they succeed, technical communication will gain respect. If they do not, efforts to add luster to the discipline will be undermined by the incompetence of its new practitioners.

We are pleased to report—admittedly with a tinge of bias!—that the early returns point toward success.

The responsibility to teach extends beyond full-time faculty, however. A significant number of professional communicators are also adjunct instructors for community colleges and universities—sometimes within the discipline of technical communication, sometimes in other communication areas such as English literature or public speaking. In all cases, their performance not only helps train the next generation of technical and business communicators, it also reflects on the present corps of technical communicators.

Technical communicators also teach colleagues through formal company in-house training programs as well as in various informal forums (see Chapter 12).

Our commitment to education should not be limited to our *own* education, nor that of our own students, or our professional colleagues. Recognizing the value of education to the community at large and, ultimately, to our society, professionals give of their time, talent, and treasures to support the educational institutions which not only have provided their own training but which are charged with the education of the next generation.

Commitment to education means the physician who purchases an ad in the high school yearbook—not because the ad is going to gain him any patients, but because the community he lives in is served by the co-curricular high school programs that help keep students off the streets. Commitment to education means the law firm that sets up a matching-funds program to encourage its attorneys to support their law schools and undergraduate colleges through alumni contributions.

Technical communicators are uniquely qualified to serve the educational community because our core skills in communication happen to be the most important thing that the schools are trying—often against all reasonable odds—to teach.

Commitment to education is the Education Committee of the Society for Technical Communication in Orlando, Florida, organizing a team of professional writers and editors to volunteer their time to work with high school instructors and help them come up with specific, practical suggestions on how to use the principles of technical communication to improve the teaching of writing across all disciplines. (In the spirit of open disclosure, one of the authors is involved with this committee. The case is cited not as a "paid political announcement"; rather, it is a legitimate example of an innovative form of community service that exploits the unique talents of our profession.)

Thus, when a high school English instructor in Eatonville, Florida, floats a simulated Request for Proposal to her exceptional education students for the design, construction, and operation of a music store specializing in rock and rap music, she will be doing so with the assistance of professionals from Lockheed Martin who deal with real RFPs for real electro-optical weapon systems. The weapon systems defend our nation from without. What the teacher's doing defends it from within.

Commitment to education is caring.

It Is Commitment to the Service Ethic

The true measure of professionalism for technical communicators, both individually and for our associations, is that we serve not just the interests of technical communicators but the interests of the public at large. The service ethic applies both to the performance of our professional duties for which we are paid and to volunteer activities of the type previously described.

In applying the service ethic to our professional activities, the STC Professionalism Committee's research found two areas of focus:

1. Ethical behavior
2. Quality performance[13]

On the former, we are obligated to protect our clients and the public at large from dishonesty, breach of contract, conflict of interest, or other forms of misconduct by an unethical practitioner. This brings us back to the need for an ethical code and a means of regulating the behavior of technical communicators to conform to the dictates of such a code (as in certification).

On the latter, we are obligated to protect our clients and the public at large from shoddy performance and poor communication products that hurt the user. Skelton and Andersen cite Juran and Gryna, who said that products or services provided by the professional technical communicator must meet "the needs of those members of society who will actually use them."[14]

This is not a particularly deep concept, when you get right down to it, but it is a very, very important one. For if we fail to serve those who use our products, we have not only fallen short in professionalism, we have failed in our most fundamental professional charter.

Commitment to the service ethic is performance.

It Is Commitment to Achieving Professional Autonomy

As previously discussed, technical communication is lagging other professions when it comes to professional autonomy. Professionalism therefore demands we demonstrate our commitment "to advancing the profession's autonomy of our work practices by striving for excellence . . . according to the following measures:

- ❑ Level of technical specialization beyond the reach of laity
- ❑ Level of commitment to the public good
- ❑ Level of monitoring necessary for controlling admission to the profession
- ❑ Level of quality in services rendered that is clearly superior to that of experienced laity"[15]

The goal, then, is to advance the status of our profession to the point that a client can contract the services of a *certified* technical communicator and know with reasonable certainty that this action brings with it a defined standard of performance, a quality product, adherence to a uniform code of ethics, and a commitment to the client's best interest.

We assume that the license on the wall in our surgeon's office means we can entrust our lives to his scalpel. We assume that the license on the wall in our psychologist's office means we can share personal information without fear she will breach our confidence.

Professionalism demands that those who hire us as technical communicators can, with a commensurate degree of confidence, assume that we will add value to their communication products, not subtract value from them or merely process them—then charge them for the privilege!

As a profession, we still have a way to go—but we *have* clearly charted our course.

Commitment to achieving professional autonomy is earning public trust.

The Image of the Profession Depends on the Performance of Its Practitioners

As we seek to enhance and advance our profession, there is no escaping one fundamental and sobering reality: *we are no better than our weakest link.*

Technical communicators add value to information. We are unique in that respect, because scarcely a field of endeavor exists where we cannot make a contribution. Whether it is the maintenance manual for a sonar tracking device or the luncheon menu for a "mom and pop" restaurant, there's always a communication product involved. The very breadth of our profession carries many opportunities to add value. The flip side is that every opportunity to impress is also an opportunity to disappoint. Thus, a very small number of incompetent technical communicators can potentially inflict a disproportionate amount of harm upon the image of our profession.

Just as the image of every dedicated, competent lawyer is tarnished by the few who place personal gain above client interest; just as the image of every dedicated, competent teacher suffers from the tenured few who are just marking time until retirement; so, too, is the reputation of every dedicated, competent technical communicator compromised by the few who edit perfectly agreeable subjects and verbs into disagreement.

We have utterly no reason to believe that the standard of professionalism among individual technical communicators is one jot lower than that in evidence among individual physicians, attorneys, teachers, or members of any other profession. In all professions, we believe integrity is the standard and unethical behavior is the exception.

The difference is that as a profession technical communication has no sanctions in place to censure the few practitioners who behave unethically. This is bad for two reasons: (1) Unless they break the law or get fired (and sometimes even then), they can continue their inappropriate practices; and (2) they can do so with relative impunity, at least within the profession. In the public eye, the second problem is even worse than the first. Sure, if news breaks of a physician being caught up in a kickback scheme, heads may nod and eyes may roll, but with the disapproval comes a sense that the violator will be punished, that the profession will purge its own ranks of the unethical—that justice will be served. Where the public good is concerned, it is axiomatic that those who can't keep their own house in order are inviting others to step in and do so for them.

We Must Uphold Professional Standards

Advancing the profession demands technical communicators adhere steadfastly to basic professional standards and refuse to compromise them. The two primary areas in which this issue is operative are quality and ethics.

Holding the Line on Quality

Chapter 7 explores the value of quality. A key concept is the fact that quality in technical communication is best defined as meeting (not exceeding) the user's requirement.

We are not serving our users if we insist on imposing a higher level of quality than the job requires, any more than a car dealer is serving the customer by insisting on selling her a Mercedes when what she wants—not to mention needs and can afford—is a Chevy. For example, if an attorney needs a clean, professional-looking flyer to explain her rate structure, a simple trifold design in black-and-white with one accent color would be appropriate; an expensive, slick four-color brochure would not (in fact, given the subject, it might even backfire).

On the other hand, what if the client insists on a quality level so low that the product reflects poorly on the preparer? For instance, suppose a schedule- or cost-driven client directs a technical communicator to take a totally unedited document, dump it into a desktop publishing application, and publish it, complete with grammatical atrocities, linguistic barbarisms, and garbled syntax? Does our commitment to our users compel us to put our name on something that is a professional embarrassment? The answer, emphatically, is no. Indeed, we pay a terrible price as professionals if we do place our imprimatur upon trash.

Okay—now that we have taken such an emphatic position, it would be sporting of us to explore the consequences. What, exactly, does "just say no" mean when it comes to technical communication? There are two perspectives: that of the individual technical communicator and that of the profession as a whole.

For independent contractors, the personal side of the conflict is fairly straightforward: If an independent technical communicator is asked to deliver a low-quality product, it becomes a tradeoff between cash in pocket and

possible harm to his reputation if the product should find its way into the hands of other prospective clients. The implications for the profession as a whole are not as straightforward.

Technical Ted in Twin Falls, for instance, might be willing to risk professional embarrassment to land a quick paycheck, but does he have a right in so doing to chip away at the reputation of Documentary Dot in Duluth?

For technical communicators in the employ of others, the quality requirement takes on another dimension: Not only is the reputation of the technical communicator and the profession on the line; so, too, is the image of the company. What if the company is willing to put out an inferior communication product in the interest of squeezing out the last dime of profit? As discussed in Chapter 7, such an approach to quality can yield short-term benefits but it generally carries long-term penalties; it's "penny-wise and pound-foolish."

Realistically, however, the technical communicator rarely has the final say in such decisions; if the program manager says the prime item specification is going out unedited, it will go out unedited. About all the technical communicator can do in a situation like this is to go on record, both within her own organization and within company channels, that she feels the document should be edited. Unless she is being asked to put her name on the document, which is unlikely, her personal reputation is not at stake and she is hardly in a position to defy the program manager, edit the document anyway, and bill the contract for her time! Such an action would be inimical with her further employment.

Insofar as the implications for the technical communication profession are concerned, that depends on whether the document in question could, even inadvertently, ever be misrepresented as an edited document. That, in turn, depends primarily on the audience. If the prime item specification is written by engineers and contract administrators for an audience of government engineers and contract specialists, the lack of editing is not likely to reflect poorly on technical communication as a profession. If, on the other hand, the document in question is an article for publication in a national journal, both the company and the communications profession could be embarrassed by the submittal of unedited copy.

Taking a Stand on Ethics

Professional communicators have an obligation both to themselves and to the profession to behave ethically. Until the profession adapts a uniform code of ethics, its practitioners will have to define this obligation in terms of available guidelines such as those contained in this book and—even more importantly—in terms of their own fundamental moral awareness. As we've said in numerous places throughout the book, it all boils down to a sense of what's right and what's wrong. Doing what's right is not always painless. Sometimes there's a hefty price to pay. This concept was developed in detail in the opening section of Chapter 4.

If technical communication is to be perceived by the public as an honorable profession, it behooves each individual practitioner to uphold its values. That means honesty, fairness, social responsibility—the whole nine yards. Here again, the "weakest link" concept applies. To strengthen the ethical "chain" within technical communication, we must be sure our membership understands and is committed to common values that govern the conduct of our profession.

Professional associations have a pivotal role to play in this process. Faced with ethical conflicts of the type illustrated throughout this book, individual technical communicators are often "overmatched" and outranked by other stakeholders. Suppose a technical editor is directed to omit critical test data that reflects negatively on product safety, including a veiled threat of layoff if he refuses to comply. Suppose, further, that the company in question is rather small and does not have an ethics office to whom the embattled editor can turn. A situation like this demands a strong professional voice—a national association with the vision to see the conflict, the courage to support its practitioner, and the resources to translate that support into legal action if need be.

That's asking a lot.

But to do less not only hangs the individual editor out to dry, it diminishes the entire profession. If a physician is pressured by a profit-oriented hospital administrator to deny vital medical care to a patient whose insurance has run out, you can bet the American Medical Association is going to come down on the institution with baseball cleats—as well it should.

Technical communicators need—and deserve—the same measure of support.

Professional Relationships Demand Mutual Respect

An important aspect of professionalism is professional relationships. Technical communicators interrelate with other professionals in many different roles. Each relationship is governed by a different set of circumstances and a different range of interests. But the core values of the profession should transcend the differences and apply to *all* professional relationships into which we enter.

One key ingredient in forming and maintaining professional relationships is the ability to focus on tasks rather than personality—to subjugate one's own interests to those of the team. One major challenge in preserving professionalism is learning how to deal with the thorny issue of incompetence in a fellow practitioner. The following sections address these two concerns.

Suspending "Ego Involvement"

Good writers must be confident. Facing an empty page or an empty screen can be intimidating. To overcome this feeling and generate fluent prose requires strong self-assurance: "I can do this. I'm a good writer. This is what I *do*. In fact, I do it better than most other people."

That's all necessary. It's part of the craft. The writer who sits in front of a computer without the conviction that she may one day win the Nobel Prize for Literature probably isn't going to write worth a lick.

That's all a matter of confidence—and confidence is good.

But there's a thin gray line between confidence and arrogance. And arrogance is bad. It's especially bad for professional relationships.

Two relationships that dwell on the fault line between confidence and arrogance are internal relationships between technical communicators and the classic love-hate alliance of author and editor.

The difference between confidence and arrogance can be defined by the presence or absence of *ego involvement*. Ego involvement means investing one's self-concept in the work to the point where criticism of the work becomes criticism of oneself. Ego involvement is counterproductive, because it creates tension in interpersonal relationships and blocks constructive criticism that is needed to improve communication products. Editors like to believe the affliction is the sole province of authors; authors swear by the inverse. As long-time editors lately turned authors, we have seen this from both sides of the fence. As usual, the truth lies in the middle:

> *To our editorial comrades:* Not all authors are wild-eyed egotists who foam at the mouth if the editor dares to change a single word of their deathless prose.
>
> *To our fellow authors:* Not all editors are anal-retentive comma punchers who aren't satisfied until they have wrung the last drop of life out of a creative passage.

The author-editor relationship can be very fruitful if both sides avoid ego involvement. As technical communicators, we often play both roles. First, let's put on our editing hat.

If we're editing a manual or proposal that does not carry a byline, we should edit as aggressively as necessary to achieve the smoothest possible

How editors see authors.

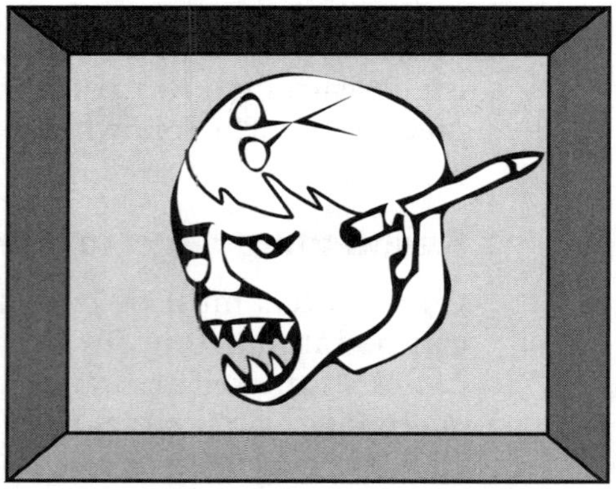

How authors see editors.

syntax, the greatest possible clarity, the nth degree of conciseness, within the constraints of budget and schedule (see Chapter 7). If, on the other hand, we're editing a bylined article, we should turn down the gain on our blue pencils (oops, sorry for the anachronism—cursors) and edit under the principle of "less is more." We still have a professional responsibility to the author to ensure mechanical correctness, to attack truly garbled syntax, even—on occasion—to spin a creative phrase that better expresses a key concept, but we should never fall into the trap of editing just to edit.

We must recognize that the author has a distinctive style; good editing means editing and polishing *within* that style, not fracturing the style and draining the life from it. The mark of a superior copy editor is the ability to perform a significant edit and have the author pronounce with a beaming smile: "He did a great job. He didn't change a thing!" Too often editors seem to fall prey to "effort justification"; that is, we feel we haven't done our jobs until the page is covered with markup, whether it needs it. Editing "six" to "half a dozen" is inexcusable; not only is it a time- and cost-waster, it also constitutes legitimate grounds for the author to take umbrage at our needless meddling.

Now, let's look at it from the author's perspective.

The hardest thing for an author to do is to "let go" of the work—to realize that once created a piece of writing is exactly that, a piece of writing—not a piece of ourselves lying out there, vulnerable, open to attack. Rarely, if ever, is any piece of writing "perfect" on the first pass. As authors, we should greet any assistance we can get from another professional with open arms. That doesn't mean we shouldn't hold our turf if the editor oversteps her bounds; it does mean that we should be receptive to a good, clean edit even if it hurts (and it does!).

The other place where ego involvement comes into play is in the relationship between two, or more, technical communicators. If you don't believe us, sit in on a professional association committee meeting when it comes time for a group of editors to draft or review a report! The important point to remember is that *there are multiple solutions to most communication challenges.*

Sometimes one solution is better than another. Professionalism means that if a colleague happens to present a better solution, we should have the grace to recognize the improvement and yield. In fact, there are times when we should yield even if the colleague's solution is no better than our own, as long as it is just as good—in the interest of professional relations and teamwork.

How to Deal with the Incompetence of a Colleague

Having waxed eloquent on the need to be flexible and tolerant in dealing with conflicts with fellow technical communicators, we now must shift gears and address the question, "What if it's not a matter of flexibility, of openness to different approaches? What if a fellow technical communicator is blatantly incompetent?"

That's not easy. Dealing professionally with the incompetence of a colleague is awkward in any profession. But deal we must. Technical communication as a profession can no more tolerate the "value subtracted" edit than medicine can accept the amputation of the wrong limb.

Okay, we're agreed in principle. But how do you commit such a value into practice? Let's say you happen to proofread a document that was "edited" by a recently hired colleague and find it laced with errors of both omission and—worse!—commission, from cover to cover. How you handle this depends on your professional relationship with the colleague. If you are the immediate superior, you must provide candid performance feedback and insist on immediate improvement. You must also institute regular product reviews and *document thoroughly* to be prepared for an unattractive possibility—that it might become necessary to fire the person. A supervisor in technical communication has this responsibility both to the company and to the profession.

What if you are on an equal professional level with the editor who made the errors? That's even harder, because you could conceivably be construed to be in a conflict-of-interest situation. In a competitive workplace and a downsizing environment, it can be hard to separate company loyalty from personal interest. You could be accused of trying to diminish someone else's job security to improve your own.

So what, then, are your options? That depends on the severity of the case. For isolated errors (unlike the previous case), the best path is to simply make the corrections and gently point them out, in the hope that the colleague's performance will improve and the problem will go away. But in a case of serious deficiency such as the one described, this approach is not adequate—the supervisor needs to know, so that something can be done to correct the situation.

A technical communicator who becomes aware that a communication product of unacceptable quality is about to go out to a client owes it to the work group, the company, and the profession to make sure that doesn't happen. If that can be accomplished through direct professional assistance, great. If it can't, the only acceptable alternative is to surface the problem to those who *can* do something about it. And as far as conflict of interest is concerned, as long as the incompetence is clearly evident (not a matter of subjective judgment), the intervention can be construed only as a legitimate effort to uphold the standards of the profession.

As technical communicators, we must never consciously denigrate a fellow professional, yet neither can we endorse incompetence or poor performance. Sitting idly by while an unacceptable product goes out to our users does, indeed, constitute an endorsement. If we do nothing, we become accomplices in what amounts to malpractice of our profession.

Avoiding the "Doormat" Syndrome: How to Earn Professional Respect

Having explored many aspects of professions and professionalism, let us return now to the question of respect. Why is it that many dedicated technical communicators who meet every criterion set forth in this chapter still have to walk around on eggshells bowing before the powers-that-be in the corporate world, reacting in knee-jerk fashion to unreasonable deadlines, suffering the indignity of being treated as "second-class citizens" with a charter to serve, yet with no franchise even to influence the conditions of that servitude?

Some might attribute the plight of such poor souls to a satiric business version of the Golden Rule: "He who has the gold, rules."

Yet with a full understanding of the politics of corporate America derived from personal experience therein, we would nevertheless venture to say that much of the lack of professional respect that technical communicators decry is at least to some degree self-inflicted. It's a matter of attitude, sort of a self-fulfilling prophecy. *Those who walk around expecting to be treated like doormats will be treated like doormats.*

While we not advocating stomping into the CEO's office and slapping a Technical Communicators' Bill of Rights on his desk, we *are* advocating a polite yet assertive insistence upon professional treatment. Every work situation is different, which makes it very difficult to provide specific operational guidelines for what we mean by "polite but assertive." What works fine in one situation could become a career-stopper in another. With this in mind, and urging you to run our suggestions through the political filter of your own organization, we offer some specific ideas on how you as a technical communicator can enhance your professional standing and treatment within your company.

1. *Publish Schedule Guidelines with Standard Turnarounds.* Resist unreasonable deadlines to the maximum extent consistent with your continued employment. Obviously, if the company president wants the job done overnight, she will get it done overnight, if it's physically possible. But you don't owe the same service to every underling (even if he outranks you) who is trying to parlay *his* poor planning into *your* crisis. This is especially true of internal jobs where a delay does not mean lost business. There's no need to become confrontational; the message comes across quite nicely if you simply explain there are sev-

eral priority jobs in work and politely decline to provide the instant turnaround requested. The person may take a few lumps for lateness; so be it. Next time maybe he'll accord you the professional respect of some lead time.

2. *Dress Consistently with the Executives.* Until a few years ago, dressing appropriately meant the customary business uniform: jacket and tie or even three-piece suit for men, comparable conservative business attire for women. Today, dress codes are gradually evolving toward informality—but professionals who are complaining about lack of respect from management would be well-advised to posture themselves on the trailing rather than the leading edge of the stampede toward casual attire! *Dress for Success* [16] remains a valid concept; we would modify it to read *Dress for Respect*!
3. *Establish First-Name Communication with Managers.* Nothing is more professionally demeaning than to be in a meeting with the Big Boss where you are the only one who has to genuflect and address him as "Mr. Bigg" while everyone else is on a first-name basis. This can't be solved at the meeting of course; the trick is to seize an opportunity while working with the boss on an individual basis or even when just talking to him in the hallway and politely work in the question, "May I call you Phil?" Nine times out of ten, the answer will be "Sure." It takes a moment of courage to make the request, but that sure beats coming off like a page in King Arthur's court while everyone else is seated at the Round Table!
4. *Aggressively Seek Invitations to Senior-Level Meetings.* It's important to be in attendance at meetings where policy pertinent to your organization is determined or where the requirements of your job are being discussed. Establish an expectation that you must routinely be involved on matters pertinent to technical communication. Make yourself indispensable.
5. *Show Initiative—Take on Communication Problems and Solve Them.* Overworked cliches notwithstanding (as in "Leslie, I have an opportunity for you to excel," followed by "Uh-oh, I remember the last opportunity!"), problems really *are* best viewed as opportunities. You may be a communication expert, but that won't help your professional stature one iota if those in power don't see you as one. Make sure they know. *Show* them. Let them benefit from your expertise, from your performance.

If, after all this, you still find yourself cast as the doormat, you can conclude that the basic problem is elsewhere, not with you—in which case it may be time to head for the door.

Summary

To improve their professional standing, technical communicators must work to bring the profession up to certain basic criteria. The Professionalism Committee of the Society for Technical Communication (STC), after extensive research, identified six criteria for a profession:

1. Specialized education and preparation
2. A body of knowledge acquired through research
3. A unique and indispensable public service
4. Autonomy in work practices
5. Ethical professional practices according to enforceable codes
6. Commitment to values of public service and social responsibility

In general, technical communication stacks up quite well on Criteria 1 through 3 and 6; on Criteria 4 and 5, there is room for growth.

In the same study, the STC committee also established six guidelines for professionalism:

1. Commitment to the profession
2. Commitment to a professional calling
3. Commitment to organizing the profession
4. Commitment to education
5. Commitment to the service ethic
6. Commitment to achieving professional autonomy

As we seek to advance our profession, we are no better than our weakest link; poor performance or unethical behavior by a very few can tarnish the image of the rest of the profession. Certification of technical communicators to a specified standard of performance and a prescribed code of ethics would improve the ability of the profession to distance itself from incompetent or unprincipled practitioners.

An important element in upholding the profession is holding the line on quality standards and taking a stand on ethics—even when doing so may cause conflict. Professional associations have a responsibility to assist those who experience conflict on behalf of the profession.

Professionalism demands technical communicators avoid ego involvement, whether they are authoring, editing, or problem-solving with colleagues.

Technical communicators should never consciously demean a fellow professional, but neither should they endorse incompetence in a colleague by ignoring it.

Technical communicators who still find themselves treated as doormats are advised to stand up and start opening doors.

Application Exercises

1. List the six criteria for a profession identified by the STC Professional Committee. Next to each, state briefly the present status of technical communication as a profession.
2. For any deficiencies in #1, outline a course of action that might improve our professional standing.
3. List the six criteria for professionalism identified by the STC Professional Committee. Next to each, state briefly the present status of technical communication as a profession.
4. For any deficiencies in #3, outline a course of action to improve our professional standing.
5. When it comes to professionalism, we are no better than our weakest link. Explain and exemplify.
6. What is "ego involvement"? How does it relate to technical communication?
7. If the customer wants a technical communicator to produce an inferior communication product, should the technical communicator do so? Consider the impact of producing the inferior product or refusing to do so—both on the individual technical communicator and on the profession.
8. Explain and exemplify "doormat syndrome."

Exploration Exercises

1. Describe one of the most "professional" people you know. What is it about this person that makes him or her a professional in your eyes?
2. Give an example of outstanding commitment to a profession that you have observed (different from #1).
3. Give an example of unprofessional behavior that you have observed which has diminished the image of a profession. How should the situation have been handled differently?
4. Read a research article in a professional association journal (e.g., IEEE *Transactions*, STC's *Technical Communication Journal)* and write a brief (1- to 2-page) reaction paper.

5. Outline a workshop you could present at a professional conference.
6. If you knew a fellow technical communicator had deliberately overcharged a client, would you do anything about it? What? Why? What are the implications of your answers for the profession?
7. Illustrate "ego involvement" from your personal or professional experience. (If need be, change the names to protect the guilty!)
8. If you knew a fellow technical communicator had botched a job to the point that a shoddy communication product was about to go out the door to a client, would you do anything about it? What? Why? What are the implications of your answers for the profession?
9. Write an R_x to cure "doormat syndrome" in technical communication—a formula for winning professional respect.

References

1. Terry Skelton, Manager STC Professionalism Committee, and Shirley Andersen, Assistant to the STC President, for Professional Development, "Professionalism in Technical Communication," Professionalism Committee Report in *Technical Communication* (Second Quarter 1993): 202-207.
2. Andrew Malcolm, "On Certifying Technical Communicators," *Technical Communication (*Second Quarter, 1987): 94-102. Benchmark STC study
3. Terry Skelton and Larry Kunz, moderators, panel discussion on professional certification, 42nd Annual STC Conference, Washington, D.C., April 1995. List combines the opinions of several experts on the panel.
4. Skelton and Andersen, "Professionalism."
5. *Ibid.*
6. *Ibid.*
7. STC, "1995 Salary Survey," http://www.clark.net/pub/stc/www/salary.html.
8. M. S. Larson, *"The Rise of Professionalism: A Sociological Analysis" (*Berkeley: University of California Press, 1977) cited in Skelton and Andersen, "Professionalism," 204.
9. Skelton and Andersen, "Professionalism."
10. *Ibid.*
11. *Ibid.*
12. *Ibid.*
13. *Ibid.*

14. J. M. Guran and F. M. Gryna, Jr., *Quality Planning and Analysis: From Product Development Through Use* (New York: McGraw-Hill, 1980) cited in Skelton and Andersen, "Professionalism," 207.
15. Skelton and Andersen, "Professionalism."
16. John T. Molloy, *Dress for Success* (New York: Warner Books, 1975). Current paperback edition, June 1993.

Part III

Applying the Values: The Cases

- How to Use the Case Study to Examine Ethical Conflicts
- Nine Case Studies in Technical Communication
- Creating Your Own Case Study

Chapter 14 Overview

Case Studies . . .

The Mission: **Case studies provide the opportunity to bridge the gap between abstract concepts and their real-world application.**

Chapter Roadmap

What Is a Case Study? Chapter 14 defines a case study as a slice of the real world relocated to the classroom, allowing experiential learning in a safe venue.

What Will You Gain from a Case Study? Chapter 14 explains the benefits of using case studies as learning tools, such as the opportunity to:

- ✓ Apply theory to realistic situations
- ✓ Confront and resolve ethical dilemmas without facing dire consequences
- ✓ Gain insight into the many variables involved in reaching decisions in business
- ✓ Enhance your ability to think through and resolve problems
- ✓ Gain insight into your own personal value system and philosophies

Making the Most of the Cases*. Chapter 1 presents tips you can use for resolving the cases, advising you to:

- ✓ Read each case thoughtfully.
- ✓ Become familiar with all the details.
- ✓ Follow the classic problem-solving methodology.
- ✓ Think about each case not in relation to common issues.
- ✓ Get involved in discussions.

Applying Value Analysis to Resolve Ethical Conflicts.* Chapter 14 guides you through the process of resolving the cases and includes both exercises and a worksheet.

References

***Chapters 15 through 23** present nine representative cases covering a broad range of ethical conflicts that can occur in technical and business communication. And last, but not least, *Chapter 24* allows you to design, develop, and document your own case study on ethics.

. . . "Because Wisdom Can't Be Told"

CHAPTER 14

Case Studies: Bringing the School of Hard Knocks to the Classroom

It can be said flatly that the mere act of listening to wise statements and sound advice does little for anyone. In the process of learning, the learner's dynamic cooperation is required.

—*Charles I. Cragg*[1]

We teach and learn in a variety of ways, ranging along a continuum from passive to active. At the passive end of the scale, we assimilate information and knowledge imparted by others, storing it away for future use. In your formal education, you know passive learning as lecture. The instructor determines the important facts and figures, theories and concepts, presenting them for you to soak up like a sponge and bring them out when you need them later. On the active end, you learn by doing. The ultimate active learning experience is the internship—the real-world application of all the knowledge you absorbed and stored.

Clearly, each type of learning has its place. The first two parts of this book, for instance, lean toward passive teaching in that we are providing information for you to absorb, offering perspectives for you to consider, and giving you a practical approach that we hope you will find useful in determining appropriate ethical behavior in situations of conflict. But we are also advocates of the school of hard knocks; we believe in the value of experiential learning. Unfortunately, creating an internship purely in ethics is unrealistic. It is easy to find documentation assignments that allow you to practice the skills you learned in the classroom; it is more difficult to track down suitable ethical dilemmas to resolve.

How, then, can we focus on experience, accenting practical application of the ideas you've absorbed so far? The answer is the case study.

What Is a Case Study?

In the most basic terms, a case study is a slice of the real world relocated to the classroom or training center. As John Brockman explains in *The Case Method in Technical Communication*, "good cases have the feel of reality."[2] In each case, the narrative presents an issue that "feels" realistic—situations and circumstances that you might actually face when you prepare for a job search or walk into the office in the morning.

We wrote or adapted these cases specifically for this text. We designed them to echo the concepts, ideas, values, and suggestions presented throughout the book. We designed them to reflect the realities and complexities of professional life. We designed them to let you wrestle with the same sort of problems technical communicators confront in the workplace every day. In short, we designed them to "embody a functional relationship between theory and practice."[3]

We gleaned the cases from a variety of environments in which technical communicators work. They represent a variety of circumstances, as well. Some cases involve issues faced by an individual employee and concern personal values; others involve departments, companies, or organizations and deal with corporate values. Some cases are isolated incidents; others are recurring issues. Finally, some cases combine these circumstances.

There are no demonstrably "correct" answers. Every dilemma has more than one "right" solution. Of course, this lack of absolutes can be frustrating when you're used to calculating the correct answer to an algebra equation or determining the best response to a multiple choice question. In some instances, for example, you may have to balance the scales between multiple possible actions, some of which are motivated by virtue, some by prudence—even some that are both or neither.

Other cases force you to consider both short- and long-term consequences. Recall the story of the CEO (see Chapter 4) who had to choose between (1) revealing that his company had inadvertently obtained sensitive pricing information from a competitor, risking possible disqualification from a major proposal, or (2) exploiting that information to land a major contract. The short-term results of choosing the ethical course of action were severe: loss of the contract and a layoff of employees. However, making the unethical choice could have resulted in even more dire consequences. Ultimately, taking the virtuous path helped build a reputation as an ethical company, which reaped long-term business benefits.

As novelist Aldus Huxley once said, "You pays your money and you takes your choice." Indeed, the cases encourage you to make choices, select courses of action, and predict the consequences of such courses of action, rather than settle on "correct" answers.

What Will You Gain from a Case Study?

Cases enable you to apply theory to realistic situations in a secure venue—to walk the high wire of ethical dilemmas with a safety net. They allow you to overcome real-world obstacles in the relative safety of the classroom or training center rather than in the office where your decisions and actions may carry more severe consequences. If you underestimate the amount of time needed to prepare a class assignment, late submission may mean you lose a letter grade on the report. If you underestimate the amount of time needed to prepare a competitive proposal, late submission could cost your company the chance to bid on a multimillion dollar contract—and cost you a job.

Confronting and resolving ethical dilemmas in the workplace can have equally grave results. Make an unethical choice by submitting a ridiculously low cost estimate and you may delight in bringing new work into your department. Unfortunately, in the process you'll also almost certainly destroy your company's reputation for dealing honestly with clients. Engaging in unethical behavior can, in fact, lead to lost contracts, damaged reputations, ruined careers, even lawsuits and criminal charges. Case studies allow you to explore all the options and play all the angles without suffering the consequences.

Solving cases will also give you insight into the many variables involved in reaching decisions in business—many of which you may have no other opportunity to encounter in school. In the classroom, for instance, you generally have sufficient information to adequately complete your assignments. In industry, however, you may have only a limited amount of information with which to evaluate a situation. So, just as in real life, when you consider cases you have to make the best decision you can based on what you know. Each case is self-contained, so evaluate each one based only on the information you are given. If details aren't provided in the case, don't invent your own; after all, none of us is omniscient. In fact, in the workplace the pressures of time and money often preclude additional information gathering, so you may be forced to make decisions without complete information and required to act without conclusive data. Case studies give you the chance to practice doing so without serious consequences.

Developing your ability to think through and resolve problems in this way will be a measure of your success in the business world. Sound judgment will serve you well in the professional arena. So, too, will the ability to analyze a situation, formulate a course of action, then follow through on it.

Finally, considering how you would behave if faced with the circumstances presented in the cases may help you gain insight into your own personal value system and philosophies. Don't be afraid to re-evaluate them, by the way; that's the hallmark of intellectual growth and education.

Making the Most of the Cases

The case studies that follow provide you with the opportunity to bridge the gap between abstract concepts and their real-world application in the workplace. As with any other opportunity, to make the most of it you have to accept the responsibility.

First, prepare to discuss the cases. It's likely that classroom or training group discussions of the cases will focus on the significant issues, raising aspects of the case you may not have considered, and sparking critical and creative thinking. To participate fully in such analyses and to exercise sound judgments, you must be familiar with all the details of the case. Read each case thoughtfully. Make sure you understand the facts. Identify all the factors involved.

Once you understand the basic situation presented in the case, follow the classic problem-solving methodology. Make inferences; that is, draw conclusions based on the facts. Having done so, ascertain and define the problem. Next, identify who the stakeholders are, discern their interests, and figure out what values are at issue. You may wish to use the 10 values we have discussed in this book as a yardstick against which to gauge the issues and conflicts in the cases. You may apply the concentric-ring value model or the hierarchical model (see Chapter 2), or you may develop and apply your own ethical framework. Next, develop options for courses of action (that is, alternate solutions). Finally, analyze, evaluate, and explore those options. Consider the short- and long-term consequences (benefits and risks) of each course of action. A good approach is to combine individual preparation and small-group discussion, followed by role-playing. This brings multiple perspectives and ethical frameworks to bear on the issues involved in the scenarios.

It will also be useful to think about each case not just in isolation, but in relation to common themes developed in analyzing and discussing other cases.

The matrix in Figure 14.1 is a useful tool for applying value analysis to the cases because it combines the nine cases and key tenets of several ethical frameworks in one place for easy reference.

In the classroom or training center, get involved in discussions. Ask questions. Offer opinions. As you've seen throughout the text, ethical behavior is a very personal and individual matter. There are rarely black-and-white answers—mostly there are *shades of gray*. Your impressions, judgments, even hunches will provide insight, spark reactions, and provoke ideas for others. Remain open-minded, too, so you can explore options you may have overlooked, revisit options you examined but discarded, or consider ideas you never even thought about.

Finally—and this is our best advice to date—enjoy yourself! Because of the myriad of alternatives, you'll find yourself involved in some lively debates. Often you and your colleagues will fail to reach unanimous accord. Hopefully, you will reach consensus. Even if you don't, one thing is certain—you are sure to find a bounty of possibilities, probabilities, and expediencies. Hopefully, you will learn to appreciate the complexity of ethics—Winston Churchill's famous comment "a riddle wrapped in a mystery inside an enigma" certainly seems applicable—and develop the capacity to discern all the shades of ethical gray.

By the way, you will notice there is no answer key for the cases. That's the whole point. The cases are designed to portray real life, and real life has no answer key. You're on your own.

Forge ahead!

Applying Value Analysis to Resolve Ethical Conflicts

Each of the case studies in Chapters 15 through 23 contains a set of four activities: value analysis, small-group discussion, role-playing, and large-group discussion. In each instance, you must refer to this chapter for the directions for completing the value analysis and large-group discussion.

In addition, this is your handy-dandy ethics "crib sheet." How appropriate! Consider it your recipe for whipping up instant, insightful, and utterly brilliant solutions to the nine appetizing cases of professional heartburn we ask you to explore in Chapters 15 through 23. To resolve each case, apply the 10 core values pertinent to technical communication, the vertical hierarchy value model, the concentric-ring value model, as well as your *personal* value model to the situation. Having done so, you should be able to develop alternative courses of action and predict the resulting outcomes. Use Figure 14.1 as a worksheet to record your own values and your suggestions for the cases.

Value Analysis Exercises

Use the following exercises for each of the case studies in Chapters 15 through 23.

Directions: Using the value analysis matrix (Figure 14.1), apply the following four (and optional fifth) ethical frameworks to this scenario to determine which values are in conflict and come up with an ethical course of action that resolves the conflict.

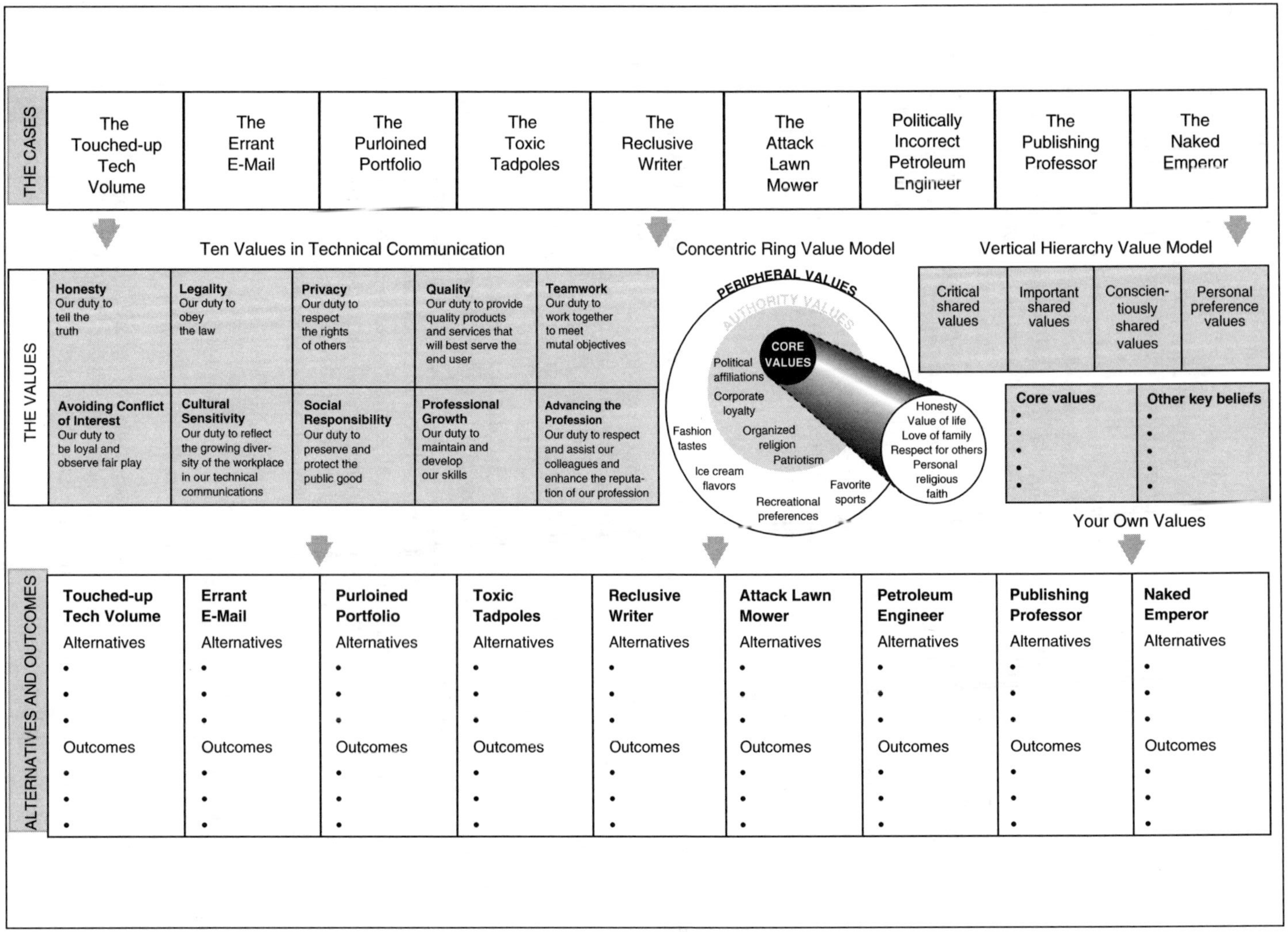

Figure 14.1 Shades of Gray: Applying Value Analysis to Resolve Ethical Conflicts in Technical Communication

Suggestion: An excellent approach is to divide into small groups, with one or two persons focusing on each of the first three frameworks and everyone applying the fourth. For purposes of subsequent role-playing, where there are three characters it's handy if you can come up with either three or six teams. The exact number of persons on each team doesn't matter; anywhere from three to six will work. After individual analysis, the group should work to a consensus on what course of action will best resolve the conflicts with the greatest good (or least harm) to stakeholders. Where possible, a compromise course of action may be considered—but not at the expense of behaving unethically.

Framework #1. The 10 basic values of technical communication as set forth in this text:

- ❑ Honesty
- ❑ Legality
- ❑ Privacy
- ❑ Quality
- ❑ Teamwork
- ❑ Avoiding conflict of interest
- ❑ Teamwork
- ❑ Cultural sensitivity
- ❑ Social responsibility
- ❑ Professional growth
- ❑ Advancing the profession

Framework #2. The concentric-ring value model:

- ❑ Core values
- ❑ Authority values
- ❑ Peripheral values

Framework #3. The hierarchical value model:

- ❑ Critical shared values
- ❑ Important shared values
- ❑ Conscientiously shared values
- ❑ Personal preference values

Framework #4. Your own personal value model.

Framework #5 (optional): An interesting supplementary activity is to have one or two group members review the ethical codes and guidelines of professional associations of technical and business communicators and apply them to the scenario. Refer to Exploration Exercise 3 in Chapter 3.

Steps: Working first individually, then as a group, complete the following steps:

1. Identify the interests (open and hidden) of each stakeholder. Consider not only those present in the scenario but anyone else who may be affected by the outcome.
2. Identify the basic values that are in conflict.
3. Determine which value(s) should prevail on an ethical basis.
4. Suggest courses of action that might be taken to resolve the conflict.
5. For each course of action, predict the likely consequences for each stakeholder. Consider short-term and long-term benefits and risks.
6. Choose the course of action that does the most good and the least harm.

Large-Group Discussion

Follow these directions to guide your large-group discussions for each of the case studies in Chapters 15 through 23.

Directions: Have a spokesperson from each group briefly summarize what the group believes to be the fundamental value conflicts in the scenario and recommend a course of action. After all groups have reported, see if the large group can arrive at a consensus on the best solution to the situation.

Note: Be sure to complete the role-playing exercise *before* the large-group discussion. There's nothing like a lively dramatic improvisation to bring ethical conflicts into sharp focus!

References

1. Charles I. Cragg, "Because Wisdom Can't Be Told," *The Case Method at the Harvard Business School,* ed. Malcolm P. McNair (New York: McGraw-Hill, 1954), 6.
2. John R. Brockman, *The Case Method in Technical Communication: Theory and Models* (The Association of Teachers of Technical Writing, 1984), 9.
3. *Ibid.*, 12.

CHAPTER 15

The Case of the Touched-Up Technical Volume

The Scenario

"Time for another cup..."

Sighing, Gina poured her fourth and final cup of double-grade, industrial-strength editorial coffee out of her insulated thermos into a foam cup already stained from several previous cups. Papers littered her desk. It was 2:30 A.M., 24 hours before the technical volume on the lightweight heat-seeking hyper-velocity air-to-air missile (Air Slam) had to hit the press to make the submittal deadline to the Third World nation whose request for proposal her company was responding to.

Gina flipped through the final draft of the document, which had been reviewed by the proposal manager earlier that evening. The changes were quite extensive; the document was heavily red-lined.

"It's going to be close," Gina muttered, to nobody in particular.

Still, with the extra graphic and word processing support she had lined up for the morning, they should make it—just barely.

Gina reached the section on test results. Most of the data contained therein was favorable to her company's marketing position. Range, velocity, warhead lethality—data on several key parameters appeared to give Allied Discount Aerospace the edge over its Air Slam competitors. The only area of concern was some negative test data identifying potential safety issues associated with the warhead igniter under cold weather conditions. Since the statement of work from the customer specifically called for performance data in extreme temperature conditions, Gina had left the results in just as the test engineer presented them—along with a plan of corrective action to ensure the problem would be resolved before the product entered full-scale production.

A thick red line ran through the section containing the negative test data. Next to it was the scribbled word "Omit!" The handwriting belonged to the program manager.

Gina reached for the telephone. Then, remembering the hour, she resolved to work the issue in the morning.

* * * * *

The next morning, a bleary-eyed Gina was waiting for the proposal manager in his office when he arrived, bright-eyed, bushy-tailed, and nattily attired after a full night's sleep.

"Bob, I'm concerned about that test data in Section 7."

"So am I," he replied brusquely, knowing exactly where the editor is headed. "That's why I took it out."

"But the statement of work specifically calls for this data."

"It's bad data. The tests are inconclusive. Besides, it will kill us."

"But if we don't let them know, and they try to test the prototype missile without proper precautions, somebody could get hurt."

"Gina, there are 500 jobs riding on this contract. One of them is mine. And one of them, in all probability, is yours. Do you get my drift?"

"But—"

"No 'but's.' Leave the data out."

"I hear you."

Gina returned to her cubicle. She reread the deleted section. The words "could pose a potentially lethal danger to test personnel" leaped out at her.

She pulled up the phone directory from e-mail and dialed Frank, the vice president for marketing, to whom Bob reported as proposal manager.

Value Analysis

Refer to Chapter 14 for value analysis exercises to apply to the case study.

Questions for Small-Group Discussion

Directions: Not every aspect of ethical dilemmas is amenable to systematic analysis. It is, therefore, useful to pursue open-ended discussion as well. Here are some jumping-off points.

1. Is a company ethically obligated to include negative information in a competitive proposal when the inclusion of such data is clearly apt to lessen its chances of winning? After all, the whole idea of a proposal is to present your product in the *best* favorable light, is it not?
2. When Gina elected to include the test data in the first place, what else did she include in order to provide "damage control"? Do you think her strategy would be effective? Why?
3. What do you think would happen to Gina if she included the data even after being directed to omit it? (See Chapter 8, the "Elephant Theory of Management Sensitivity.")
4. What would you do if you were in her place?
5. What position would you expect Frank, the marketing director, to take when Gina brings the problem to his attention? Why?
6. Could it conceivably be in Allied's best interest to *include* the information? Explain.
7. Reach into your personal experience. Can you come up with any conflicts, either on the job or elsewhere, resembling this one? If so, share them and discuss possible alternatives.

Role-Playing

Directions: Assign each group one of the three characters in the scenario. If there are six groups, simply enact each scene a second time with a different set of players. (Indeed, this can be particularly interesting!) In no instance, however, should a group be assigned more than one character. Whatever character the group is assigned, it must choose an actor or actress to portray the assigned role and coach or coax that person into an Academy Award-winning performance. In the role-playing scenes, characters should act in accordance with their own interests and values, not according to a prescribed ethical "script." The name of the game is creative improvisation!

Note: This exercise is gender-friendly. Gender changes may be made as follows:

- Gina-Gene
- Bob-Barbara
- Frank-Francis

Scene #1
Replay the scene between Gina and Bob.

Scene #2
Dramatize a meeting involving Gina, Bob, and Frank, the marketing director.

Scene #3
Play scenes involving just Bob and Frank, immediately before and after their meeting with Gina.

Variations:

- ❑ Assume Gina puts the test data back in and the proposal is delivered that way. Enact the scene between Gina and Bob immediately after Bob discovers what she has done.
- ❑ Same as above, but have Frank be the one to make the discovery instead of Bob.

Epilogue (optional):
Assume the company does include the negative test data, along with a plan for correcting the situation. Assume, further, that Allied's Air Slam missile is superior to the other competitors' missiles in every other respect. Enact a scene in which three military leaders in the Third World country are evaluating the proposals, trying to make a final decision.

Large-Group Discussion

Refer to Chapter 14 for large-group discussion questions to apply to the case study.

CHAPTER 16

The Case of the Errant E-Mail

"...and this one called me an idiot–transfer him to International Falls in January. That one called me a buffoon–fire her!"

The Scenario

"Well, if you'll just stop being indignant long enough to calm down," Julia crooned in her most soothing tone, patting her colleague on the back, "you'll see that there might be a certain benefit in this for our department. You know—the proverbial silver lining."

Kristen wasn't buying it. "How can there possibly be a silver lining associated with the fact that the division director has been reading my e-mail without my knowledge? It's unfair. It's unethical. It's sneaky. Hell, it's probably even dishonest! And why aren't you just as angry as me? He's probably reading everyone's mail—it's just that I'm the only one who got caught sending a document he didn't approve of. At least, I'm the only one so far."

Kristen returned to her pacing while Julia speculated. "Now let's see if I have this straight. You e-mailed Kevin to review your first cut at the white paper?"

"Yes—silly me to think a rough, rough draft would never be seen by anyone I didn't intend it for!" Kristen couldn't resist adding the editorial comment.

"OK, so old man Jefferson took offense at several controversial comments. . . "

Before Julia could finish her sentence, Kristen turned on her, "What do you mean controversial? Just because Jefferson doesn't like something doesn't make it controversial."

"Geez, I'm sorry. To use your words, he didn't like a few of your assertions. So he put the kibosh on the whole project?"

"Hell, no! Just on my involvement in it," Kristen exploded. "He had Jim turn it over to Eric—like always, he had our supervisor do his dirty work."

"Well, there's not much you can do about it now, so let's look at the bright side. If he can read your mail, why couldn't he read everyone's mail? He's probably doing it already, anyway. Don't look so horrified, Kristen. Think about it . . . our department has lost a lot of work ever since that dumb policy change that allows the engineers to bid their publications projects to outside vendors. That can't be making Jefferson look good. His budget has got to be hurting." A wicked grin was beginning at the corners of Julia's mouth.

Bewilderment tempered Kristen's anger. "What on earth does one have to do with the other?"

"Think about it. All those wretched engineers who can't write a sentence are mad at us for insisting on changes to their reports, so they're sending work outside and not even giving us a chance to bid on it. Right?"

"Right, Julia, and your point is?"

"If Jefferson kept tabs on the e-mail of the engineering departments and knew what they were working on, he could twist some arms up there on executive row to send some work our way!" Julia was beaming very much like the proverbial cat that swallows the mouse.

Kristen's demeanor was decidedly different, however. "You can't be serious. E-mail should be private. We can't do something unethical just to get more business."

Julia jumped in, "You mean, we can't take every measure possible to keep our department going and stay out of the unemployment line?"

"You know what I mean, Julia. Don't put words in my mouth."

Undaunted, Julia forged ahead. "You wouldn't object if someone just happened to overhear a conversation in the hall and got us more work that way. Don't be so stuffy. I'm not putting words in your mouth, but I will put ideas in people's heads—starting with Jefferson."

"He'll never go for it," Kristen said smugly, assuming she was getting in the last word.

"Wanna bet?" was Julia's tart response as she waltzed out of Kristen's office.

Surely, Julia must be kidding. She sounded so serious, though. I'll fight her, and anyone else, every step of the way, Kristen thought.

Value Analysis

Refer to Chapter 14 for value analysis exercises to apply to the case study.

Questions for Small-Group Discussion

Directions: Not every aspect of ethical dilemmas is amenable to systematic analysis. It is therefore useful to pursue open-ended discussion as well. Here are some jumping-off points for such discussion.

1. Is it ethical for a company to read employees' e-mail? Why or why not? Can you think of any circumstances in which such an action might be justified? Explain.
2. Given that bosses (or at least one boss) *already* read others' e-mail in this company, is there anything wrong with Julia's idea?
3. What would you do in Kristen's place?
4. What position would you expect Jim, Kristen, and Julia's supervisor, to take?
5. Was Jefferson justified in taking Kristen off the project? Discuss the purpose of rough drafts. Who should review them? What should be the intent of early reviews?
6. Reach into your personal experience. Have you encountered a similar situation in which your privacy was invaded? What happened? How did you handle the situation? What, if anything, would you do differently now?

Role-Playing

Directions: Assign each group one of the three characters in the scenario. If there are six groups, simply enact each scene a second time with a different set of players. (Indeed, this can be particularly interesting!) In no instance, however, should a group be assigned more than one character. Whatever character the group is assigned, it must choose an actor or actress to portray the assigned role and coax that person into an Academy Award-winning performance. In the role-playing scenes, characters should act in accordance with their own interests and values, not according to a prescribed ethical "script." The name of the game is improvisation!

Note: This exercise is gender-friendly. Gender changes may be made as follows:

- Julia-Jules
- Kristen-Kris
- Jim-Jenny

Scene #1
Replay the scene between Kristen and Julia.

Scene #2
Dramatize a second meeting between Kristen and Julia after both have had time to reconsider.

Scene #3
Dramatize the meeting in which Julia presents her idea to Jefferson.

Variations:
Assume Kristen is so concerned that Julia will go through with her plan, she attempts a preemptive strike by going to their supervisor, Jim. Enact the meeting between Kristen and Jim.

Epilogue (optional):
Role-play a company meeting with both management and employees discussing the adoption of a policy regarding reading of private e-mail.

Large-Group Discussion

Refer to Chapter 14 for large-group discussion questions to apply to the case study.

CHAPTER 17

The Case of the Purloined Portfolio

"To cheat or not to cheat, that is the question. Whether 'tis nobler..."

The Scenario

Now Sally wished she'd paid more attention at that STC meeting on campus instead of flirting with that cute guy in the back row. But at the time, her spring semester internship had just led to a job offer, which she had accepted, so she was more concerned about getting a date than in learning how to prepare a portfolio for a job search.

Unfortunately, she was becoming acquainted all too soon with corporate downsizing. Only five months into her first real job, Sally had learned just two days ago that her entire department was being eliminated. Now, as she perused the classified ads over her first cup of Sunday-morning coffee, she felt confident about her ability to put together a pretty solid resume and cover letter. Thank heaven for those mandatory seminars in Career Services! But that darned portfolio. . .

Finally, after a second cup of coffee and a croissant, Sally decided she might as well tackle the portfolio dilemma. First, she dug through her files for copies of papers from classes over the last four years. Let's see—Freshman Composition? No way—Sally couldn't believe she'd ever written so poorly!

She kept digging. Ah, these were better—certainly the final projects from hardware and software documentation classes would be impressive. Lucky she'd kept the electronic files. She'd incorporate the professors' suggestions, print clean copies, and have some pretty good stuff. Maybe this wasn't so bad, after all!

Her soaring spirits crashed all too soon. A search of the remaining files didn't turn up anything else that Sally considered portfolio-quality material. Well, she certainly needed another croissant before she could figure out what to do—maybe even one with cherry jelly.

"OK," Sally thought, after finishing off her second pastry and third coffee, "there has to be something here I can use. Four years of school and eight months on a job—I must have *something* to show for it."

Maybe it was the sugar high, but her next pass through her files produced more possibilities, *and* some perplexing questions. Like the 20-page research paper from her senior capstone course. It was very well written, contained sound research, and it was on a timely topic—the significance of the Americans with Disabilities Act to technical communicators. In a portfolio, it would showcase her research skills, her ability to develop a well-organized argument, and her familiarity with issues within the discipline. Just one small problem—the paper had been a collaborative project with three other classmates. Sally felt kind of guilty about including it. But, after all, if she created a new title page with her name as sole author, who would ever know?

Then there was the copy of the department newsletter. She had word processed the articles; someone else had actually finished the layout. But her name was listed first on the masthead as part of the production crew. It was the closest she could come to a design piece for her portfolio. And who would ever know the difference?

And there was that journal article she had edited during her first few months on the job. She'd done her best, but the senior editor had found a number of errors and improvements that Sally had missed. But if she included "before" and "after" versions of the article, it would point out how well edited it was. No one would ever know she'd had a little help.

The little pang of guilt returned. She couldn't remember any classroom discussion of this particular issue from the lectures on ethics. She did remember something about personal values. Well, wasn't personal employment an important value? Certainly she'd weaken her competitive posture in the job market if she didn't have more than just a few pieces in her portfolio. And how did she know other applicants would play totally fair? How did she know *they* wouldn't use collaborative material in *their* portfolios? After all, she'd received top marks in school and earned praise on her short-lived job. She'd be an asset to some company—if she could just convince someone to hire her.

What was so wrong with including those items in her portfolio? After all, taking credit for them was only a *partial* untruth. No one would care.

Still, that nagging little pang of guilt . . .

Value Analysis

Refer to Chapter 14 for value analysis exercises to apply to the case study.

Questions for Small-Group Discussion

Directions: Not every aspect of ethical dilemmas is amenable to systematic analysis. It is therefore useful to pursue open-ended discussion as well. Here are some jumping-off points for such discussion.

1. Aren't employers fostering unethical behavior by requiring submission of a portfolio as part of an interview? How can employers ensure that applicants are truthful in their submissions? Should they do so?
2. Do you have any collaborative projects that would enhance your portfolio? How would you handle their inclusion?
3. How would you react if several collaborators included your work in their portfolios without giving you credit?
4. How would your professors encourage you to deal with collaborative projects when it comes to portfolios?
5. Is it possible to prepare an impressive portfolio that contains numerous collaborative projects? How?

Role-Playing

Directions: Assign two groups the role of Sally/Sal. Assign the other characters to the remaining groups. (Indeed, this can be particularly interesting!) In no instance should a group be assigned more than one character. Whatever character the group is assigned, it must choose an actor or actress to portray the assigned role and coax that person into an Academy Award-winning performance. In the role-playing scenes, characters should act in accordance with their own interests and values, not according to a prescribed ethical "script." Begin with two interpretations of Scene #1; then proceed to the other scenes. The name of the game is improvisation!

Note: This exercise is gender-friendly. Gender changes may be made as follows:

- Sally-Sal
- The other characters may be of either gender

Scene #1
Dramatize a monologue expressing Sally's thoughts as she finalizes her portfolio.

Scene #2
Dramatize a meeting that Sally requested with her former professor to seek advice in this matter.

Scene #3
Bring the four collaborators of the senior project together to discuss how best to handle this and similar projects in a portfolio.

Variations:
Assume you ask a former professor to evaluate your portfolio and suggest improvements. You have not credited collaborators. Role-play the professor's response.

Epilogue (optional):
Assume the role of an employer interviewing a job applicant. The applicant's portfolio contains a software manual that he claims to have written. However, you just happen to know that a colleague on your former job actually wrote it. Would you confront the applicant? If so, how? If not, how would you continue with the interview?

Large-Group Discussion

Refer to Chapter 14 for large-group discussion questions to apply to the case study.

CHAPTER 18

The Case of the Toxic Tadpoles

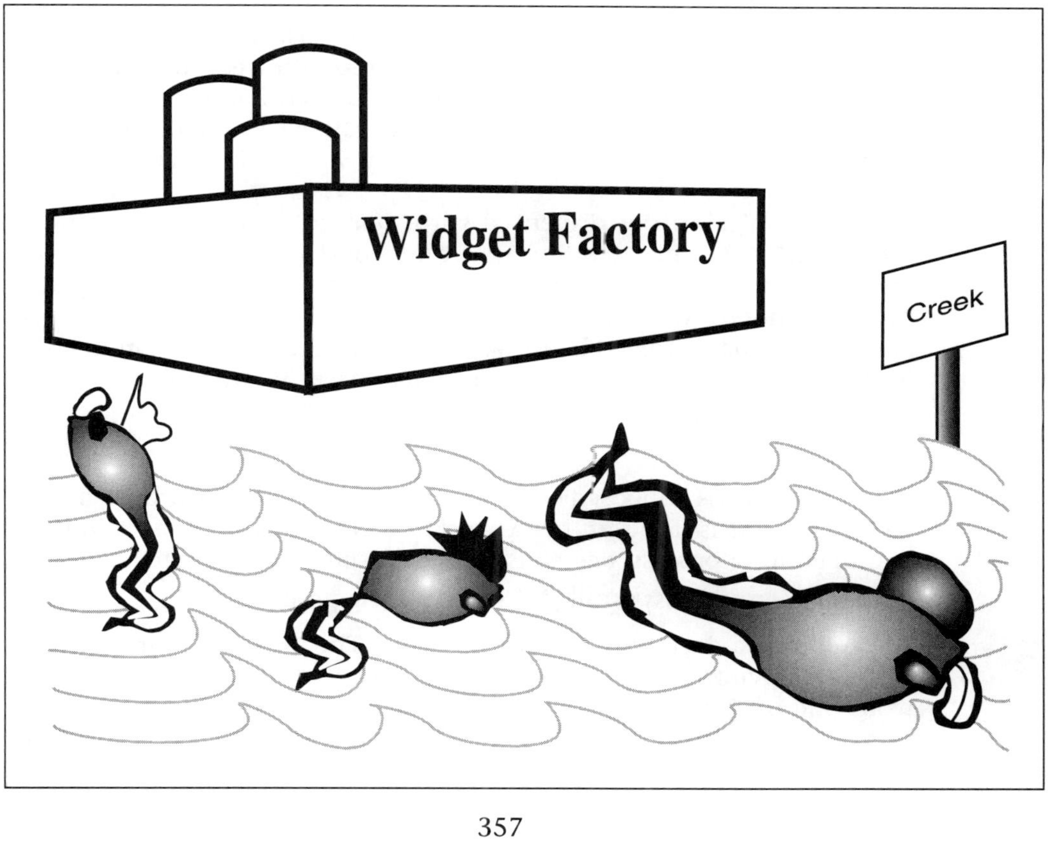

The Scenario

Maybe what he needed was a cup of coffee. Ken had already read the paragraph three times and it still made so little sense he couldn't even begin to edit it.

He'd been doing this job—and doing it quite well—for four years, but this was the first time he was totally stumped. Maybe it was just the lateness of the hour and a sleepy reaction to that huge lunch. A good caffeine jolt and a walk around the office would surely rejuvenate him and let him get a handle on this document.

Unfortunately, 15 minutes, a cup of strong coffee, and a brisk walk didn't help. When he read the paragraph the fourth time, he still didn't know where to begin his edit. The words simply didn't make sense. It was jargon, gobbledygook, and doublespeak. He hoped that the other 38 pages of this environmental report were written more clearly. Well, it was nearly quitting time—too late to tackle the rest now. Maybe it would look better in the morning.

* * * * *

Oh, no! Not only did the opening paragraph not look any better in the morning, the rest of the report was equally confusing. No sense wasting time on a draft that needed a serious rewrite instead of the edit that was requested.

He'd just talk to Melissa, the author, first. Ken had worked with Melissa before and he was just a little puzzled. She really wasn't a bad writer—not for an environmental engineer! This certainly wasn't the quality he'd expect from her.

All the pleasantries dispensed with, Ken got right to the point. "Look Melissa, it's my understanding that this report is intended for stakeholders and the general public who are up in arms about the proposed expansion of our Phosphorescent Frisbee fabrication facility. They say our plant is already spilling too many toxic chemicals into Pristine Creek as byproducts of our Frisbee manufacturing. They want us to cut back. Meanwhile, we're planning to expand *and* triple Frisbee production. Right so far?"

"You nailed it. Go on."

"So this report is supposed to calm them down. To reassure them that our current process is safe and the expansion will make it even safer: improve the manufacturing process, decrease byproduct spill, etc., etc., etc."

"Exactly, Ken," Melissa smiled. "So what's the problem?"

Ken took a deep breath and started in. "The problem is that I've been an editor of environmental publications for four years and I don't understand word one of this report. Now, call me overzealous if you'd like, but I don't think our audience will understand it, either, and they can hardly be persuaded by what they can't comprehend. Just look at this sentence. . . ." Ken randomly opened to a page and read to Melissa:

> *". . . the firm, analytical, studied, and well-documented opinion after a significant and major year long study that the use, release, elimination, accumulation and disposal of flow through into Pristine Creek falls within accepted 3Σ industry standards for 3-year rolling volume, enabling the continuation of capacity maximization at performance ranges of greater than 86.5 percent resulting in near optimal organic filtration processes permitting anti-eutrophificative cleansing of aquatic resources within accepted levels for neonatal amphibious host occupied environments. . ."*

"Just what does that mean? In English, please," Ken concluded.

"Hallelujah! It was tough, but it looks like I've done my job!"

Ken was more puzzled than ever, given Melissa's delighted response and Cheshire-cat grin.

"Look," she chirped, "I just followed orders. You and I both know that some major dollars and heavy careers are on the line with this expansion. If the people in this town and the Sierra Club go through with their threats to stop the expansion and make us clean up their creek, heads will roll. Mine isn't going to be one of them."

"And your point is . . . ?" Ken questioned, still unclear what all of this had to do with the report.

"I had orders from the third floor. They seem to think that if this report sounds official and scientific enough, it won't be questioned. 'Uneducated masses' and 'fanatic, eco-freak do-gooders' are the phrases I heard being tossed around up there."

At least Melissa had the decency to blush as she reported the deception to Ken.

"So what are you telling me?" Ken was almost afraid to hear the answer.

"It stays as it is. Whether I agree with the strategy is irrelevant—I'm not in a position to argue and I may well be in a position to be promoted."

Melissa was suddenly brusque and abrupt.

"Come on, Melissa," Ken felt as though he were whining. "We can't let this go out."

"I can. If you don't like it, take it up with my supervisor and his manager and her director. You can't win this one, Ken. May as well be a team player and make yourself look good."

Was it Stan Laurel who used to say something like "what a fine mess"? That's exactly how Ken felt. Should he simply produce what he knew was a poorly written—and very likely dishonest— report? Or should he go to Melissa's boss? Or maybe his own boss?

Value Analysis

Refer to Chapter 14 for value analysis exercises to apply to the case study.

Questions for Small-Group Discussion

Directions: Not every aspect of ethical dilemmas is amenable to systematic analysis. It is therefore useful to pursue open-ended discussion as well. Here are some jumping-off points for such discussion.

1. Are there other approaches Ken might take in dealing with Melissa? What are they?
2. What course of action is likely to have the most dire consequences for Ken?
3. Ken assumes the report's content is dishonest. Do you agree with his assessment? Why or why not?
4. What might be the response of the public to this report?
5. What might be the response of the Sierra Club to this report?
6. What is likely to be the response of upper management if Ken pursues his concerns?

Role-Playing

Directions: Assign each group one of the characters in the scenario. If there are more groups than characters, simply enact each scene a second time with a different set of players. (Indeed, this can be particularly interesting!) In no instance, however, should a group be assigned more than one character. Whatever character the group is assigned, it must choose an actor or actress to portray the assigned role and coax that person into an Academy Award-winning performance. In the role-playing scenes, characters should act in accordance with their own interests and values, not according to a prescribed ethical "script." The name of the game is improvisation!

Note: This exercise is gender-friendly. Gender changes may be made as follows:

- Melissa-Mel
- Ken-Kendra
- The supervisors may be either gender

Scene #1
Dramatize the meeting in which Ken tries to persuade Melissa to go with him to talk to their supervisors about the report.

Scene #2
Dramatize the meeting in which Ken expresses his concerns to his supervisor.

Scene #3
Dramatize the meeting in which Ken expresses his concerns to Melissa's supervisor.

Variations:
Assume Ken's supervisor concurs with his concerns and requests a meeting with Ken, Melissa, and Melissa's supervisor. Dramatize the meeting.

Epilogue (optional):
Assume the report is distributed to the public. Role-play a meeting in which three members of the company's upper management meet with the people of the town in an open forum.

Large-Group Discussion

Refer to Chapter 14 for large-group discussion questions to apply to the case study.

CHAPTER 19

The Case of the Reclusive Writer

The Scenario

Dina stared glumly at the e-mail message blinking menacingly from her screen:

> *To: Dina Miller, Communications Task Leader, Technical Publications*
>
> *From: Leigh M. Oph, Human Resources*
>
> *cc: Carlos Mendoza, Manager, Communication Department*
>
> *Re: Attendance at Professional Development Workshops*
>
> *Our records indicate that the personnel on your team have failed to maintain a satisfactory attendance record at the mandatory professional development workshops in the first quarter of this year. Your overall attendance rate is 73 percent—well below the required minimum of 85 percent. These workshops are an essential part of our horizontal growth initiative within the teaming concept. Please take immediate corrective action to ensure that your personnel attend at the required rate or higher, effective immediately.*

"Or we shall confiscate your home, your car, and your first-born," Dina mumbled. "Someone oughta teach these yo-yo's a lesson on how to win friends and influence people through e-mail."

The electronic rebuke from Human Resources was ill-timed. Dina's annual salary review was coming up the following month, and the low participation of her overtime-weary team in after-hours professional development workshops had been an area marked "Needs Improvement" last year. To be certain, Carlos would have something to say about this.

Actually, it wasn't really a *team* problem at all. Of the 15 people on Dina's team, 14 of them had maintained a record of 82 percent—close enough to the mandated minimum to keep the heat off Dina, given the heavy overtime her team had to work.

The driving factor was one recreant—Dr. Alec Trode. Dr. Trode's attendance rate at the workshops was zero. An eccentric and curmudgeonly old-school editor with a Ph.D. in Medieval Literature, Dr. Trode was without question one of the department's finest wordsmiths. He was not, however, well-suited to the teaming environment. He preferred to work alone, in his ill-lighted, paper-strewn office in the back of the balcony, away from the rest of the team.

He was keenly distrustful of management in general, younger management in particular, and female management above all.

"Three strikes and I'm out," Dina mused.

Moreover, Dr. Trode possessed a deep and abiding skepticism about computers. He had only recently made a grudging truce with a basic word processing program, reluctantly trading in his blue pencil for a mouse three years from retirement, but he took no stock whatsoever in what he acidly termed "new-fangled twaddle" such as hypertext, online documentation, and interactive multimedia—areas that naturally were often the subject of professional development workshops.

"Don't bother me with these silly workshops about these damned computers!" he would bark. "Just give me the documentation, and I'll fix it."

And fix it he would. No matter how garbled the syntax, how tangled the message, how incoherent the development, copy sent to Dr. Trode came back fluent, correct, and cohesive.

That's why he was still around.

But if Dina couldn't figure out some way to get him to start attending professional development workshops, maybe *she* wouldn't be.

Taking a deep breath, she prepared herself mentally and walked to the back of the balcony where Dr. Trode's office was located. He was in. She could see the unruly rug of silver hair. He was seated facing the computer, back to the door. As usual, his desk was littered with paper, empty coffee cups, and telltale cigarette ashes. The company had instituted a No Smoking policy three years ago, but Dr. Trode never missed an opportunity after hours to sneak a cigarette. If he was really stressed, sometimes he even flouted the policy openly. Wisps of smoke had been reported to waft over the eight-foot partitions of his office.

All that was missing to make the scene complete was a sign by the door bearing the dire admonition from the *Inferno* in Dante's *The Divine Comedy*:

All hope abandon, ye who enter here.

It was now or never. With much trepidation, Dina rapped her knuckles on the door.

Value Analysis

Refer to Chapter 14 for value analysis exercises to apply to the case study.

Questions for Small-Group Discussion

Directions: Not every aspect of ethical dilemmas is amenable to systematic analysis. It is, therefore, useful to pursue open-ended discussion as well. Here are some jumping-off points.

1. Should participation in after-hours professional development workshops be mandatory? Why?
2. How much iconoclastic behavior should be tolerated by management or teammates on the basis of the offender's talent and work output?

3. Should Dina take a conciliatory or hard-line approach? Which has the best chance of success?
4. Is it still possible for a technical writer or editor to be a valuable contributor to a Communications Department with little or no computer literacy? If so, what if the person is *not* nearing retirement as Dr. Trode is? Is there a viable career path in technical communication without keeping current with technology and computer tool skills?
5. How would you encourage colleagues who are already burned out on overtime to attend after-hours professional development workshops?
6. Do you share Dina's opinion of the tone of the e-mail message from Human Resources? If so, how better might the message have been worded?
7. If you were Carlos, what advice would you give Dina for handling Dr. Trode?
8. Drawing upon your own experience, have you been involved with any situations involving a cantankerous colleague like Dr. Trode? If so, how did you handle the situation? Compare notes.

Role-Playing

Directions: Assign each group one of the three characters in the scenario. If there are six groups, simply enact each scene a second time with a different set of players. (Indeed, this can be particularly interesting!) In no instance, however, should a group be assigned more than one character. Whatever character the group is assigned, it must choose an actor or actress to portray the assigned role and coach or coax that person into an Academy Award-winning performance. In the role-playing scenes, characters should act in accordance with their own interests and values, not according to a prescribed ethical "script." The name of the game is creative improvisation!

Note: This exercise is gender-friendly. Gender changes may be made as follows:

- Dina-Dean
- Dr. Alex Trode-Dr. Alice Trode
- Carlos-Carlotta

Scene #1
Play out the confrontation between Dina and Dr. Trode. Then bring in new actors and try it again.

Scene #2
Dramatize a meeting between Dina and Carlos following the scene with Dr. Trode.

Scene #3
Now bring Carlos and Dina together back to Dr. Trode's office. Or, alternatively, play the trilogue scene in Carlos' office.

Variations:

- ❑ Introduce Leigh (or Lee) from Human Resources. Various combinations exist: Leigh and Dina; Leigh, Dina, and Carlos; Leigh, Dina, and Dr. Trode, etc.; improvise.
- ❑ Invent three or four teammates from the Tech Pubs team and enact a staff meeting where Dina reports the team's low attendance at professional development workshops. First play the scene without Dr. Trode present (which would probably be typical!). Then try it with him present.

Epilogue (optional):
Dramatize Dr. Trode's retirement party (premature or otherwise).

Large-Group Discussion

Refer to Chapter 14 for large-group discussion questions to apply to the case study.

CHAPTER 20

The Case of the Attack Lawn Mower

"Hey Dad, good shot! The mower nailed me again!"

The Scenario

"Holy smokes!"

Quality engineer Shawna Karing sifted through the safety reports. There were at least five documented cases of a faulty rivet on the Iago II Super Mower giving way and being slung from the whirling blades at high velocity—a deadly projectile. Fortunately—indeed, almost miraculously—no one had been in the path of the rivet. So far, the damage had been confined to windows, shrubs, and, in one case, a Lamborghini tire.

"This must be fixed immediately!"

Shawna's fingers raced across the keyboard.

"Level 1 Safety Alert..."

* * * * *

Troy Trooper was putting the finishing touches on the introduction to the new user's manual for the Iago II when he saw the urgent e-mail message blinking on his computer terminal.

"Now what's *this*?"

It had already been a bad day, starting with a second overdue notice on his mortgage payment.

Troy read the message.

"Oh, brother..."

* * * * *

"You're going to have *to insert a warning, Troy," Shawna insisted. "There's no question about it."*

"I see where you're coming from," Troy admitted, "but it may not be that easy. This thing could really hurt our sales next quarter. Besides, I believe the rivet is supplied by a disadvantaged minority vendor, which would mean more bad press." He paused dramatically for the coup de grace. "And you know that Ms. Cuda has to approve any changes."

Shawna cringed. The vice president for domestic sales, Beri Cuda, was better known for her political savvy and her commitment to the bottom line than for her dedication to the public good.

* * * * *

"I want you to lose those safety reports on your desk until after we print." Beri Cuda leaned forward over her large oak desk and pointed at Shawna for emphasis.

Shawna held her ground. "I can't do that. You have *to put some kind of warning in the manual."*

"Right. And tell our faithful customers we're selling a combination mower-machine gun that will not only mow down their grass, but possibly one or two of their children? Not to mention making our minority partner look like a total idiot and blowing our government contract to bail out their business."

Shawna skewered Troy with a stare. He wriggled in his seat.

"Well, Ms. Cuda, Shawna does have a point. Maybe if we word the warning appropriately..."

"No. We'll retrofit the rivet as soon as possible. But the manuals go as is. Am I making myself clear?"

Shawna and Troy looked at each other.

Clarity was not the problem.

Value Analysis

Refer to Chapter 14 for value analysis exercises to apply to the case study.

Questions for Small-Group Discussion

Directions: Not every aspect of ethical dilemmas is amenable to systematic analysis. It is, therefore, useful to pursue open-ended discussion as well. Here are some jumping-off points.

1. Of the three participants, who appears to have the greatest concern for the user? The least? Why?
2. Why is Troy so reluctant to confront Beri Cuda? What item dominates his personal agenda? Is it a legitimate concern?
3. Do you agree with Beri Cuda's contention that exposing the rivet problem with a safety warning would inflict irrecoverable harm upon the disadvantaged minority vendor that supplies the part? Explain.
4. What options, if any, do Shawna and Troy have after Beri Cuda's final directive? What would you do if you were in their place?
5. Forget the user for a moment and consider only the company's bottom line. From this perspective, is Beri Cuda's decision wise? Explain.
6. Do you know of any real-world incidents (or even dramatized incidents, as in a movie) involving the same issues that are at stake in this scenario? If so, share them and compare the outcomes with those discussed above.

Role-Playing

Directions: Assign each group one of the three characters in the scenario. If there are six groups, simply enact each scene a second time with a different set of players. (Indeed, this can be particularly interesting!) In no instance, however, should a group be assigned more than one character. Whatever character the group is assigned, it must choose an actor or actress to portray the assigned role and coach or coax that person into an Academy Award-winning performance. In the role-playing scenes, characters should act in accordance with their own interests and values, not according to a prescribed ethical "script." The name of the game is creative improvisation!

Note: This exercise is gender-friendly. Gender changes may be made as follows:
- Shawna-Shawn
- Troy-Troyla
- Beri-Barry

Scene #1
Play out the pow-wow between Shawna and Troy before the big meeting with Beri Cuda.

Scene #2
Enact the meeting in Beri Cuda's office with all three characters.

Scene #3
Directly after the scene in Beri Cuda's office, dramatize an ensuing discussion between Shawna and Troy.

Variations:

- ❑ Bring Shawna and Troy back to Beri Cuda's office, along with a corporate vice president.
- ❑ Involve OSHA (the Federal Occupational Safety and Health Administration).
- ❑ Involve the local news media.

Epilogue (optional):
Dramatize the confrontation between the husband and wife in the chapter-opening cartoon, starting from the part where the wife finds little Billy dangling from the eaves by a lawn-mower rivet.

Large-Group Discussion

Refer to Chapter 14 for large-group discussion questions to apply to the case study.

CHAPTER 21

The Case of the Politically Incorrect Petroleum Engineer

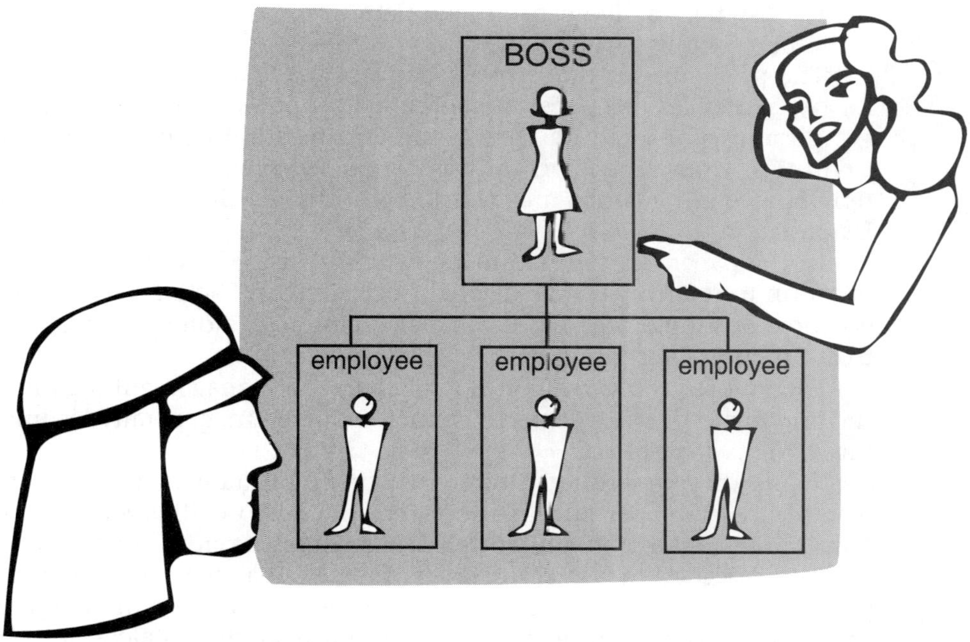

"No offense, Madam, but there are very limited opportunities for females in oil field management."

The Scenario

"Uh-oh."

This was going to be a problem.

Ken flipped back through the article on "The Application of Teaming Strategies to Increase Productivity in Petroleum Refining and Delivering Operations." Then he flipped through the visual aids for the accompanying briefing.

"Uh-oh."

It wasn't that the article and briefing weren't *good*. As a matter of fact, they were excellent. Not only was the argument cogent and the material lucidly presented, the writing style was, well, exceptional. The author, brilliant petroleum engineer and rising management star Kristen Schmidt, had left him little room to add value through wordsmithing.

Good thing, too. Based on everything he had heard about Kristen from other editors in the department—he had not worked with her before personally—she was inclined to take umbrage at any editorial tinkering with her deathless prose. Her posture was consistent with her overall reputation as an extremely aggressive, talented, and upwardly mobile businesswoman. Fortunately, her resistance to editing was not an insurmountable problem, because her fluid syntax rarely needed much help.

What, then, was the problem?

The problem was that Kristen had not made any allowance whatsoever for her audience. The paper was intended for publication at an international symposium sponsored by OPEC—the Organization of Petroleum-Exporting Countries, whose membership was comprised predominantly of Arabic nations. Indeed, the conference was to be hosted by the Syrian government in Damascus.

In the paper, Kristen had made extensive use of the third person female pronoun in referring to management decisions and actions, as in "The refinery director should organize *her* management team along product lines rather than functional responsibilities."

Even more visibly, the conceptual charts on management organization in the briefing frequently depicted women involved in, generally leading, meetings and discussions.

While they were unquestionably a ringing testament to diversity in the workplace, the paper and briefing were bound to collide with the cultural sensibilities of the predominantly Islamic, almost exclusively male audience.

And, for good measure, Kristen had tossed in a couple of mild profanities (for example, "Teaming does a helluvalot more for morale than Type X management") and capped off the paper with a reference to alcohol in her dramatic closing ("So when the CEO hoists a glass of Chardonnay to toast a 40 percent increase in productivity this time next year, she will have teaming to thank for the company's success"). Word was, Kristen was particularly sensitive about any changes to her dramatic closings.

Would it play in Philadelphia? Maybe. Would it play in Damascus? Or Riyadh? Or Cairo? No chance.

Ken had no choice. The culturally insensitive material *had* to be changed. The only question was whether to go ahead and make the changes, incurring Kristen's wrath in the final review cycle, or to talk to her immediately and explain why he felt the changes needed to be made. Or perhaps he should involve a representative from the EEO Office to help him explain the cultural issues involved.

Not normally a coward, Ken found himself gravitating toward the former strategy—except his conscience told him avoiding the confrontation was not the right thing to do. Finally, he tossed the draft of the paper and the roughs for the visuals into a manila folder and set out for Kristen's office.

It was a corner office, not quite executive size, but with a window.

It was a lot bigger than Ken's cubicle, which had no window.

Ken sighed.

"May as well give it the old college try," he figured. "If that doesn't work, and I'm still alive, I can always come back with the EEO representative."

He knocked on the door.

Value Analysis

Refer to Chapter 14 for value analysis exercises to apply to the case study.

Questions for Small-Group Discussion

Directions: Not every aspect of ethical dilemmas is amenable to systematic analysis. It is, therefore, useful to pursue open-ended discussion as well. Here are some jumping-off points.

1. Is Ken correct in his judgment that the culturally questionable material must be changed? How far should authors and editors in Western society modify their works to accommodate the cultures of Eastern or Middle Eastern audiences?
2. Suppose Kristen, known to have feminist leanings, digs in her heels on including a significant female management presence in her pitch. Does she have a point? And if she prevails, what will likely be the communication outcome?
3. Suppose Ken prevails and an all-male depiction of management is substituted. Would such a change represent a strategic communication

decision, showing respect for host cultures; or a moral cop-out on diversity, selling out on the rights of females in the workforce?

4. What about the mild profanity for emphasis? Does that have any place in business communication? If so, should it be removed for this particular audience? Or is the presenter sovereign in matters of style, doing what is most comfortable for *her*?
5. How about the alcohol reference? Could such an oblique reference to alcohol in the context of a celebratory toast really be offensive?
6. Assume Ken and Kristen wind up at loggerheads and the EEO representative must be involved. What position would you expect him or her to take? Why?
7. Can you recall any communication dilemmas involving differences between cultures? If so, share your experiences and compare them with the case at issue.

Role-Playing

Directions: Assign each group one of the three characters in the scenario. If there are six groups, simply enact each scene a second time with a different set of players. (Indeed, this can be particularly interesting!) In no instance, however, should a group be assigned more than one character. Whatever character the group is assigned, it must choose an actor or actress to portray the assigned role and coach or coax that person into an Academy Award-winning performance. In the role-playing scenes, characters should act in accordance with their own interests and values, not according to a prescribed ethical "script." The name of the game is creative improvisation!

Note: This exercise is gender-friendly. Gender changes may be made as follows:

- Ken-Kenia
- Kristen-Kris
- EEO representative: Karl-Karla

Scene #1
Play out the confrontation between Ken and Kristen. Then bring in a new pair of actors and try it again.

Scene #2
Reenact the scene adding a third character, the EEO representative (Karl or Karla).

Variations:

- ❑ Vary the scenario so it becomes a real estate marketing pitch to the Board of Directors of an exclusive country club whose token minority membership can be counted on the fingers of one hand. In this case, the roles are reversed; the editor (Kenia) has incorporated cultural diversity into the slides while the marketeer (Kris) wants them redone to be lily white.

Large-Group Discussion

Refer to Chapter 14 for large-group discussion questions to apply to the case study.

CHAPTER 22

The Case of the Publishing Professor

The Scenario

Mark trudged to Dr. Ernstein's fourth-floor office barely in time to catch the professor before the end of his afternoon office hours. He could easily have avoided this last-minute meeting. The paper he had to discuss with Dr. Ernstein was actually finished last night. But Mark had procrastinated as long as possible because he just didn't know how to pose his concerns.

Taking the stairs instead of the elevator was one more way for Mark to put off the confrontation, if, in fact, there was to be one. He was still undecided. Mark recalled how flattered and thrilled he'd been at the start of the last semester when Dr. Ernstein had asked him to assist in the final stages of an ongoing research project. Sharing a byline with such a respected professional would certainly improve his application for doctoral programs! More important to Mark was the opportunity to conduct research first-hand with a master—he'd learn so much!

Unfortunately, he was learning something other than what he'd expected. His job had been to document the final stages of the experiment, compile Dr. Ernstein's earlier notes, and scrutinize and update the professor's initial literature review. Having accomplished all that, Mark had given Dr. Ernstein an initial draft of the paper three weeks ago. He was confident the draft was good, even though it was covered with Post-it notes full of questions for the professor. Dr. Ernstein had returned the edited paper last week with instructions for Mark to complete the rewrite by today.

"It's probably just a mistake—an oversight by Ernstein," thought Mark. "After all, who am I to question *him*? He's one of the most frequently published physicists in the field. He has an international reputation. I'm just a second-year master's degree student," he sighed, halfway to the second floor.

But Mark knew better. If *he* could spot the flaws in the paper, certainly the brilliant Dr. Ernstein already had. Mark was afraid it wasn't an oversight at all, but an intentional disregard for the basic tenets of research. Certainly if *he* was aware of the theories of Dr. Charles Darword, Dr. Ernstein had read them, too. Granted, the journals in which Darword had published were somewhat obscure. And his findings were too controversial to be held in very high esteem among physicists. In fact, they had at first been ignored. Nonetheless, now six years after their publication, portions of Darword's theories were beginning to be revisited with some interest by younger, more innovative physicists, which meant they were getting some discussion in the discipline.

Darword's area of research was clearly akin to Ernstein's. Of more concern to Mark, was the fact that Darword's findings seemed to contradict Dr. Ernstein's. Mark had included yellow, sticky notes questioning Dr. Ernstein about the need to refer to Darword's findings—and refute them—on at least half a dozen pages in the draft. The professor's hand-written response in every instance was a resounding "IGNORE THE QUACK!"

That wasn't all. Several pieces in Ernstein's notes had struck a familiar chord with Mark, as if he had read them elsewhere first. But the Post-it queries he'd left for Dr. Ernstein had been pulled from the draft without a response, leaving Mark with the uncomfortable suspicion that they would not be giving credit where it was, in fact, due.

So what was he supposed to do now? Accuse the greatest physicist Southern U. had ever employed of plagiarism and shoddy research procedures? The least that could happen was that Dr. Ernstein would unceremoniously remove Mark from the project, leaving him publicly humiliated and ruining his chances for admission to a first-rate doctoral program—probably even threatening his chances to complete his master's degree.

And if Mark continued to pursue this, he'd be forced to escalate his accusations to the department chair and dean, at the least. Certainly they had a vested interest in supporting Ernstein—who brought in thousands of dollars each year in research grants.

Mark groaned. His career was definitely ruined—his life was certainly over if he mentioned his suspicions.

Of course, the alternative wasn't too pleasant either. He could keep his mouth shut, finish his degree, and reap the glory of publishing with Dr. Ernstein. But that could backfire, as well. Surely, if he was right, somewhere down the road—and it may not even be *too* far, since he felt that the flaws in the paper were rather glaring—a discerning and ethical physicist would point out the weaknesses in their work. Another form of public humiliation!

Well, here he was in front of his mentor. Might as well bite the bullet and approach his concerns as diplomatically and delicately as possible.

"Uhm, doctor, I'm afraid that perhaps I didn't communicate some of my questions adequately. Or maybe the notes fell off. I'm sure you intended for me to document this data that I thought sounded familiar, if you could just steer me to the correct authors. And I probably didn't make my concerns about Darword's theories clear, because if I had I'm *sure* you'd agree that we need to address them . . ."

"Mark, you made yourself perfectly clear, as, I assume, did I. When I tell you to ignore something, *I mean it*. Do *you* dare to question *me*? A second-year grad student? *Certainly not.*"

No one, not even a very nervous Mark, could not miss the menacing tone in Ernstein's remarks.

Clearly the professor did not intend to be questioned any further by a student.

Now what should Mark do?

Value Analysis

Refer to Chapter 14 for value analysis exercises to apply to the case study.

Questions for Small-Group Discussion

Directions: Not every aspect of ethical dilemmas is amenable to systematic analysis. It is therefore useful to pursue open-ended discussion as well. Here are some jumping-off points for such discussion.

1. Is a researcher ethically obligated to consider contradictory data when publishing research? Why or why not? If so, how should Mark address this issue with the professor?
2. Clearly Dr. Ernstein has been published many times before. If, in fact, he *does* make a habit of plagiarizing information, what are the responsibilities of the publishers of the journals in which his papers appear? What about the publishers of his books?
3. Are there other approaches Mark might try in dealing with Dr. Ernstein? What?
4. What course of action is likely to have the most dire consequences for Mark?
5. Mark assumes how the department and dean would react to his accusations. Do you agree with his assessment?
6. Reach into your own personal experience. Can you come up with any conflicts, either during your education or on the job, resembling this one? If so, share them, and discuss possible alternatives.

Role-Playing

Directions: Assign each group one of the characters in the scenario. If there are more people than groups, simply enact each scene a second time with a different set of players. (Indeed, this can be particularly interesting!) In no instance, however, should a group be assigned more than one character. Whatever character the group is assigned, it must choose an actor or actress to portray the assigned role and coax that person into an Academy Award-winning performance. In the role-playing scenes, characters should act in accordance with their own interests and values, not according to a prescribed ethical "script." The name of the game is improvisation!

Note: This exercise is gender-friendly. Gender changes may be made as follows:

- Mark-Mary
- The professor, department chair, and dean may be either gender

Scene #1
Replay the scene between Mark and Dr. Ernstein.

Scene #2
Dramatize a meeting in which Dr. Ernstein is more open to Mark's concerns.

Scene #3
Dramatize a casual lunch at the faculty club the next day in which Dr. Ernstein tells the department chair about his confrontation with Mark.

Variations:

- ❑ Assume Mark does go to the department chair with his concerns. Enact the scene between them, role-playing several possible outcomes.
- ❑ Assume the department chair gives credence to Mark's concerns. Role-play his meeting with the dean. You might also dramatize a subsequent meeting involving the department chair, the dean, and Dr. Ernstein.

Epilogue (optional):
Assume that Mark's concerns are disregarded at all levels of the college administration. Mark may still have the option of approaching the publishers of the journal in which the paper is to appear. Enact a scene or scenes between Mark and both a junior and senior editor at the journal.

Large-Group Discussion

Refer to Chapter 14 for large-group discussion questions to apply to the case study.

CHAPTER 23

The Case of the Naked Emperor

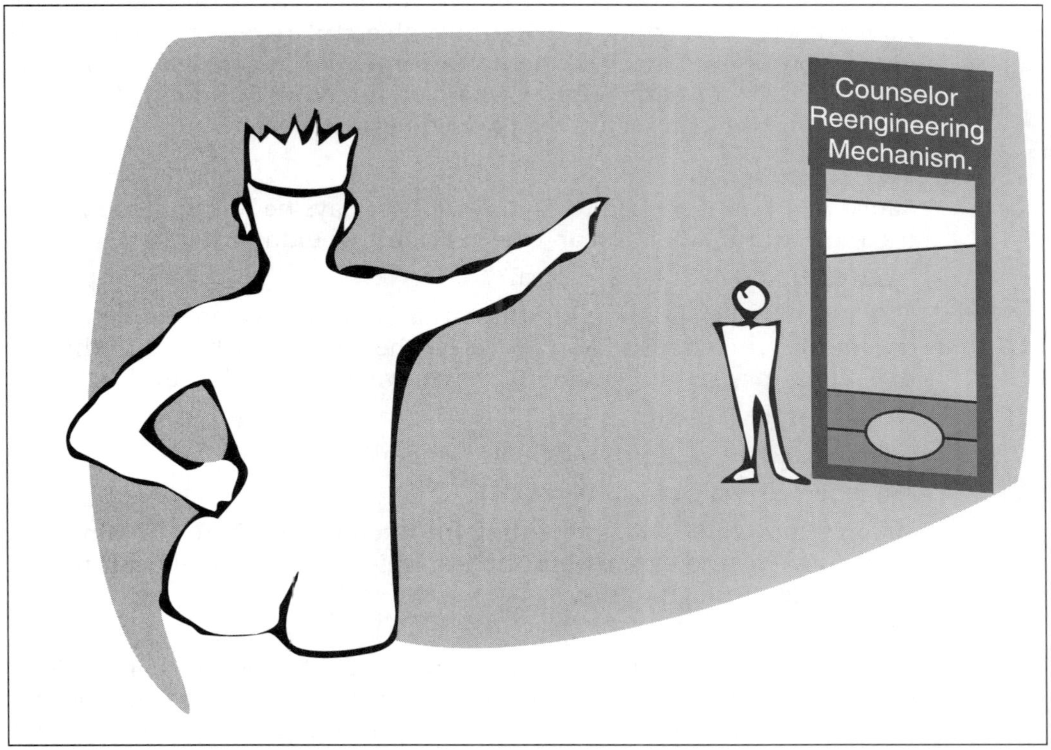

The Scenario

"So you see, Becky, everyone tried to tell the Emperor he had no clothes on, but he just wouldn't listen."

"So what happened, Daddy?"

"It's 'to be continued' tomorrow," Travis said. "It's past your bedtime."

"But–"

"No 'buts'," Travis said firmly, closing the children's book and turning out the light.

He had a tough day ahead of him tomorrow as a video producer at Acme Widgets. A major marketing pitch he was producing had been unreasonably schedule-compressed by the marketing director to the point of severely compromising quality. Every attempt Travis had made to warn the director had fallen upon deaf ears. The threadbare product was due for a final review at 7 A.M.

* * * * *

"Star-six and hurry!" Ears still ringing, Julie frantically entered the "Stop Message" sequence for her voice mail. Too late. The shrill tones of Bert Bigman's Level 3, Maximum-Irritation voice still reverberated in her head: "Julie, you've *really* screwed me up this time. What garbage! I want the producer fired! And I want to know what you intend to do about..."

Sighing, Julie picked up the phone to call Travis.

* * * * *

"But I told him if we had to do this in three days he was going to get trash! I used that word, Julie, I swear—trash! He just wouldn't listen."

"And so that's what he got, trash?"

"Well, not really. Considering the turnaround, the video was actually fairly decent. But it obviously didn't have all the bells and whistles of the interactive pitch we gave him last month when we had enough time."

"And he *wanted* them."

"He *expected* them. Hell, he *demanded* them!"

"So where do we go from here?"

"I don't know, Julie. He's presenting the pitch today. There's no time to do all that stuff. We knew that from the beginning. I tried to tell him that. I just couldn't get through to him."

* * * * *

Julie knocked on the 10-foot mahogany door—slightly ajar—of Bert Bigman's huge corner office. Travis stood next to her, shifting from foot to foot on the deep pile carpet. From past experience, Julie could almost script the impending confrontation. But this time there was an added wrinkle for her.

For all his arrogance, Bert had generally favored Julie and boosted her career. Now she was in line for a major promotion. To face him down on the video could jeopardize her chances. But the only other option would be to let Travis take the fall. Maybe even lose his job.

* * * * *

"So you see, Becky, the wise servant held up the mirror so the Emperor could see he had no clothes on."

"So what happened to the servant, Daddy?"

ENDING 1: "He was promoted to Wise Counselor and everyone lived happily ever after."

ENDING 2: "He was beheaded."

* * * * *

"Julie, come on in..."

Value Analysis

Refer to Chapter 14 for value analysis questions to apply to the case study.

Questions for Small-Group Discussion

Directions: Not every aspect of ethical dilemmas is amenable to systematic analysis. It is, therefore, useful to pursue open-ended discussion as well. Here are some jumping-off points.

1. Are you familiar with the child's story that Travis is reading to Becky? If not, read it.
   a. How does the original child's story end?
   b. Which character in the scenario resembles the emperor? In what sense does he have no clothes on?
   c. Which character represents the counselor? Explain how his efforts parallel those of the counselor in the story.
   d. Which character in the scenario has no counterpart in the story? Invent one.
2. Could Travis have done anything differently to avoid this problem? Is there any "magic mirror" he could have held up to Bert to get his message across?
3. What mistake did Travis make? How could he have lessened his personal risk, even if he could not find a magic mirror?
4. What is Julie's personal agenda? How might this affect her handling of the situation? How does it relate to the risk to Travis?

5. Is there any way Julie can serve her own interests *without* selling out Travis?
6. Is there any way Travis can protect himself at this point if Julie does sell him out?
7. Share your experiences. Have you encountered any "naked emperors" in your personal and business travels? If so, have you found any magic mirrors that are particularly effective for conducting reality counseling with the egotistically myopic?

Role-Playing

Directions: Assign each group one of the three characters in the scenario. If there are six groups, simply enact each scene a second time with a different set of players. (Indeed, this can be particularly interesting!) In no instance, however, should a group be assigned more than one character. Whatever character the group is assigned, it must choose an actor or actress to portray the assigned role and coach or coax that person into an Academy Award-winning performance. In the role-playing scenes, characters should act in accordance with their own interests and values, not according to a prescribed ethical "script." The name of the game is creative improvisation!

Note: This exercise is gender-friendly. Gender changes may be made as follows:

- Travis-Travette
- Julie-Jules
- Bert Bigman-Bertha Bigwoman

Scene #1
Play out the scene in Bert Bigman's office with just Julie and Bert.

Scene #2
Directly after the scene in Bert's office, dramatize an ensuing meeting between Julie and Travis.

Scene #3
Replay the scene in Bert Bigman's office with all three characters (adding Travis). There are two possible versions of this. First, play one where Julie brings Travis with her for the initial meeting with Bert. Then, if time permits, play one which follows up Scenes #1 and #2.

Variations:

- ❑ Enact one of Travis' earlier attempts to communicate with Bert.
- ❑ Involve a fourth character (your choice).

Other Applications (Optional):
Drawing from your collective experience (see Question #7), create and enact other scenarios involving "naked emperors" and frustrated counselors.

Large-Group Discussion

Refer to Chapter 14 for large-group discussion questions to apply to the case study.

CHAPTER 24

Ethical Dilemmas—Your Way

The Scenario

Now that you've become expert—or at least more "practiced"— in resolving thorny ethical dilemmas, you have the opportunity to create one for your classmates or training colleagues to resolve.

In *The Case Method in Technical Communication: Theory and Models*, John R. Brockman identifies the six features of cases. According to Brockman, cases must:

- ❑ Have verisimilitude
- ❑ Have fullness of . . . content
- ❑ Illustrate how communication is part of the problem-solving process
- ❑ Focus on technical communication
- ❑ Embody a functional relationship between theory and practice
- ❑ Be difficult and open to divergent solutions[1]

Let's examine each of Brockman's guidelines in more detail. By verisimilitude, Brockman refers to the realism of the cases. Not only the underlying situation, but all of the details of everyday life—both in the narrative and the dialogue—must ring true. *The best way to achieve this is to place your case study in a situation and location you know well, so you can make it true-to-life*. It's difficult to take the ethical conflicts seriously if the scenario feels forced.

Fullness of . . . content refers to the *rhetorical* context of cases according to Brockman, but the same advice holds for ethical cases. To be useful for exploring values in technical communication, your case must have "fullness" of ethical content. In other words, the case should revolve around ethical conflicts. *Develop a case that poses a conflict between at least two, preferably more, of the 10 core values presented in this book.*

Although Brockman's remarks apply to communication cases, your cases will illustrate how *value analysis*, rather then communication, is part of the problem-solving process. In other words, *your case should lend itself to the use of the six-step value analysis process as a means of resolution.* To ensure a close fit, refer to Chapter 2 as you develop it.

Brockman also advises us to focus our cases on technical communication. Clearly, we all face ethical dilemmas every day that bear no relation to our profession. However, this is, after all, a book about ethics in technical communication, so *focus your case on an academic- or career-related situation*, not on family, personal relationships, and so on.

As readers work through your case, they should be able to identify and apply ethical theories that are applicable across the board to any ethical dilemma. *Carefully arrange the case to demonstrate the application of value analysis using the 10 core values.*

Finally, there should be no single answer to your case—no "right" or "wrong" solutions. *Your case should allow, even force, readers to make choices, select courses of action, and predict the consequences of such courses of action, rather than settle on "correct" answers.* If you'd like, use our exercises following each case as a guide. Be sure to leave plenty of room for discussion.

The Assignment

Using Brockman's guidelines and our suggestions, create a detailed case study that:

1. Presents a realistic, true-to-life situation
2. Poses a conflict between two or more of the 10 core values presented in this book
3. Requires the use of the six-step value analysis process as a means of resolving the conflicts
4. Focuses on an academic- or career-related situation in technical communication
5. Defines the relationship between theory (value analysis using the core values) and practice
6. Allows readers to make choices, select courses of action, and predict consequences of such courses of action, rather than settle on "correct" answers

7. Sets up role-playing scenes wherein readers can breathe dramatic life into conflicts
8. Provides questions to spark and guide small- and large-group discussion of ethical issues and potential consequences

For further instructions, see your instructor or training facilitator.

Some Tips

We can offer some specific tips as a result of experience creating our own case studies.

1. Decide which values you want to bring into conflict and design your scenario around that central conflict(s). This will give you a much more natural-flowing scene than you can get by whipping up a scenario, then trying to force value conflicts into it.
2. In selecting a case study topic, stick with what you know; that will help you achieve the verisimilitude Brockman talks about. On the one hand, avoid oversimplifying; try to incorporate a number of conflicts among the stakeholders' values and interests. On the other hand, provide focus; avoid making conflicts so complex nobody can figure them out!
3. Pay particular attention to characterization; create interesting characters that will help sustain interest in your scenario. Consider characters like Dr. Ernstein and Dr. Trode; not only were they fun to create, but they also provide grist for the ethical mill!
4. Make your character names gender-friendly (e.g., Paul/Paula) to avoid creating artificial constraints when assigning roles for classroom or training session dramatization.
5. Have fun. Loosen up. Wake up your muse. Don't be afraid to do something a little bit different. Think "outside the box." When it comes to creating case studies, originality is half the battle.

References

1. John R. Brockman, ed., *The Case Method in Technical Communication: Theory and Models* (The Association of Teachers of Technical Writing, 1994), 1-13.

Bibliography

Albert, Rosita D. and Harry Triandis. "Intercultural Education for Multicultural Societies: Critical Issues." In *Intercultural Communication: A Reader,* edited by Larry A. Samovar and Richard E. Porter. Belmont, CA: Wadsworth Publishing, 1994.

American Heritage Electronic Dictionary. 3d ed. WordStar International, Inc. New York: Houghton Mifflin, 1992.

Amernick, Barton A. *Patent Law for the Nonlawyer: A Guide for the Engineer, Scientist, Technologist, and Manager*. 2d ed. New York: Van Nostrand Reinhold, 1991.

Andersen-Hancock, Shirley. "Presenting the New STC Ethical Guidelines for the Technical Communicator." *Intercom*, (July-August 1995): 6-7.

Anderson, Paul V. *Technical Writing: A Reader-Centered Approach*. New York: Harcourt Brace Jovanovich, 1991.

Apollo 13. Universal City Studios MCA Home Video, 1995.

Augustine, Norman R. "The Cost of Being Ethical." *Executive Excellence* (February 1994): 5.

Bacard, Andre. *The Computer Privacy Handbook*. Berkeley, CA: Peachpit Press, 1993.

Bigelow, Barbara, Liam Fahey, and John F. Mahon. "Political Strategies and Issues Evolution: A Framework for Analysis and Action." In *Contemporary Issues in Business Ethics and Politics,* edited by Karen Paul. *Mellon Studies in Business*, Vol. 7. Lewiston: The Edwin Mellon Press, 1991.

Boiarski, Carolyn. *Technical Writing: Contexts, Audiences, and Communities*. Boston: Allyn & Bacon, 1996.

Branscomb, Anne W. *Who Owns Information?: From Privacy to Public Access*. New York: Basic Books, 1994.

Brecht, Bertolt, and Kurt Weill, composer. *The Three-Penny Opera*. Act II, Scene I.

Brockman, John R., ed. *The Case Method in Technical Communication: Theory and Models*. The Association of Teachers of Technical Writing, 1984.

Brown, Marvin T. "Discovering the Right Decision." In *Working Ethics: Strategies for Decision Making and Organization Responsibility*. San Francisco: Jossey-Bass Publishing, 1990.

Bryan, John. "Responding to Ethical Dilemmas When the Mortgage Is Still Due." Presented at 15th Annual Practical Conference on Communication, Oak Ridge, TN, October 1991.

Brydon, Steve R. and Michael D. Scott. *Between One and Many: The Art and Science of Public Speaking*. Mountain View, CA: Mayfield Publishing, 1994.

Byham, W.C. and J. Cox. *ZAPP! The Lightening of Empowerment: How to Improve Quality, Productivity, and Employee Satisfaction*. New York: Fawcett, 1988.

Carr, Albert Z. "Is Business Bluffing Ethical?" *Harvard Business Review* 46 (January-February 1968): 143-153.

CBS Evening News, April 3, 1996.

Chodorow, Jeoffrey. *Cases, Research, and Issues in Technical Communication*. New Jersey: Humanities Press International, 1990.

Class Action. Twentieth Century Fox, 1991. Fox Home Video, 1995.

Cragg, Charles I. "Because Wisdom Can't Be Told." In *The Case Method at the Harvard Business School*, edited by Malcolm P. McNair. New York: McGraw-Hill, 1954.

Crosby, Philip B. *Quality Without Tears: The Art of Hassle-Free Management*. New York McGraw Hill, 1984.

DeGeorge, R.T. "Professions, Business, and Ethical Codes of Behavior." *Business Ethics*. 3rd ed. New York: MacMillan, 1990.

Dead Poets Society. Touchstone Films.

Dickens, Charles. *Martin Chuzzlewit*. London: Collins, 1953.

Dolan, Edward. *Your Privacy: Protecting It in a Nosy World*. New York: Cobblehill Books, 1995.

Dombroski, Paul. "Can Ethics be Technologized? Lessons from *Challenger*, Philosophy, and Rhetoric." *IEEE Transactions on Professional Communication* 38 no. 3. (September 1995): 146.

Don, Robert. *Protecting Trade Secrets, Patents, Copyright and Trademarks*. New York: John Wiley, 1995.

Doublespeak. Films for the Humanities. Princeton, NJ, 1991.

Drucker, Peter. "The Responsibilities of Management." In *Ethics, Leadership, and the Bottom Line*, edited by Charles Nelson and Robert Cavey. Croton-on-Hudson, NY: North River Press, 1991.

Eagleson, Robert D. *Writing in Plain English*. Canberra, Australia: Australian Government Publishing Service, 1991.

Elias, Stephen. *Patent, Copyright, and Trademarks: A Desk Reference to Intellectual Property*. Berkeley, CA: Nolo Press, 1996.

Ellison, Ralph. *Invisible Man*. New York: Vintage Books, 1952.

Etzkorn, I.A. "Why Government Has Difficulty Communicating." In *Plain Language: Principles and Practices*, edited by E. R. Steinberg. Detroit: Wayne State University Press, 1991.

Fenwick, Ellen. "Ethics of Helping with Student Papers: Summary." TECHWR-L electronic discussion. January 2, 1996.

Fishman, Stephen. *The Copyright Handbook: How to Protect and Use Written Works*. Berkeley, CA: Nolo Press, 1992.

Flynn, Laurie. "They're Watching You: Electronic Surveillance of Workers Raises Privacy Concerns." *San Jose Mercury News* (June 13, 1993): 1F. Cited in Gini Graham Scott, Ph.D. *Mind Your Own Business: The Battle for Personal Privacy*. New York: Plenum Books, 1995.

Foltz, Raymond. *Handbook for Protecting Ideas and Inventions*. Cleveland, OH: Penn Institute, 1992.

Frankel, Mark. "Professional Codes: Why, How, and with What Impact?" *Journal of Business Ethics* 8 (1989): 109-115.

Friedman, Milton. "The Social Responsibility of Business to Increase Its Profits." In *Ethics, Leadership, and the Bottom Line*, edited by Charles Nelson and Robert Cavey. Croton-on-Hudson, NY: North River Press, 1991.

Gilliam, John. *Surveillance, Privacy, and the Law: Employee Drug Testing and the Politics of Social Control*. Ann Arbor, MI: University of Michigan Press, 1994.

Gorlin, R. *Codes of Professional Responsibility*. Washington, D.C.: Bureau of National Affairs, 1986.

Grolier Electronic Encyclopedia, Version 7 04, 1995.

Grove, Laurel. "Speaking for the Environment." *IEEE Transactions on Professional Communication* 38 no. 3. (September 1995).

Gunning, Robert. *Fog*. Chicago: Dartnell Press, 1964.

Guran, J. M. and F. M. Gryna, Jr. *Quality Planning and Analysis: From Product Development Through Use*. NY: McGraw-Hill. Cited in Skelton and Anderson, "Professionalism in Technical Communication." *Technical Communication* (Second Quarter 1993): 202-207.

Guy, Mary. *Ethical Decision Making in Everyday Work Situations*. New York: Quorum Books, 1990.

Harcourt, Jules, "Buddy" Krizan, and Patricia Merrier. *Business Communication*. 3d ed. Cincinnati: Southwestern Educational Publishing, 1996.

Hatch, Elvin. "The Evaluation of Culture." In *International Communication: A Reader*. 7th ed. Edited by Larry A. Samovar and Richard E. Porter. Belmont, CA: Wadsworth Publishing, 1994.

Hemingway, Ernest. *Death in the Afternoon*. New York: Scribner, 1932.

Henderson, Verne E. "The Ethical Algorithm." In *What's Ethical in Business*. New York: McGraw-Hill, 1992.

Hodgson, Kent. *A Rock and a Hard Place: How to Make Ethical Decisions when the Choices Are Tough*. New York: American Management Association, 1992.

Hornung, Mark. "Honesty's valued, but transgressions abound." *Crain's Chicago Business* 10(34):1; 1987: 61-63.

Horowitz, Renee . "Beyond Plagiarism: Ethical Issues in the Technical Communication Classroom." *Proceedings of the 41st Annual Conference*. Washington D.C.: Society for Technical Communication, 1994.

Horton, William K. *The Book of Icons*. New York: John Wiley, 1994.

"How Ethical Is American Business?" *Working Woman* (September 1990): 113.

Hudson Institute. *Workforce 2000: Work and Workers for the 21st Century*. Washington, D.C.: U.S. Department of Labor, 1987.

Huff, Darrell. *How to Lie with Statistics*. New York: W.W. Norton, 1954.

James, Stewart. *Calculus*. 2d edition. Brookes/Cole Publishing, 1991.

Jones, Dan. *Technical Writing Style*. Boston: Allyn & Bacon. Forthcoming Fall 1997.

Joseph, Chief. Upon surrender of Nez Perce Indians to federal forces, northern Idaho, September 30, 1877.

Josephson, Michael. *On Ethics*. Washington, D.C.: Josephson Institute of Ethics, 1992. Videotape.

Kale, David W. "Peace as an Ethic for Intercultural Communication." In *Intercultural Communication: A Reader,* edited by Larry A. Samovar and Richard E. Porter. Belmont, CA: Wadsworth Publishing, 1994.

Kaplan, Linda. "Corporate Mentoring." In *Proceedings of the 38th Conference*. Washington, D.C.: Society for Technical Communication, 1991.

KCNC Morning News, Denver, CO, April 4, 1996.

Killingsworth, M. Jimmie. *Information in Action: A Guide to Technical Communication*. Boston: Allyn & Bacon, 1996.

Kim, Young Yun. "Intercultural Personhood : An Integration of Eastern and Western Perspectives." In *Intercultural Communication: A Reader,* edited by Larry A. Samovar and Richard E. Porter. Belmont, CA: Wadsworth Publishing, 1994.

King, Martin Luther. "I Have a Dream." Washington, D.C. August 28, 1963.

Lannon, John. *Technical Writing.* 6th ed. New York: Harper Collins, 1994.

Larson, M.S. "The Rise of Professionalism: A Sociological Analysis." Berkeley: University of California Press. Cited in Skelton and Anderson, "Professionalism in Technical Communication." *Technical Communication* (Second Quarter 1993): 202-207.

Lutz, William. *Doublespeak.* New York: Harper & Row, 1989.

Lutz, William. "The First Casualty." *Quarterly Review of Doublespeak* XVII:4 (July 1991): 1.

MacKenzie, Gordon. Keynote address at the 41st Annual Conference of the Society for Technical Communication, Minneapolis, 1994.

Malcolm, Andrew. "On Certifying Technical Communicators." *Technical Communication* (Second Quarter 1987): 94-102.

Manley, Walter. *Critical Issues in Business Conduct: Legal, Ethical, and Social Challenges for the 1990s.* New York: Quorum Books, 1990.

Manley, Walter. *Executive's Handbook of Model Business Conduct Codes.* Englewood Cliffs, NJ: Prentice-Hall, 1991.

Mannering, Dennis. Closing address at the 40th Annual Conference of the Society for Technical Communication, Dallas, June 1993.

Markel, Mike. *Technical Communication: Situations and Strategies.* 4th ed. New York: St. Martin's Press, 1996.

McLean, Deckle. *Privacy and Its Invasion.* Westport, Connecticut: Praeger, 1995.

Molloy, John T. *Dress for Success.* New York: Warner Books, 1975.

Mount, Eric. *Professional Ethics in Context: Institutions, Images, and Empathy.* Louisville, KY: Westminster/John Knox Press, 1990.

Nash, Laura, "Why Business Ethics Now?" In *Managing Business Ethics,* edited by John Drummond and Bill Bain. Oxford: Butterworth-Heineman, 1994: 9.

National Council of Teachers of English, 61st Annual Meeting, 1971.

Nelson, Charles, and Robert Cavey, eds. *Ethics Leadership and the Bottom Line.* Croton-on-Hudson, North River Press, 1991.

New York Times, December 12, 1973.

New York Times, February 6, 1996.

Newman, Edwin. Keynote address at the 42nd Annual Conference of the Society for Technical Communication, Washington, D.C., April 1995.

"Pair sues, claims ad defamatory." *The Denver Post* (January 20, 1996): 3A.

Payne, Lynn, PhD. "Hierarchical Model for Value Analysis." University of Virginia, 1991.

"Pentagon stands tall in doublespeak." *The Denver Post* (October 5, 1994).

Piller, Charles. "Bosses with X-Ray Eyes." *MacWorld* (July 1993): 118-123.

Pinella, T. E., R. O. Barclay, M. L. Keene, J. M. Kennedy, and L. F. Hecht. "From Student to Entry-Level Professional: Examining the Role of Language and Written Communications in the Reacculturation of Aerospace Engineering Students." *Technical Communication* (Third Quarter 1995).

Redish, Janice, and Judith Ramey. "Measuring the Value Added by Professional Technical Communicators." *Technical Communication* (First Quarter 1995).

Reed, Ishmael. "The Multinational Society." In *The Bedford Reader*. 5th ed. edited by X. J. Kennedy, Dorothy Kennedy, and Jane E. Aaron. Boston: Bedford Books of St. Martin's Press, 1994.

Robin, Donald P. and R. Eric Reidenbach. *Business Ethics: Where Profits Meet Value Systems*. Englewood Cliffs, NJ: Prentice-Hall, 1989.

Rodgers, Richard, and Oscar Hammerstein II. "You Have to Be Carefully Taught." *South Pacific*. New York: Williamson Music, 1949.

Rose, Mark. *Authors and Owners: The Invention of Copyright.* Cambridge, MA: Harvard University Press, 1993.

Samovar, Larry A. and Richard E. Porter. *Communication Between Cultures*. Belmont, CA: Wadsworth Publishing, 1991.

Samuels, Jeffrey M., ed. *Patent, Trademark and Copyright Laws*. Washington, D.C.: Bureau of National Affairs, 1993.

Scott, Gini Graham, Ph.D. *Mind Your Own Business: The Battle for Personal Privacy*. New York: Plenum Books, 1995.

Shatner, William, and Chris Kreski. *Star Trek Memories*. New York: Harper Collins, 1993.

Simply Stated. "Gas Utilities Switch to Plain Language." Washington, D.C.: American Institutes for Research, 1986: 2. Cited in Karen A. Shriver, "Quality in Technical Communication." *Technical Communication*, 2d Quarter, 1993: 250–51.

Skelton, Terry, and Shirley Anderson. "Professionalism in Technical Communication." *Technical Communication* (Second Quarter 1993): 202-207.

Skelton, Terry, and Larry Kunz, moderators. Panel discussion at the 42nd Annual Conference of the Society for Technical Communication, Washington, D.C., April 1995.

Smart, Karl L., Kristie K. Seawright, and Kristen Bell DeTienne. "Defining Quality in Technical Communication: A Holistic Approach." *Technical Communication* (Third Quarter, 1995): 474–481.

Smith, H. Jeff. *Managing Privacy: Information Technology and Corporate America*. Chapel Hill, NC: University of North Carolina Press, 1994.

Sniffen, Michael J. "Brother did agonize before going to FBI." *Denver Post.* April 5, 1996: 5A.

Star Trek VI: The Undiscovered Country. Paramount Pictures, 1992.

Terra Pacific. San Diego and Los Angeles, CA. "Training Guide." Handouts from Internationalization Seminar, Denver, CO, 1994.

The Net. Columbia Pictures, 1995.

"Today." NBC Broadcasting, February 7, 1996.

Troyka, Lynn Quitman. *Handbook for Writers*. 3d ed. Englewood Cliffs, NJ: Prentice-Hall, 1993.

Tuleja, Tad. *Beyond the Bottom Line: How Business Leaders Are Turning Principles into Profit*. New York: Facts on File, 1985.

Voss, Daniel. "Editorial Training." Martin Marietta Corp. (now Lockheed Martin). In *Proceedings of the 37th Annual Conference*. Washington, D.C.: Society for Technical Communication, 1991.

Voss, Daniel. "Editorial Training II." Martin Marietta Corp. (now Lockheed Martin). In *Proceedings of the 38th Annual Conference*. Washington, D.C.: Society for Technical Communication, 1992.

Voss, Daniel. "Sharpening the Focus: A Workshop on Ethics and Technical Communication." Martin Marietta Corp. (now Lockheed Martin). In *Proceedings of the 41st Annual Conference*. Washington, D.C.: Society for Technical Communication, 1994.

Voss, Daniel. "A Slippery Slope: Using Value Analysis to Resolve Ethical Conflicts." Martin Marietta Corp. (now Lockheed Martin). In *Proceedings of the 40th Annual Conference*. Washington, D.C.: Society for Technical Communication, 1993.

Voss, Daniel. "Three Whiz-Bang Online Ways to Promote Professional Development in a Whiz-Bang Online World." Martin Marietta Corp. (now Lockheed Martin). In *Proceedings of the 42nd Annual Conference*. Washington, D.C.: Society for Technical Communication, 1995.

Wall Street Journal (May 5, 1983): 1.

Weber, Richard C. "The Group: A Cycle from Birth to Death." *NTL Reading Book for Human Relations Training*. NTL Institute, 1982.

When Harry Met Sally. Castle Rock Entertainment, 1989.

White, Fred D. *Communicating Technology: Dynamic Processes and Models for Writers*. New York: Harper Collins, 1996.

Winch, Peter. *Ethics and Action*. London: Routledge & Keyan Paul, 1972.

Zelezny, John D. *Common Laws: Liberties, Restraints, and the Modern Media*. Belmont, CA: Wadsworth Publishing, 1993.

Index

D

E

Q

R

S

T

U

V